INTRODUCTION TO ENGINEERING DESIGN

BOOK 9, SECOND EDITION

ENGINEERING SKILLS AND HOVERCRAFT MISSIONS

James W. Dally
Kevin Calabro
William L. Fourney
Wesley G. Lawson
Gary A. Pertmer
Guangming Zhang

University of Maryland, College Park

College House Enterprises, LLC
Knoxville, Tennessee

The manuscript was prepared using Word 2002 with 11 point Times New Roman font. Publishing and Printing Inc., Knoxville, TN printed this book from pdf files.

College House Enterprises, LLC.
5713 Glen Cove Drive
Knoxville, TN 37919, U. S. A.
Phone   (865) 558 6111
FAX   (865) 558 6111
Email   jwd@collegehousebooks.com
http://www.collegehousebooks.com

10 Digit ISBN 0-9792581-0-3
13 Digit ISBN-978-0-9792581-0-7

PREFACE

This book is the second edition of the ninth textbook in this series dealing with Introduction to Engineering Design. Jim Dally, working with College House Enterprises, has prepared eight previous books in this series—a new one almost every academic year—for the first-year engineering students of the University of Maryland at College Park. Several other Colleges of Engineering have adopted one or more of the books in this series to introduce design and engineering skills for their first or second year students.

PROCESS AND CONTENT

The procedure followed in offering a design experience to first or second year engineering students is to:

- Teach the class in moderate size sections.
- Divide the class into product development teams with five or six members per team.
- Assign a project entailing the development of a prototype that will require the entire semester.
- All student teams develop the same product in this instance a hovercraft.
- In the product realization process, the students:
  - Design
  - Manufacture or procure parts for the prototype.
  - Assemble the prototype.
  - Test and evaluate the prototype.
- In developing the prototype, the students have the opportunity to learn:
  - Communication skills.
  - Team building skills.
  - Engineering graphics.
  - Software applications including CAD.
  - Design methods and procedures.

The textbook is used to support the students during a semester-long project. Some of the material may be covered in class or in a computer laboratory. Additional material is covered with reading assignments. In other instances, the students use the text as a reference document for independent study. Exercises, provided at the end of each chapter, may be used for assignments when the demands of the project on the students' time are not excessive.

The book is organized into six parts to present many topics that first year engineering students should understand as they proceed through a significant portion of the product realization process. Product and system development processes is introduced in Part I. Information on team skills and the importance of the product development process is covered in Chapter 1 and 2. The hovercraft missions are also presented in Chapter 2 together with a description of the design concepts involved in hovercraft development. By assigning a demanding project, a holistic approach is employed in the student's first engineering experience that motivates them. Design of a hovercraft enables the instructor an opportunity to integrate a spectrum of knowledge about many topics. The student's hands-on participation in a design, building and testing a hovercraft significantly enhances their learning process. The theoretical background needed to conduct design analyses for the hovercraft is presented in Chapter 3. Chapter 4 describes basic electric circuits to provide technical background helpful for the design of the hovercraft. Sensors used to provide feedback signals for control of the hovercraft are discussed in Chapter 5. Finally, an extensive and detailed description of programming in ROBOLAB is included in Chapter 6.

Part II presents three chapters on engineering graphics. This coverage includes a relatively complete treatment of three-view drawings in Chapter 7. Pictorial drawings including isometric, oblique, and perspective are covered in Chapter 8. Pro/Engineer a computer aided design (CAD) program is described in detail in Chapter 9. The use of tables and graphs in communicating engineering information using Microsoft Excel is presented in Chapter 10.

Part III treats the very important topic of communications. A chapter on technical reports describes many aspects of technical writing and library research. The most important lesson here is that a technical report is different than a paper for the History or English Departments. An effective professional report is written for a predefined audience with specific objectives. The technical writing process is described, and many suggestions to facilitate composing, revision, editing and proofreading are given. In the chapter on design briefings a distinction is drawn among speeches, presentations and group discussions. Emphasis is placed on the technical presentation and the importance of preparing excellent visual aids. PowerPoint slides are employed as visual aids in a design briefing.

Part IV introduces the students to engineering. Chapter 13 describes engineering disciplines and on-the-job activities. A useful student survival guide is also included in Chapter 14.

Part V contains three chapters dealing with engineering and society. A historical perspective on the role engineering played in developing civilization and on improving the lives of the masses is presented in Chapter 15. In this chapter, we move from the past into the present and indicate the current relationship between business, consumers and society. Chapter 16 discusses the balance among safety, risk and performance. It includes a listing of hazards, which is important in identifying the many different ways users of a product can be harmed. Chapter 17 on ethics, character and engineering includes a large number of topics so the instructor can select from among them. A description of the Challenger and Columbia accidents is also given because both of these fatal crashes provide excellent case histories covering safety related conflicts between management and engineers.

ABET AND EDUCATIONAL OUTCOMES

ABET's Criteria 2000 was considered in writing this book. It is believed that a first course in engineering design could include the educational outcomes listed below:

Communication Skills
1. Engineering Graphics
   - Understand the role of graphics in engineering design.
   - Understand orthographic projection in producing multi-view drawings.
   - Understand three-dimensional representation with pictorial drawings.
   - Understand dimensioning and section views.
   - Demonstrate capability of preparing drawings using both manual and computer methods.
   - Demonstrate understanding of engineering graphics by incorporating appropriate high-quality drawings in the design documentation.
2. Design Briefings:
   - Within a team format, present a design review for the class using appropriate visual aids.
   - Each team member demonstrates briefing skills.
3. Design Reports:
   - The team's design is documented in a professional style report incorporating time schedules, costs, parts list, drawings and an analysis.

Team Experience
- Develop an awareness of the challenges arising in teamwork.
- Demonstrate teamwork in the product realization process through a systematic design concept selection process involving participation of all team members.
- Demonstrate planning from conceptualization to the evaluation of the prototype.
- Understand and demonstrate sharing responsibility among team members.
- Demonstrate teamwork in preparing design reports and presenting design briefings.

Software Applications
- Demonstrate entry-level skills in using spreadsheets for calculations and data analysis.
- Show a capability to prepare graphs and charts with a spreadsheet.
- Show a capability to prepare professional quality visual aids.
- Understand entry-level skills in a feature-based solid modeling program.
- Demonstrate these computer skills in preparing appropriate materials for design briefings and design reports.

Design Project:
- The design project is the overarching theme of the course.
- Utilize all the skills listed above to assist in the product development process.
- Demonstrate competence in defining design objectives.
- Generate design concepts that meet the design objectives.
- Understand the basis for design for manufacturing, assembly and maintenance.
- Manage the team and the project effectively.

ACKNOWLEDGEMENTS

Acknowledgments are always necessary in preparing a textbook because so many people are involved in many helpful ways. First, it is important to recognize the contributions of many dedicated faculty members and teaching fellows at the University of Maryland. To date over 60 different faculty members have taught this introductory engineering course on at least one occasion. Many of them have made suggestions for improving the material in this book. In particular, thanks are due to Dr. Gary Pertmer and Dr. William L. Fourney for their commitment in providing the leadership necessary to operate a multi-section offering of this course and in providing support for the many different instructors and teaching fellows involved each semester. We appreciate the review and changes to Chapters 3 suggested by Ken Kiger. Thanks are also due to Drs. Christopher Cadou, Sheryl Ehrman, Bruce Jacob, Glen Moglen and Guangming Zhang for helpful suggestions made during a recent study of the content in this Introduction to Engineering Design course. We also appreciate the support of Dean Nariman Farvardin who has provided the significant resources necessary for a hands-on course like this one to be successful.

Several chapters in this text were either revised or written entirely by others. Dr. Guangming Zhang authored Chapter 9 on Pro/ENGINEER in its entirety. Kevin Calabro edited and revised Chapters 1, 2 and 5 and combined two previous versions of chapters dealing with Tables and Graphs and Excel to produce a new updated version in Chapter 10. Wes Lawson reviewed and suggested changes to Chapter 4 on Basic Electrical Circuits. I also enlisted the help of two undergraduate students, McKenzie C. Primerano and Emi P. DiStefano to edit Chapter 14—A Student Survival Guide—and to check its content from an undergraduate viewpoint.

Thanks are also in order for Anne S. McClain from the University of Alabama at Birmingham who shared experiences and class notes with us. A hovercraft project involving lift capabilities and speed was used in a first year engineering course at the University of Alabama in the spring 2003 semester.

Jim Dally, Kevin Calabro, Bill Fourney, Gary Pertmer and Guangming Zhang
University of Maryland at College Park
June 2007

CONTENTS

PART I PRODUCT AND SYSTEM DEVELOPMENT PROCESSES

CHAPTER 1 DEVELOPMENT TEAMS

CHAPTER 2 PRODUCT DEVELOPMENT AND HOVERCRAFT MISSIONS

CHAPTER 3 FLUID MECHANICS AND DESIGN ANALYSIS

CHAPTER 4 BASIC ELECTRIC CIRCUITS

CHAPTER 5 SENSORS

CHAPTER 6 PROGRAMMING IN ROBOLAB

PART II ENGINEERING GRAPHICS

CHAPTER 7 THREE-VIEW DRAWINGS

CHAPTER 8 PICTORIAL DRAWING

CHAPTER 9 COMPUTER AIDED DESIGN (CAD)

CHAPTER 10 MICROSOFT EXCEL

PART III COMMUNICATION

CHAPTER 11 TECHNICAL REPORTS

CHAPTER 12 DESIGN BRIEFINGS

PART IV   ENGINEERING AND SUCCESS SKILLS

CHAPTER 13 THE ENGINEERING PROFESSION

CHAPTER 14 A STUDENT SURVIVAL GUIDE

PART V ENGINEERING AND SOCIETY

CHAPTER 15 ENGINEERING AND SOCIETY

CHAPTER 16 SAFETY, RISK AND PERFORMANCE

CHAPTER 17 ETHICS, CHARACTER AND ENGINEERING

APPENDIX A: FORMS FOR DESIGN TEAMS

PART I

PRODUCT AND SYSTEM DEVELOPMENT PROCESSES

CHAPTER 1

DEVELOPMENT TEAMS

1.1 INTRODUCTION

The development of quality, leading edge, high-performance products requires the coordinated efforts of many skillful individuals from different disciplines over an extended period of time. This group of individuals is organized within a company to act in an integrated manner to successfully complete the development process. The organizational structure employed varies from one corporation to another; it also depends on the size of the company engaged in the development. For relatively new firms in the entrepreneurial stage, team structure is usually not an issue. These firms are small with a single product and only ten to twenty individuals are involved in the development process. Everyone on the payroll is deeply committed to this product, and communication is usually accomplished around the lunch table. On the other hand for very large corporations, where the number of employees may exceed 100,000, the company is often organized into divisions or operating groups. These large companies have many products or services that are offered by each division, and in some instances one division may actually be a direct competitor of another division. For example, General Motors with its five automotive divisions produces many different models of each of its lines that compete for the same market segment.

In a large corporation, the total number of new products that are introduced each year may exceed 100. Communication is often very difficult because hundreds or even thousands of miles sometimes separate personnel involved in the development of a given product. Sometimes different divisions are in different countries, and business is conducted in different languages. Clearly, in a large firm, organization of the product development team and the physical location of its team members becomes extremely important.

In the past decade, there have been many examples [1] demonstrating that a cross disciplinary team provides a very effective organizational structure for product development. The idea of forming a cross disciplinary team to design a new product appears simple until one examines the existing organizational structure of all but the smaller of the product oriented corporations. Larger corporations are usually organized along functional lines. The organization chart presented in Fig. 1.1 shows many of the functional departments that provide personnel needed on a typical product development team. Each department has a functional manager. Also, several departments are often grouped to form a division, and of course, each division has a director. Titles for the division leaders vary from firm to firm—but Director or Vice President are commonly employed.

The functional organization with its leadership structure and its ongoing operations presents a management problem when forming product development teams. As a team member, do you report to the team leader (product manager) or to the functional manager? When important decisions must be made, are they made by the product manager, by one or more of the functional managers or by the division directors? Clearly, the establishment of a product development team within a company that is organized along functional lines creates several operational problems for management. Functional organizations may also create problems for working engineers attempting to serve several different managers. In the next section, these problems are addressed, and a popular team structure employed in many corporations that are organized along functional lines is described.

Fig. 1.1 Typical organization chart showing the disciplinary functions. The organization includes three divisions and three levels of management.

1.2 A BALANCED TEAM STRUCTURE IN A FUNCTIONAL ORGANIZATION

There are several organizational arrangements frequently used in forming teams within companies organized along functional lines. They include:

- Functional teaming
- Modified functional teaming
- Balanced team structure
- Independent team organization

Each of these techniques for forming product development teams in a functional organization has advantages and disadvantages. A complete discussion of these four different team structures is given in reference [2]. To introduce you to the concept of forming teams within a functional organization, the balanced team structure will be described.

The balanced team structure is represented in the organization chart presented in Fig. 1.2. The team is formed with members drawn from the functional departments; however, in most situations, the team members are located in close proximity in a separate room or a building wing reserved for the team. The team members are usually dedicated to a single project, and their responsibilities are coordinated and focused. The team members retain their reporting relationship to the functional manager, but the contact is less frequent than in functional teaming arrangements, and the relation is less intense. Members of a balanced team often consider the product manager to be more important than their functional manager. The product manager is frequently a senior technical administrator with more experience, status and rank than the typical functional manager. The status of the product manager is dependent on the product development costs and the size of the development team. The product manager reports to the general manager, and usually is equal in status to the division directors in the organization. The senior program manager, with significant status and clout, leads the team and insures that resources will be available as required to meet schedules. The team members recognize the freedom from functional constraints that the balanced team structure implies. They assume ownership of the product specifications and commit fully to the success of the development activities. Communication is accomplished through the use of division and department coordinators and daily meetings. The team makes most of the routine decisions and only the higher-level decisions require the approval of the appropriate division directors.

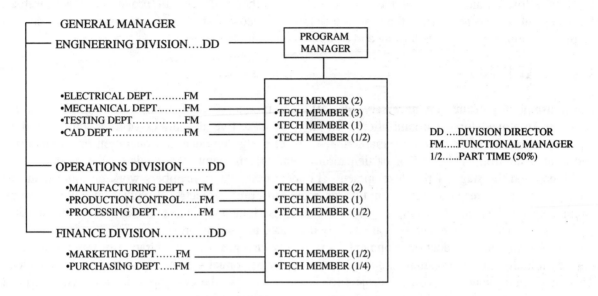

Fig. 1.2 Organization with a balanced team structure.

The primary disadvantage of the balanced team structure is the dedication of the team members to a single project. Many talented team members are needed by the functional managers to support design activities across several product lines. With some of the functional resources dedicated to a single product development, other products may not receive adequate attention. The balanced team structure enhances the development capability of a given product at the cost of reducing the technical capability that can be applied across the company's entire product line. A second disadvantage is that some team members from a given department or division may not have sufficient expertise to perform adequately. With loose and distant functional supervision and review, this fact may not be apparent. The new design may not be at the cutting edge of technology.

1.3   DEVELOPMENT PLANNING

As the development team becomes more independent and moves away from the functional departments, planning becomes more important. Team planning documents are often prepared by a small group of senior team members with talents in business, marketing, finance, engineering design and manufacturing. These planning documents incorporate detailed descriptions of the product, market analysis, business strategy for capturing market share, product performance specifications, development schedules and budgets, team staffing requirements, material selections, design strategies, technology prerequisites, manufacturing processes, tooling requirements, inspection procedures, test specifications, facility availability and distribution capabilities. An element of the plan covers techniques and personnel that are to be employed to insure quality and product reliability. Finally, the plan includes a section specifying the deliverables and other sections on methods to measure performance of the product and the productivity of the team.

The development planning documents are extremely important since they outline strategies for achieving the company's goals and objectives in developing a new product. Management may use some of these documents during periodic design reviews to monitor progress and to judge the team's performance. The plan is a well-defined contract between the team and the divisional managers involved in the product development. The team members are expected to sign-on; to commit to the level of effort required to meet the schedule, achieve the goals and to produce the deliverables. Management is expected to fund the program, to provide adequate staffing and to provide the necessary facilities, computers, software, tooling and equipment in a timely manner.

1.4 STAFFING

Cross-disciplinary teams are necessary in staffing a product development team regardless of the structure employed by the organization. Usually, at least five disciplines or functions should be represented on the team including: marketing, finance, design engineering, operations/manufacturing, and quality control. The level of participation of each team member depends on the product being developed and the stage of the development. In some cases, team members work on more than one product and have responsibilities on two or more different projects. In other instances, several engineers from the same discipline are needed to complete long and complex tasks to meet the schedule for a single product. These members are usually dedicated to a single project.

Staffing on a product development team usually changes over the duration of the development period. Initially the team is small and staffed with senior personnel with demonstrated skills and talent. This small cadre may stay with the project for its duration. As the development proceeds, additional design engineers are required as technical strategies are converted into design concepts, then to design proposals and finally into detailed engineering drawings. Manufacturing engineers are needed in larger numbers when the early prototypes are being produced. After release of the design by engineering, the number of designers is reduced, and the emphasis is shifted to production where staffing from operations (production control, quality control, purchasing and plant maintenance) increases sharply. Clearly, the staffing of the development team is dynamic. Usually only a small fraction of the total team is committed to the project on a full time basis from its initiation to its completion. In some companies, individuals are committed to the development team for the duration of the project, but at varying levels of participation. This approach enhances ownership of the project by the individual team members.

1.5 TEAM LEADERS

The experience and leadership skills required of the product manager are strongly dependent on the team structure employed as well as the size and cost of the product development. Some products can be developed with a relatively small commitment of staff (say less than 10 members). In these circumstances, the team established is often structured along functional lines, and the product manager is often one of the functional managers. The functional managers are experts in their respective disciplines, skilled in managing personnel and knowledgeable (but not expert) in the other disciplines. They are also familiar with the company's entire product line.

Other product developments require a larger commitment of personnel (say a team numbering 10 to 50), and the balanced team structure shown in Fig. 1.2 is probably the most suitable organization. The product manager is a seasoned staff member well-known for his or her expertise and respected for performing at a level that exceeds expectations. The appointment as a product manager (equal in rank and status to the functional department managers) often represents a promotion for the individual involved and a first experience in managing a technical group. The functional managers and the divisional directors provide close support through periodic reviews that monitor progress and reveal problems early in the development process.

As the size of development team becomes even larger (100 or more members), team leadership assumes major importance. The product manager in this situation is a senior member of the company's executive management with a well-established record of accomplishments. He or she is senior to the managers of the functional departments, and is at least equal in status to the division directors. The authority and responsibility given the product manager is significant, and his or her career depends strongly on the success of the product in the market and the productivity of the development team.

1.6 TEAM MEMBER RESPONSIBILITIES

Members of a product development team have two different sets of responsibilities—one to his or her disciplinary function and the other to the team. A team member, from say the electrical engineering department, provides the technical expertise in his or her specialty and ensures that the product is correctly designed with leading edge, state-of-the-art technology. The team member acts to bring all of the important functional issues that affect product performance to the attention of the team. The team member also represents the functional department to ensure that the product development team maintains the principles, goals and objectives of the functional discipline.

The most important team activity for the individual member is to share responsibility. Highly effective teams that routinely succeed have team members that hold themselves mutually accountable for the overall success of the product. Next, the team member must recognize and understand all of the product features, and fully participate in the techniques employed to meet the design objectives. Individual team members must assess team progress and participate in improving team performance. A team member must cooperate in establishing all of the reporting relationships required to maintain the communication among the team and functional departments. In many of the team structures, an individual may report to three different managers—the Division Director, the Department Manager and the Program Manager. The matrix organization structure (Fig. 1.2), inherent in forming development teams, creates complex reporting relations. Flexibility and cooperation on the part of the individual team members are required for the team structure to be effective.

Finally, the team members must be able to communicate in a clear concise manner in all three modes: **writing, speaking and graphics**.

1.7 TEAM MEMBER TRAITS

When a group of individuals work together on a team to achieve a common goal, they can be extremely effective [3]. The team interaction promotes productivity for several different reasons. First, meeting together is synergistic in that one's ideas freely expressed stimulate additional ideas by other team members. The net result is many more original ideas than would have been possible by the same group of individuals working independently. Another advantage is in the breadth of knowledge available in the team. The team is cross-disciplinary and the very wide range of skills necessary for the product development process is included within the team. Each member of the team is different with some combination of strengths and weaknesses. Acting together the team can build on the strengths of each member and compensate for any weaknesses. The grouping of individuals provides social benefits that are also very important to the individuals and to the corporation. There is a bonding that occurs over time and a support system develops that team members appreciate. The team develops a sense of autonomy valued by each of its members. The team develops solutions, implements them and then monitors the results with little or no direction from management. Empowered teams assume ownership of the project. They commit to the product development process and consistently perform beyond expectations.

While cross disciplinary teams have been employed in many different companies for at least two decades, little has been done by the educational system to develop team skills. On the contrary, both the secondary school and many college systems tend to encourage competitive attitudes and independence. Students compete for the "A" grades that might be given in a course and are discouraged from cooperating on assignments. This system prepares students to work independently, but team-building skills are not addressed.

The industrial workplace is much different. Team members compete, but not within the team. They compete with a similar development team from a rival corporation. Team members bond, cooperate and consistently help each other by sharing assignments. Competition between team members is discouraged. Recognition and rewards go to the team as a whole much more often than to select individuals.

There is a set of characteristics that describe a good team member, and another set that depicts an individual who can destroy the efforts of a team.

The characteristics of a good team member are:

1. Treat every team member with respect, trust their judgment and value their friendship.
2. Maintain an inquiring attitude free of predetermined bias so that team members will all participate and share knowledge and opinions in an open and creative manner. Listen carefully to the other team members.
3. Pose questions to those hesitant team members to encourage them to share their knowledge and experience more fully. Share your experiences and opinions in a casual, easy-going manner. Help other members to relax and enjoy the interactive process involved in team cooperation. Participate but do not dominate.
4. Observe the body language of the other team members because it may indicate lack of interest, defensive-attitudes, hostility, etc. Act with the team leader to defuse hostility and to stimulate interest. Disagree if it is important, but with good reason and in good taste.
5. Emotional responses will occur when issues are elevated to a personal level. It is important to accept an emotional response even when you are opposed. It is also important to control your emotions and to think and speak objectively. You should be self confident in your discussions, but not dogmatic.
6. Allocate time for self-assessment so the team can determine if it is performing up to expectations. Make suggestions during these assessment periods to build team skills.
7. Be comfortable with your disciplinary skills but continue to study to improve your capabilities. Communicate effectively by speaking clearly, writing clearly and concisely, and illustrating effectively with modern graphics.

The characteristics of a destructive team member are:

1. No member of the team should ever participate in a conversation that is derogatory about a person on the team or in the corporation. If you cannot make complimentary remarks about a person, keep the negative thoughts to yourself. Respect, trust and friendship are vital elements in the founding of a successful product development team. Derogatory remarks destroy the foundation necessary for respect, trust and friendship.
2. Arguments among team members should be avoided. You are encouraged to introduce a different opinion and participate in a discussion with a different viewpoint, but the discussion should never degenerate into an argument. Every member on the team is responsible for quickly resolving arguments that arise among the team members.
3. All members of the team are responsible for on-time attendance at their meeting. If a member is absent, the leader should know the reason beforehand and explain it to the team. You must be certain that all members appreciate and respect the reason for everyone's role on the team and the importance of every member attending every meeting.

4. Team progress is hindered when one member dominates the meeting. These people are often overly critical, intimidating and stimulate confrontations. If you have these characteristics, work hard to suppress them.

5. The team leader must be extremely careful about intervention. If the team is moving and working effectively, the leader should be quiet because intervention in this instance is counter productive. When the team is having difficulty, intervention may be necessary depending on the problem. In the event a team member becomes hostile, intervention should be quick, and the disagreement producing the hostility should be dealt with immediately. When a team is seeking consensus, the process may require time and the leader should wait 5 to 10 minutes before intervening.

6. No member of the team can be opinionated including the leader. He or she is not a judge determining the correct solution. The leader seeks to facilitate so that the team can reach a consensus on the correct solution. It is only when the team accepts the solution that implementation can begin.

Compare your traits with those listed above. You will probably exhibit both good and bad characteristics. To be effective team members, you must work to enhance the favorable traits and to suppress those that are destructive to the efforts of the team.

1.8 EVOLUTION OF A DEVELOPMENT TEAM

If a number of workers or students are assigned to a new team, it is possible to observe several behavioral phases as the team bonds, melds and matures [4]. In the very beginning (Phase I), the team members usually exhibit both excitement and concern. They are excited about working on a high-profile project with a new group of talented people. It should be an opportunity to learn, make new friends, advance in position and/or stature and have fun in the process. At the same time, the team members exhibit signs of concern. They are worried about meeting and understanding other team members. The tasks assigned to the team are probably extremely vague at this stage of the development, and vagueness leads to uncertainty. The skill levels and areas of expertise of fellow team members have yet to be determined. The personality of the team leader is unclear. There is worry over the new reporting relations that new team membership implies. In this orientation phase, team members are searching for their role in the team and evaluating their possibilities for success or failure. It is a tense time.

Dissatisfaction is the next phase in the evolution of the team (Phase II). You have met your fellow team members, and the news is not good. You recognize the differences in personalities and in work habits. A few of the team members do not understand how to tell time and never arrive on time when attending team meetings. Member schedules make it very difficult for the entire team to meet. Some team members lack social graces. Some members are inexperienced and unable to cope with their assignments. You have met the team leader and he or she is very difficult (stressed, demanding, exacting, impatient and abrupt). The schedule calls for the team to leap over mountains on a daily basis. The budget for the project is totally inadequate. Surely you are in a lose—lose situation. Your thoughts are dominated by schemes to transfer to a better team. During this phase, the progress of the team toward scheduled milestones is exceedingly slow.

With time progress is made, and the team members start learning how to work together. This is the resolution phase (Phase III). Everyone agrees to attend meetings and arrive on time. Personality conflicts are smoothed over. Team members may never be true friends, but everyone manages to get along and show mutual respect. Team members have come to terms with the schedule, and have committed to the extra effort that it requires. The manager has mellowed, or team members have learned how to handle his or her impatient demands. Solutions for reducing development costs have been found. There appears to be a good possibility of meeting the product development plan. The team is beginning to perform well, and more members are committed.

The team has melded into an efficient unit. Conflicts rarely arise. Lasting friendships begin to be formed. The strengths of each team member are fully utilized. The manager recognizes the talents of the team. Tasks are completed ahead of schedule and under budget. Upper level managers recognize the accomplishments of the team. You now find yourself in a win—win situation. This is the production phase of the development team (Phase IV).

The project has been completed and the ribbons have been cut. The team prepares to disband. In this termination phase (Phase V), you reflect on your experiences, good and bad, over the duration of the project. You examine your individual performance and evaluate various factors that improved the effectiveness of the team. You meet with the manager for a performance evaluation. The team members share a sense of accomplishment in achieving the goals and objectives of the development.

1.9   A TEAM CONTRACT

Teams are effective because they meet together to focus their wide range of disciplinary skills and natural talents toward the solution of a set of well-formulated problems. The team meeting is the format for synergistic efforts of the team. Unfortunately, not all teams are effective because some members are so disruptive they destroy the cohesiveness of the team. Bonding of team members is vital if the team is to successfully solve the multitude of problems that arise in the product development process.

Teams **fail** because of three main reasons. First, they deviate from the goals and objectives of the development, and cannot meet the milestones on the development schedule. Second, the team members become alienated and the bonding, trust and understanding critical to the success of the team never develops. Third, when things begin to go wrong and adversity occurs, "finger pointing" begins and an attempt to fix blame on someone replaces creative actions.

Clearly, many problems will arise to impair the progress of a development team. It is recommended that each team establish a set of simple rules it considers necessary to govern the team. Once established, each team member should sign and date this contract. Suggestions for your team to consider are provided below:

- The information discussed by team members will remain confidential
- Members will acknowledge problems and deal with them
- Members will be supportive rather than judgmental
- Members will respect differences
- Members will provide responses directly and openly in a timely fashion
- Members will provide information that is specific and focused on the task and not on personalities
- Members will be open, but will respect the right of privacy
- Members will not discount the ideas of others
- Members are each responsible for the success of the team experience
- The team will have the resources needed to solve the problems that arise
- Every member of the team will contribute to its success
- Members recognize the importance of team meetings to the success of the group, and accordingly agree to:

    1. Always attend the scheduled team meetings
    2. When attendance is impossible, notify the team leader and as many other members as possible in advance
    3. Use meeting time wisely

4. Start on time
5. Limit breaks and return from them on time
6. Keep focused on the goals
7. Avoid side issues, personality conflicts and hidden agenda
8. Share responsibility for briefing members when they miss a meeting
9. Avoid making phone calls or engaging in side conversations that interrupt the team

1.10 EFFECTIVE TEAM MEETINGS

Leadership is important to ensure that the team remains focused on the overall project goals/objectives and on the agenda at each meeting. Creating the correct team environment or atmosphere is essential so that the team members cooperate, share ideas and support each other to achieve solutions [5]. It is important for the team leader to assess the performance of the entire team at each meeting. When the team fails to make progress, it is critical for the team leader or possibly an individual member to make changes as necessary to enhance the team's effectiveness. Some typical difficulties encountered during team meetings are identified in the following paragraphs, along with suggested corrective actions team members can take to enhance the team's productivity.

One common problem impairing team productivity occurs when the goals and objectives of the meeting are not clear or they appear to be changing. While in many cases team meetings are planned well in advance and are supposed to follow an agenda, new topics are introduced and the meeting drifts from one item to another. This team behavior indicates that either some of the team members do not understand the goals, or they do not accept them. Instead, they are marching to the beat of their own drum and trying to take the team with them. Unless corrective actions are taken to focus the team on the original goals, this team meeting will fail and time and effort will be lost. A focused team stays on the agenda, and its team members accept the meeting objectives. Team discussions should generate a wide set of solutions to the problem being considered, and should continue until team members reach a consensus solution. A meeting is successful when an issue has been resolved and all of the team members accept the solution. This allows the team to move forward by organizing its talents to address the next problem.

Leadership is an essential element for team effectiveness. The most effective teams are usually democratic with a large amount of shared leadership. While there is a recognized leader, with a considerable degree of responsibility and authority, the team leader is supported by each team member. The team member with the appropriate expertise (relative to the problem being considered) will usually lead the discussion and effectively act as leader of the team during this period. However, in some instances, the team leader maintains strict control of the meeting and never shares the leadership role with any team member. Some team members resent this style of leadership and may not participate as fully as possible. The result is that complete utilization of the resources of the team does not occur.

Attitude of the team members is an important element in the success of the team. Are the team members committed? This commitment is evidenced in their attitude. If they come to the meeting table exhibiting interest and willingness to participate, they are committed and will make a positive contribution. However, if they are bored and indifferent, they will not become engaged in a meaningful way. In fact, if they are sufficiently disinterested, they may initiate side conversations or arguments and destroy the atmosphere of trust necessary for an effective meeting. It is important to evaluate each team member, and to secure their commitment to the goals and objectives of the project.

As the meeting progresses, it is important to assess the discussions that are occurring. For a meeting to be successful, the discussion should include everyone on the team. A discussion should stay focused on an agenda item with few if any deviations to unrelated topics. The team members must listen carefully to one another and give all of the ideas presented a serious hearing. Teams tend to accomplish less when the discussion is dominated by only a few of the participants. These members tend to intimidate others in their efforts to control and dominate the team. The result is disastrous because they essentially eliminate the contributions of other team members.

Teams do not operate with total agreement on every issue. There must be accommodation for disagreements and for criticism. All of the ideas or suggestions made by every team member at any meeting will not be outstanding. Indeed, some of them may be ridiculous; criticism of these ideas must occur. However, the criticism should be frank and without hostility. Personal attacks must not be a part of the critique of an idea. If the criticism is to improve an idea or to eliminate a false concept, then it is of benefit to the progress of the team. The criticism should be phrased so that the team member advancing the flawed idea is not embarrassed. When disagreements occur, they should not be suppressed. Suppressed disagreements breed hostility and distrust. It is much better for the team to deal with disagreements when they arise. The root causes of a disagreement should be ascertained, and the team should take the actions necessary to resolve them. On some occasions voting is a mechanism used to resolve conflict. This procedure must be used with care particularly if the vote indicates the team is split almost equally in their opinion. A better practice is to discuss the issues involved and attempt to reach a consensus. Reaching a team consensus may require more time than a simple vote, but the results are worth the effort. Consensus implies that all members of the team are in general agreement and willing to accept the decision of the team. When the team votes, a simple majority is sufficient to resolve an issue. However, the minority members may become resentful if they are always on the short end of the vote. In a short time, they will not accept the outcome, and they will not commit to the actions necessary to implement the team's decision.

The team meetings must be open. The agenda should be available in advance of the meetings and subject to change with added topics introduced as "new business". A sample agenda is shown in Table 1.1. Team members should believe that they have the authority to bring new topics before the team. They should feel free to discuss procedures used in the team's operation. Secret meetings of an inside group should be carefully avoided. When the fact leaks that secret meetings are being held and that issues are prejudged by a select few, the effectiveness of the team is destroyed.

At least once during a meeting it is useful to evaluate the progress of the team. Is the team following the agenda? Is someone dominating the meeting? Is the discussion to the point and free of hostility? Is everyone properly prepared to address his or her agenda items? Are the team members attentive, or are they bored and indifferent? Is the team leader leading or is he or she pushing? If a problem in the operation of the team is identified during the pause for self-appraisal, it should be resolved immediately through open discussion.

Finally, the team must act on the issues that are resolved and the problems that are identified. To discuss an issue and to reach a consensus is part of the process, but not closure. Implementation is required for closure, and implementation requires action. When decisions are made, team members are assigned **action items**. These action items are tasks to be performed by the responsible individual; only when the tasks are completed is an issue considered closed. It is important that the action items are clearly defined, and the role of each team member in completing their respective tasks is understood. A realistic date should be set for the completion of each action item. The individual responsible should be clearly identified; he or she must accept the assignment without objection or qualification. A checking system must be employed to follow up on each action item to insure timely completion. If there is a delay, the schedule for the entire product development cycle may be at risk. It is important to deal with delays immediately and for the team to participate in the development of plans to eliminate the cause of the delay. A form has been provided in Appendix A to facilitate the assignment of action items and the follow-up on each item.

Table 1.1:
Sample Agenda for a 1-hour weekly meeting
Weekly meeting of the Mighty Terrapin Team
'September 21, 2007 7:59 am

Weekly status report	Team Leader
Review of outstanding action items	
Action item #14	Member responsible
Action item #15	Member responsible
Action item #N	Member responsible
Report on progress	
Subsystem – Power	Member responsible
Subsystem – Electrical	Member responsible
Subsystem – Mechanical	Member responsible
Subsystem – Control	Member responsible
Market study results	Marketing Representative
Identify new problems	All member participate
Assignment of action items	Team Leader and volunteers
New business	All member participate
Summary	Team Leader
Adjourn at 8:59 am	

1.11 PREPARING FOR MEETINGS

Effective team meetings do not just happen. It is necessary to prepare for the meeting and to execute post-meeting activities to insure success. The preparation usually involves selecting, arranging and equipping the meeting room, scheduling the meeting so that the necessary personnel are in attendance, preparing and distributing an agenda in advance of the meeting and conducting the meeting following rules established by the team. Equally important are actions taken by the team members following the meeting to implement the decisions that have been reached.

The meeting room and its equipment also affect the team's progress. The room should be sized to accommodate the team and any visitors that have been invited. Rooms that are too large permit the members to scatter and the distance between some members becomes too long for effective and easy communication. The seating arrangement should be around a table so that everyone can observe each other's face and body language. It is very difficult to engage anyone in a meaningful conversation if they have their back to you. Water, coffee, tea or soft drinks should be available if the meeting duration exceeds an hour. Smoking is strictly prohibited. The equipment that will be needed for presentations should be available, and its operation should be checked prior to the start of the meeting. A computer with projection capability is of growing importance in a well-equipped meeting room. The availability of a computer during the meeting permits one to draw from a large database and to modify the presentation in real time. Additionally, it allows results of analyses or experiments to be clearly displayed in graphical form for the entire team to review.

Prior to the scheduled meeting, it is important to make careful preparations. Minutes from the previous meeting should be distributed with sufficient time for review before the next scheduled meeting. A detailed agenda, similar to the one shown in Table 1.1, is distributed to inform team members of the topics and/or problems that will be addressed in the next meeting. Individual team members responsible for specific agenda items are identified. The details are clear to all and responsibility shared by individual members is defined. In some instances, information will be needed from corporate employees, suppliers or visitors that are not members of the team. In these cases, arrangements must be made to invite these people to the meeting so that they can provide the necessary expertise and respond to questions from all of the disciplines represented on the team.

In conducting the meeting, it is important to start on time. It is very annoying to the majority of the members to wait five or ten minutes for a straggler or two. The team leader must make certain that everyone involved understands that 8:00 am means 8:00 am and not 8:08 or 8:12 am. The objectives of the meeting should be clearly understood and the time scheduled for each agenda item should be estimated. A team member should be assigned as the timekeeper, and another should act as a secretary to record notes and to prepare the minutes of the meeting. If team meetings are frequent and held over a long time (several months), the duties of the timekeeper and the secretary should be shared with others on the team. The meeting should follow a set of rules that govern the behavior of individual members.

As the meeting draws to a conclusion, the team leader should take a few minutes to summarize the outcome of the discussions. This summary gives an ideal opportunity to insure that assignments are understood, responsibility accepted and completion dates established. The meeting must be completed on time and everyone's schedule should be respected. If the meeting is not periodic (e.g. every Tuesday at 8:00 am), then it is important that the time and place for the next meeting be scheduled.

After the meeting, it is productive for the team leader to make certain that the complete minutes have been prepared and distributed. If any member was not able to attend the meeting, the team leader should brief that person on the team's progress. Finally, the team leader should follow up on each action item to ensure that progress is being made, and that new or unanticipated problems have not developed. It is clear that effective meetings do not happen by accident. Many members of the team work diligently before, during and after the meeting to make certain that the goals and objectives are clearly defined, that the issues and problems are thoroughly discussed and that the solutions developed result in assignments that are executed with dispatch.

1.12 SUMMARY

Arguments for the importance of development teams in industry have been presented. Experience has shown that effective development teams are vital if a corporation is to develop highly successful products and introduce them in time to win a major share of the market.

Functional organizations that exist within most corporations have been described. Reasons why a functional organizational structure often interferes with rapid low-cost product development have been given. The balanced team structure commonly employed in establishing development teams within large functionally organized corporations has been described.

The importance of location of team members in either enhancing or inhibiting communication has not been considered. However, you should be aware of the important role distance plays in effective and timely communication. The rule is simple. To enhance communication, minimize the distance between those needing to communicate with each other. The ideal situation is to place the entire development team in the same room.

Development planning and staffing of the development team has been covered. The success of the team depends strongly on generating a comprehensive plan. The plan is used initially to justify the development budget and its schedule. During the course of the development, the plan provides guidance for monitoring the progress and for judging both the team and product performances. Team staffing is cross-disciplinary with adequate representation from both the engineering and finance divisions. Engineers interact with the business personnel from marketing, sales and purchasing. Responsibility is shared between disciplines and functions.

The team's organizational structure and the team leader's management status have been discussed. This relationship is important because of executive authority that corresponds with a higher level of management. Higher-level managers carry more authority, many decisions can be made more rapidly, communication is more effective and team progress is often enhanced if the leader has status and clout associated with executive management.

Next, team member responsibilities and team member traits were discussed in considerable detail. Team members work in a matrix organization and often report to two or more managers. This arrangement is sometimes difficult particularly when you get mixed signals from different managers. It is important to be flexible and remain cooperative with the program manager and the functional managers. Team member traits are very important to your career. Evaluate your traits in an honest self-assessment. This is a critical first step in building team skills.

Most design teams evolve over the duration of a development project. Five phases usually experienced by the team in this evolution have been described. Not all of the phases provide pleasant experiences. Early in the evolution, a team encounters the dissatisfaction phase with very slow progress, low productivity and member gloom. To minimize the time in this phase, it is suggested that the team prepare and execute a contract that provides guidelines for member behavior.

Finally, guides for conducting successful team meetings have been provided. You will spend more time than you can imagine in team meetings. From the outset, learn the techniques for a successful meeting. Reasons for the failure of some meetings and the success of others have been presented. The role of the team leader and the behavior of the team members are described. The meeting room is more important than most engineers imagine. Features and furniture arrangements in a room affect the ability of a team to function effectively. Recommendations are made for meeting room size and for arrangements to enhance team productivity during meetings. The progress of the project is dependent on the team's ability to control its meetings.

REFERENCES

1. Smith, P. G., D. G. Reinertsen, Developing Products in Half the Time, Van Nostrand Reinhold, New York, NY, 1991.
2. Schmidt, L. C. and Zhang, G., Dieter, G., Cunniff, P. F., and Herrmann, J. W., Product Engineering and Manufacturing, 2nd Edition, College House Enterprises, Knoxville, TN, 2001.
3. Barczak, G. and Wilemon, D., "Leadership Differences in New Product Development Teams," Journal of Product Innovation Management, Vol. 6, 1989, pp. 259-267.
4. Lacoursiere, R. B. The Life Cycle of Groups: Group Development State Theory, Human Service Press, New York, NY, 1980.
5. Barra, R. Putting Quality Circles to Work, McGraw Hill, New York, NY, 1983.
6. Foxworth, V., "How to Work Effectively in Teams," Class Notes ENME 371, University of Maryland, College Park, September 5, 1997.
7. Clark, K. and Fujimoto, T., Product Development Performance, Harvard Business School Press, Boston, MA, 1991.
8. Wheelwright, S. C. and Clark, K. B., Revolutionizing Product Development, Free Press, New York, NY, 1992.

EXERCISES

1.1   Write an engineering brief describing why team structure in a large corporation is so important in the product development process. Add a second paragraph indicating why team structure is much less important in a very small company.

1.2   Prepare an organization chart, like the one shown in Fig. 1.1, for the University or College that you are attending. Describe the logic, as you see it, for this organizational structure.

1.3   List the advantages and disadvantages of the balanced team structure.

1.4   Outline a development plan that you would prepare if you were a senior team member representing an engineering function (discipline) on a newly formed development team. The product to be developed is a new portable digital music player to compete with iPod.

1.5   Write an engineering brief describing staffing required for the development team for the digital music player. Give the reasons for changing the staff as the digital music player evolves during the development process.

1.6   Describe the experience and status of the product manager for the balanced team structure.

1.7   Describe a matrix organization structure. Explain why this organization impacts an engineer assigned to a product development team with a balanced structure. If you were this engineer, how would you respond to inquires from your functional manager?

1.8   You (a male/female figure) are attending weekly team meetings and are always seated next to Sally/Bill who is young and very attractive. Sally/Bill is apparently interested in you because she/he frequently involves you in side conversations that are not related to the ongoing team discussions. Write a plan describing all of the actions you will take to handle this situation.

1.9   Brad, the team leader, is a great guy who believes in leading his team in a very democratic manner. He encourages open discussion of the issue under consideration for a defined time period, usually 5 or 10 minutes. At the end of the time period, he intervenes and calls for a team vote to resolve the issue. Write a critique of this style of leadership.

1.10   You together with Zack and Mary are senior members on a development team with 14 other members. The three of you are really very knowledgeable and experienced. You get along famously, and have developed a habit of gathering together at the local watering hole the evening before the scheduled team meeting. During the evening you discuss the issues on the agenda and arrive at some design decisions. Write a brief describing the consequences of continuing this behavior.

1.11   Sue has a wonderful personality, and as a team leader is skillful in promoting free and open discussion by all of the team members. After the team reaches a consensus, she calls for volunteers to implement the required action. When two or more people volunteer, she assigns them as a group to handle the action item. When no one volunteers, she assigns the least active person on the team to the action item and moves promptly to the next item on the agenda. Write a critique of this leadership style.

1.12   You are the leader of an eight-member team meeting for the first time to develop a new hair dryer. During this initial meeting the following events occur.

- The temperature in the meeting room is 88 °F.
- The room is set up for a seminar speaker.
- The meeting is scheduled to begin at 9:30 am and at 9:35 am only five of the eight members have arrived.
- During the meeting Horrible Harry begins to verbally abuse Shy Sue.
- At 9:50 am Talkative Tom begins to describe his detailed positions on the world situation and is still going at 10:10 am.
- The team breaks for coffee at 10:15 am and has not returned at 10:30 am.

- Sleepy Steve begins to snore.
- Procrastinating Peter refuses to accept an action item after leading the discussion on the issue for 15 minutes.

Describe the approach that you, acting as the team leader, would follow in handling each of these situations.

1.13 Prepare a list of ground rules that should be followed by all the members of a team in conducting an effective meeting.

1.14 You are a member of a team with an ineffective leader. What can you do to improve the productivity of the team? In responding to this question consider the irresponsible actions on the part of certain team members that are listed in Exercise 1.16. Remember in your response that you are not the leader and the most that you can do is to share the leadership role from time to time.

1.15 Meet with your team and discuss a list of rules that will be used in guiding the team's conduct during the semester. Incorporate the team's list of rules in the form of a contract. As a final step, each member of the team is to sign the contract indicating his or her commitment to the rules of conduct.

Notes:

CHAPTER 2

PRODUCT DEVELOPMENT AND HOVERCRAFT MISSIONS

2.1 THE IMPORTANCE OF PRODUCT DEVELOPMENT

Hundreds of products are used to perform many functions every day. Products surround us. This morning I prepared breakfast by toasting bread, frying eggs and making tea. How many products did I use in performing this simple task? The products used included a toaster (Phillips), a frying pan (Calphalon), a microwave (General Electric) to heat the tea and a gas cook top (Kitchen Aid) to heat the frying pan. These are all products that are designed, manufactured and sold to customers both in America and abroad. Some products are relatively simple, such as the frying pan with only a few parts, yet some are much more complex, such the microwave with over a hundred parts.

Corporations worldwide continuously develop their product lines with minor improvements introduced every year or two and with more major improvements every four or five years. Product development involves many engineering disciplines and is a major responsibility of engineers entering the work place. An examination of Table 2.1 shows that designing, manufacturing, and product sales represent more than 80% of the responsibilities of mechanical engineers in their initial position in industry. This distribution of activities is also typical of many of the other engineering disciplines.

Table 2.1
Responsibilities of Mechanical Engineers in their first position [1]

ASSIGNMENT	% of TIME	
Design Engineering	40	
Product Design		24
Systems Design		9
Equipment Design		7
Plant Engineering / Operations / Maintenance	13	
Quality Control / Reliability / Standards	12	
Production Engineering	12	
Sales Engineering	5	
Management	4	
Engineering		3
Corporate		1
Computer Applications / Systems Analysis	4	
Basic Research and Development	3	
Other Activities	7	

The financial success of many corporations depends on the introduction of a steady stream of winning products to the marketplace. A successful product must meet the customer needs and provide robust and reliable service at a competitive price.

2.1.1 Development Teams

Interdisciplinary teams develop products with members drawn from engineering, marketing, manufacturing, production, purchasing, etc. A typical organization chart for a team developing a relatively simple electro-mechanical product is shown in Fig. 2.1. The team leader coordinates and directs the activities of the designers—mechanical, electronic, industrial, etc. A marketing specialist supports the team to ensure that the evolving product meets the needs of the customer. Financial, sales and legal assistance is usually provided on an as-needed basis by corporate staff or outside contractors.

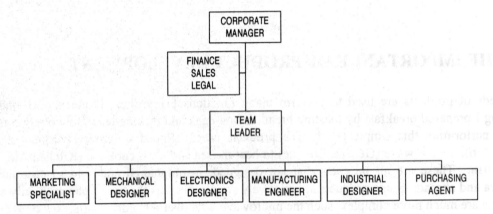

Fig. 2.1 Organization of a product development team for a relatively simple product.

The team leader is also supported by a purchasing agent who establishes close relations with the suppliers who provide materials and components used in the final assembly of the product. In the past decade, the relationship between suppliers and end-product producers has changed significantly. Suppliers become part of the development team; they often provide a significant level of engineering support that is external to the internal (core) development team. In some developments, these external (supplier funded) development teams—when totaled—are larger than the internal development team. For example, the internal development team for the Boeing 777 airliner involved 6,800 employees and the external teams totaled about 10,000 persons from many suppliers [2].

2.2 DEVELOPING WINNING PRODUCTS

There are two primary aspects involved in developing successful products that win in a competitive marketplace. The first is a combination of quality, performance and price of the product. The second involves cost and time. The time and the cost of a product development, and the manufacturing costs incurred during the production cycle are vitally important.

2.2.1 Quality, Performance and Price

Let's discuss the product first. Is it attractive and easy to use? Is it durable and reliable? Is it effective and efficient? Does it meet the needs of the customer? Is it better than the products now available in the marketplace? If the answer to all of these questions is an unequivocal **yes**, then the customer **may** want to buy the product if the price is right. Next, you need to understand what is implied by product cost and its relation to the price actually paid for the product. Cost and price are distinctly different quantities. Product cost clearly includes the cost of materials, components, manufacturing and assembly. The accountants also include less obvious costs such as the prorated costs of capital equipment (the plant and its machinery), the price of the tooling, the development cost, and even the expense of maintaining the inventory in establishing the **total cost** of producing the product.

Price is the amount of money that a customer pays to buy the product. The **difference** between the **price** and the **total product cost** is **profit**, which is usually expressed on a per unit basis.

$$\textbf{Profit = Product Price – Total Product Cost} \qquad \textbf{(2.1)}$$

This equation is the most important relation in engineering and in business. If a corporation cannot make a profit, it soon is forced into bankruptcy, its employees lose their positions and the stockholders lose their investment. It is this profit that everyone employed by a corporation seeks to maximize while maintaining the strength and vitality of the product lines. The same statement can be made for a business that provides services instead of products. If a business is to make a profit and prosper, the price paid by the customer for a specified service must be more than the cost to provide that service.

2.2.2 The Role of Time in the Development Process

Let's now discuss the role of the development process in producing a line of winning products. Developing a product involves many people with technical expertise in different disciplines. Also, a product development takes time and requires significant sums of money. Let's first consider development time. Development time, as it is used in this context, is time to market—the time from the product development kickoff to the introduction of the product to the market. This is a very important target for a development team because of the many significant benefits that follow from being first to market with a new product. A corporation realizes many competitive advantages with a rapid development capability. First, the product's market life is extended. For each month cut from the development schedule, a month is added to the life of the product in the marketplace with an additional month of sales revenue and profit. The benefits of being first to market on sales revenue are shown in Fig. 2.2. The shaded area between the two curves to the left side of the graph is the enhanced revenue due to the longer sales life.

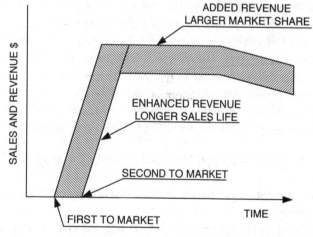

Fig. 2.2 Increased sales revenue due to extended market life and larger market share.

A second benefit of early product release is increased market share. The first product to market has 100% of market share in the absence of a competing product. For products with periodic development of "new models," it is generally recognized that the earlier a product is introduced to compete with older models—without sacrificing quality and reliability—the better chance it has for acquiring and retaining a large share of the market. The effect of gaining a larger market share on sales revenue is also illustrated in Fig. 2.2. The shaded area between the two curves at the top of the graph shows the enhanced sales revenue due to increased market share.

A third advantage of a short development cycle is higher profit margins. If a new product is introduced prior to availability of competitive products, the manufacturer is able to command a higher price for the product, which enhances the profit. With time, competitive products will be introduced forcing price reductions. However, in many instances, relatively large profit margins can still be maintained because the company that is first to market has additional time to reduce manufacturing costs. Their employees learn better methods for producing components and reduce the time needed to assemble the product. The advantage of being first to market, with a product where a learning curve for manufacturing exists, is shown graphically in Fig. 2.3. The learning curve reflects the reduced cost of manufacturing and assembling a product with time and experience in production. These cost reductions are due to many innovations introduced by the workers after mass production begins. With time and manufacturing experience, it is possible to drive down production costs.

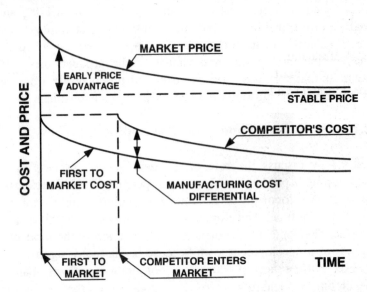

Fig. 2.3 The development team bringing a product to market first enjoys a higher initial price and cost advantages from manufacturing efficiencies.

Let's next consider development costs that represent a very important investment for the companies involved. Development costs include the salaries of the members of the development team, money paid to subcontractors, costs of pre-production tooling, expense of supplies and materials, etc. These development costs are significant, and most companies must limit the number of developments in which they invest. The size of the investment may be appreciated by considering the development cost of a new automobile is estimated at $1 billion, with an additional expenditure of $500 to $700 million for the new tooling required for high-volume production [2].

This discussion on time and cost of product development has been included to help you begin to appreciate some of the business aspects of developing winning products. Companies involved in the sale of products depend completely on their ability to continuously introduce winners to the marketplace in a timely manner. To win market share, the development team must bring a quality product to market that meets the needs of the customer. The development costs must be minimized while maintaining a development schedule that permits an early (preferably first-to-market) introduction of the product.

2.3 LEARNING ABOUT PRODUCT DEVELOPMENT

The best way to learn about product design is to work on a development team and build a prototype of some product. For this reason, your instructor is encouraged to form student development teams that will engage in the design, construction and testing of a prototype during the semester. A prototype is the first working model of a product. In some instances, companies develop several prototypes before finalizing a particular product design. However, time available during this semester will limit each team to a single prototype of the product selected by the class instructor.

The development of a prototype of a new product is an intimidating task, particularly when the assignment is given so early in the engineering curriculum. However, experience with several thousand students beginning their studies of engineering indicates that students gain significantly from a multidisciplinary product development experience early in their education. In this class you will learn to work effectively within a team structure, to make difficult design decisions, and to design and build successful prototypes. Extensive surveys from students completing this course clearly indicate that they spent many hours each week on the project, had fun in the process, learned a lot about engineering and appreciated the lessons learned in working as a member of a student development team.

2.4 HOVERCRAFT

The product described in this textbook is a hovercraft, which is a vehicle that moves on a cushion of air. As a hovercraft is without wheels, it is propelled by thrust usually generated by auxiliary fans. A photograph of a large commercial hovercraft is presented in Fig. 2.4.

Fig. 2.4 A large commercial hovercraft used to transport people at high speed over water.

Let's examine this hovercraft to identify some of its design features. First note the black rubber tube (skirt) that encircles the hovercraft, shown in Fig. 2.4. This tube serves two purposes—it provides buoyancy to support the hovercraft on water and it provides the walls of a plenum chamber that contains a cushion of air which supports the hovercraft while it is in motion. The two large axial fans located aft provide propulsion and steerage. If both fans are operating at the same speed, the thrust developed propels the hovercraft forward. Operating only one fan causes the hovercraft to turn either to port or starboard. The deck and cabin form the top of the plenum chamber. The attachment of the skirt to the deck and cabin must have an airtight seal because the air cushion in the plenum chamber is pressurized. The centrifugal blower that supplies pressurized air to the plenum chamber is hidden in Fig. 2.4.

The design of a hovercraft that operates on water differs from one that operates only on land. For operation on water, the skirt is made from a hollow tube (usually fabricated from reinforced rubber). The volume of this tube is sized to provide a buoyancy force sufficiently large to keep the hovercraft from sinking when the air pressure in the plenum chamber goes to zero, as shown by the top illustration in Fig. 2.5. However, when the hovercraft is moving under normal operations, the plenum pressure is increased to the point where the skirt is lifted until the hovercraft is skimming over the surface of the water, as shown in the bottom illustration of Fig. 2.5. The advantage of operating with plenum pressure is that the drag force is significantly reduced, thus allowing higher speeds to be reached.

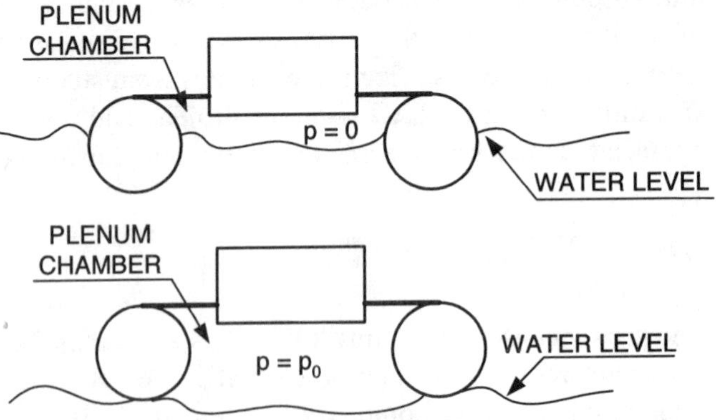

Fig. 2.5 Effect of plenum pressure on a hovercraft operating in water.

When the plenum chamber is pressurized, the rubber tube does not provide buoyancy. The lift required to keep the hovercraft from sinking is provided by the upward force due to the plenum chamber pressure. Some of the pressurized air may escape around the bottom of the skirt; however, a centrifugal blower supplies a sufficient quantity of air to maintain adequate plenum chamber pressure.

Hovercraft Operation on Land

The primary difference between hovercraft operating on land and water is in the design of the skirt. On water relatively large tubes are required to provide buoyancy when the plenum pressure goes to zero. For a hovercraft operating on a solid surface, this design requirement vanishes. A skirt is required as part of the plenum chamber, but it does not necessarily have to be tubular. As an example of a skirt suitable for land based hovercraft, examine the small radio controlled model shown in Fig. 2.6, which was designed and fabricated by students at California Institute of Technology. The plenum chamber for this model was fabricated from a Frisbee. Note the small axial fans used for lift, propulsion and steering.

Fig. 2.6 A small model hovercraft which incorporates a Frisbee as a plenum chamber.

2.5 HOVERCRAFT MISSIONS

The design of a hovercraft will depend on its mission. Clearly, the design will depend upon whether it is to operate on land, on water, or on both surfaces. However, the design will also depend upon its intended application. Is it a military hovercraft with missions on both land and water? Is it a commercial model used to carry passengers rapidly over water for relatively short distances? Is it a toy to be used by middle school children? Is it a small hovercraft intended for high speed racing by sports enthusiasts? The designs for each of these applications will differ significantly.

We recognize that in a first year engineering design class, time, resources and facilities do not exist for building a full scale hovercraft for any of these applications. However, several missions are described that involve the design and construction of relatively small models. All of these missions require autonomous operation of the hovercraft. This requirement implies that you press a start button to begin the mission, but from that instant onward the operation is totally automatic.

2.5.1 Land Race Mission

The land race mission is basically a time trial where a team's hovercraft traverses the track defined in Fig. 2.7. The track is laid out on a smooth surface such as a parking lot. The boundaries of the track are formed using a light colored tape. The exact geometry of the track is dependent upon local conditions and your instructor may change the dimensions and the configuration shown in Fig. 2.7.

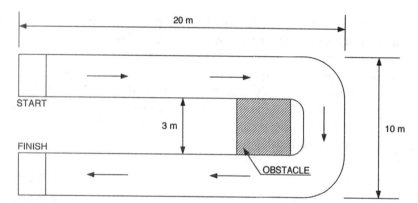

Fig. 2.7 Suggested track for the hovercraft land race.

An obstacle is shown near the turn in the track. This is a rigid obstacle and if the hovercraft strikes it at speed damage could occur. Consequently, a penalty of 15 seconds will be imposed on the team if its hovercraft strikes the obstacle. At the conclusion of the race, no penalty will be imposed if the hovercraft overshoots the finish line; however, if the hovercraft stops with 0.2 m of the line a bonus of 15 seconds will be given to the team. A 10 second penalty will be imposed each time the entire hovercraft moves outside the boundaries of the track.

2.5.2 Water Race Mission

The water race[1] is also a time trial, but it is conducted in a swimming pool. In this instance, the track is not defined. Instead the hovercraft starts from a defined position circles around a float at the far end of the pool and then returns to the finish position. The navigation and steering of the hovercraft is autonomous, although navigational aids deployed along the route are allowed. A suggested arrangement of the track is illustrated in Fig. 2.8. No dimensions have been provided in this illustration

---

[1] The water race entails significant danger to electronic equipment mounted on the hovercraft. If the hovercraft sinks, the electronic equipment may be destroyed. Also it is important that the electrical equipment be shielded from water due to splashing or wave action.

because they depend on the size of the pool available for the time trials; however, an Olympic size pool is recommended. An Olympic pool is 25 m wide with a minimum depth of 2.0 m. It is 50 m in length between touch panels. The 25 m width allows for 8 lanes, each 2.5m wide, with 2 spaces each 2.5m wide outside lanes 1 and 8.

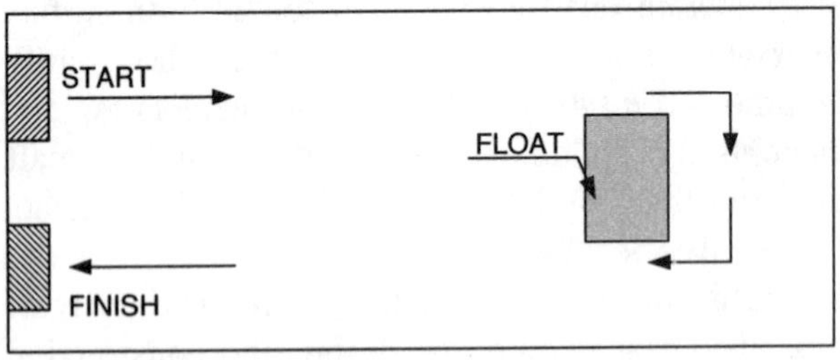

Fig. 2.8 Suggested track for the water race
mission.

A float is placed near the far end of the pool to serve as a turning point. This is a relatively rigid obstacle and if the hovercraft strikes it at speed damage might occur. Consequently, a penalty of 10 seconds will be imposed on the team if its hovercraft impacts the float. Also a 10 second penalty will be imposed each time the hovercraft strikes the far wall of the pool behind the float. At the conclusion of the race no penalty will be imposed if the hovercraft misses the finish mark; however, if the hovercraft stops with 0.4 m of the mark a bonus of 5 seconds will be given to the team.

2.5.3 The Navigation Mission

The navigation mission entails steering the hovercraft around a number of square obstacles that are arranged in a 3 by 3 array, as shown in Fig. 2.9. The obstacles are large (3 to 4 m square) and the passages between them are relatively wide (about 2 m). The surface is level and relatively smooth. The prescribed course involves ten 90° turns that take the hovercraft around three of the obstacles as it proceeds from the stating position to the finish location.

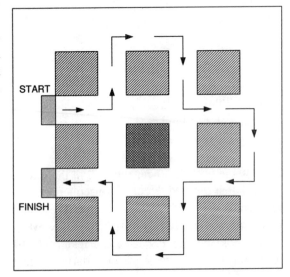

Fig. 2.9 Suggested track for the navigation mission.

This mission can be conducted with either one or two vehicles. If one hovercraft is employed, the time required to traverse the course from start to finish is the basic criterion to judge the design team's performance. If two vehicles are employed, it is suggested that one runs the standard course while the other runs the same course but starting from the opposite side. The hovercraft with the better performance will complete the entire track and return to where it started the fastest.

The center obstacle, shown with double hatching, must be avoided. If a hovercraft strikes this obstacle a penalty of 15 seconds will be imposed on the team. At the conclusion of the race, no penalty will be imposed if the hovercraft overshoots the finish line; however, if the hovercraft stops within 0.2 m of the finish line a bonus of 12 seconds will be given to the team. A 10 second penalty will be imposed each time the entire hovercraft moves outside the boundaries encircling the track.

2.5.4 The Cargo Mission

The cargo mission involves a hovercraft designed to carry freight from a starting position to an unloading dock located about 50 to 60 m from the starting line as shown in Fig. 2.10. The track for the cargo mission is laid out on a parking lot with a smooth surface. The boundaries of the track are formed using a light colored tape. The exact geometry of the track is dependent upon local conditions and your instructor may change the dimensions and the configuration shown in Fig. 2.10.

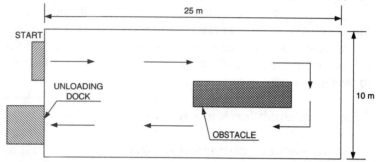

Fig. 2.10 Suggested track for the cargo mission.

The unloading dock is fixed in place and is 3 m wide, 1 m deep and 100 mm high. Your mission is to deliver cargo to this unloading dock. You may place the cargo on the deck of the hovercraft at the starting position, but it must be unloaded automatically at the loading dock. The size and weight of the cargo is decided by the instructor. The time taken to complete the delivery is the main judging parameter, with the handling of the cargo as a secondary parameter.

An obstacle is shown near the turn in the track. This is a rigid obstacle and if the hovercraft strikes it at speed damage to the cargo could occur. Consequently, a damage penalty of 15 seconds will be imposed each time the hovercraft strikes the obstacle. At the conclusion of the race, a 10 second penalty will be imposed if the cargo does not remain on the unloading dock and a 30 second penalty will be imposed if the cargo is damaged during the mission. Also a time penalty of 15 seconds will be imposed each time the entire hovercraft moves outside the boundaries of the track. Note that your instructor may decide to modify the criteria for judging the performance of the cargo hovercraft.

2.5.5 The Land and Water Mission

The land and water mission is also a time trial, but in this mission the hovercraft must traverse two different surfaces—smooth plywood and water. The track, shown in Fig 2.11 is fabricated using 4 × 8 ft sheets plywood and 2 × 4 studs. The plywood surfaces are elevated by about 4 inches above the ground; hence, the hovercraft will be disabled if it runs over the edge of the track.

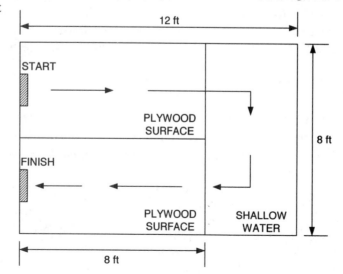

Fig. 2.11 Suggested test arrangement for the land and water mission.

The water portion of the track is formed with a 4 × 8 sheet of plywood with 2 × 4 studs as sides. The joints are sealed with caulking compound to provide a water tight tray. When filled the surface of the water in this tray is flush with the plywood surface of the solid part of the track. The water in the tray is about 3.5 inches deep.

There are no obstacles on this track except for the boundaries which are elevated about 4 inches. Clearly, the hovercraft must avoid going over the edge. We do not believe that this will occur because as soon as the lip of the skirt goes over the edge, the air pressure in the plenum chamber will be lost and the hovercraft will be grounded. If a hovercraft does go over the edge it is disqualified. At the conclusion of the race, no penalty will be imposed if the hovercraft overshoots the finish line if it does not go over the edge. If the hovercraft stops within the finish area a bonus of 15 seconds will be given to the team.

2.5.6 The Demolition Mission

As the name implies, the demolition mission is about survival of your team's hovercraft. These vehicles (three are suggested) are placed in a walled area about 3 × 4 m in size, as indicated in Fig. 2.12. The hovercrafts are to engage one another until a failure occurs. A hovercraft is declared a failure when it cannot move. During the demolition event, each hovercraft is required to be in constant motion. They can move in a circle, back and forth or side to side, but their motion cannot stop. Loss of plenum pressure even for a short period of time is also a cause of failure.

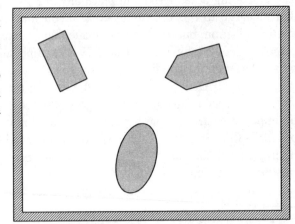

Fig. 2.12 Walled enclosure for the demolition mission.

The area of engagement is constructed from 2 × 4 studs nailed together to form a suitable wall around a rectangular area like that shown in Fig. 2.12. Placing the enclosure on a parking lot provides a relatively smooth and level surface. There is no penalty for striking the wall as the damage that may be imparted to the hovercraft is a sufficient detriment. It is anticipated that the design of the fighting hovercraft will involve both offensive and defensive devices, as well as programming for engagement or retreat. The winner of the demolition mission is the sole surviving hovercraft.

2.5.7 Summary

The six missions described above are only brief sketches of the potential design challenges involved in developing a prototype hovercraft. It is expected that the instructor will provide additional material describing the details of the track and the restrictions (such as weight or size) placed on the development of the hovercraft for a specific mission. The instructor is responsible for designing the track and arranging for its construction. The instructor can make it larger or smaller, add obstacles, add or remove turns, make the surface rough, etc. Also, the parameters used to judge the adequacy of the design of the hovercraft will depend upon the instructor. The instructor is free to change the parameters, consider others and to modify the penalties and bonuses. Regardless, final product specifications should be provided very early in the design stage to ensure sufficient time to complete the selected mission.

2.6 HOVERCRAFT DESIGN CONCEPTS

Lift

The operation of a hovercraft over land or water depends on a cushion of air generated by a fan to lift and support the vehicle. As indicated in Fig. 2.13, this cushion of air is supplied by a fan that produces a high volume of air at a relatively low pressure. The pressure acts on the deck of the hovercraft and provides an upward lift force that is sufficiently large to overcome the weight of the structure, fans and other equipment. Under ideal conditions, the hovercraft lifts up and a gap forms between the skirt and the surface, as illustrated in Fig. 2.13. Air flow through the gap is considered leakage. At an equilibrium state, the airflow into the plenum chamber equals the air flow due to leakage out of the plenum chamber and the lift force equals the hovercraft weight. This is the equilibrium condition for both forces and flow rate. When a gap forms around the entire perimeter of the hovercraft, the structure is floating on air and the hovercraft moves without the frictional constraints that occur with normal sliding.

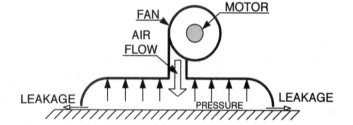

Fig. 2.13 Representation of an equilibrium state for operation of a hovercraft on a solid surface.

The height of the gap around the periphery of the plenum chamber is an important parameter. Under ideal conditions, the gap height is a constant about the periphery of the plenum chamber. However, if the hovercraft is not in perfect balance, it will tip and the skirt may contact the surface at some point on its periphery. When contact with surface is made, a friction force develops and the hovercraft will begin to drag and rotate about this point. To pilot the hovercraft in a straight line, it is essential to either avoid contacting the skirt with the surface of the course over which the hovercraft is moving or to control the propulsion forces and correct for the adverse affects of the friction force on the hovercraft's intended path.

Fan Characteristics

The fan that delivers the air upon which the hovercraft floats is an essential component of the vehicle. Thus, it is important to understand the relationship between the flow rate and the pressure of the air delivered by the fan. A typical curve that depicts this relationship is presented in Fig. 2.14. Examination of this curve shows that the pressure is a maximum when the flow rate is zero (the outlet of the fan is blocked), while the flow rate is a maximum when the pressure is zero and the fan is exhausting unobstructedly into the atmosphere.

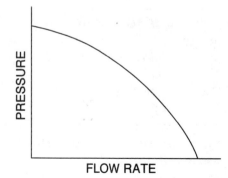

Fig. 2.14 The general shape of the fan characteristic curve showing flow rate as a function of output pressure. The exact shape of this curve is either determined experimentally or found on the web as manufacturer's data.

It is clear from Fig. 2.14 that there are many different pressure/flow rate combinations at which the fan can operate. The operating point is established by examining the rate of air flow (leakage) through the gap formed between the skirt and the surface. As the pressure in the plenum chamber increases, the flow rate through the gap increases, as illustrated in Fig. 2.15. The operating point is established at the intersection of the fan characteristic curve and the plenum chamber's leakage curve. At the operating point, the plenum pressure is p_0 and the flow rate is Q_0.

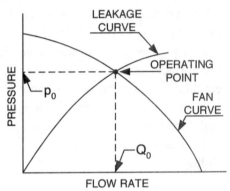

Fig. 2.15 The intersection of the fan and leakage curves gives the operating point for the fan. The leakage curve is dependent on the flow channel and the size of the gap.

Gap Size

When the plenum pressure p is sufficient to develop a lift force equal to the total weight of the hovercraft, the hovercraft lifts off of the ground and a gap with a height h_{gap} is formed between the skirt and the ground. The height h_{gap} of the gap is dependent on the total weight of the hovercraft and the fan's characteristic curve. The leakage curves, shown in Fig. 2.16, illustrate the effect of gap height. When the gap is small, the flow rate through the reduced area between the skirt and the ground is relatively low and the fan is able to maintain a higher plenum pressure. However, when the gap is large, the area between the skirt and the ground increases, the flow rate increases and the plenum pressure maintained by the fan decreases. The controlling factor in establishing the gap height is the weight of the hovercraft. The plenum pressure developed by the fan will produce a lift force equal to the hovercraft's weight if the fan is capable of developing this pressure.

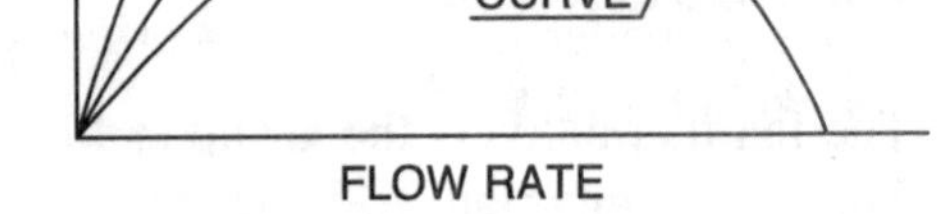

Fig. 2.16 The influence of gap size on the operating point of the fan.

Steering and Braking

Under ideal conditions the hovercraft floats on a cushion of air. It does not have wheels or brakes. How do we steer it? Let's consider a boat as we explore this question of steering. A boat floats on water and does not have wheels or brakes. A boat is steered with a rudder or with its propulsion unit. Let's consider each of these methods. The action of a boat's rudder is illustrated in Fig. 2.17. The rudder is located beneath the boat and is usually oriented along its axis. When turning the boat to the right (starboard) the rudder is rotated to the position shown in Fig. 2.17. The flow of water is interrupted by the rudder and a component of force F develops that acts on the rudder at a position some distance d from the axis of the boat. This force creates a moment M = (F)(d) that turns the boat to the right (starboard).

The rudder on an airplane acts in a similar manner except that the fluid is air and the velocity of the airplane is significantly higher than that of a boat. Will a rudder be effective on a model of a hovercraft? Note that the rudder would be located above the deck of the hovercraft and would interfere with the flow of air not water. Also be aware that the velocities of the hovercraft your team will build for these missions will be significantly lower than that of an airplane.

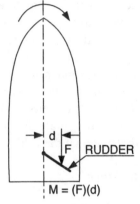

Fig. 2.17 The mechanics of steering a boat with a rudder.

Boats are also steered with thrust directional control or thrust vectoring. The propeller on an outboard motor is rotated in the water to provide a thrust force F that makes an angle with the axis of the boat. This force creates a moment to turn the boat to the right (starboard), as illustrated in Fig. 2.18.

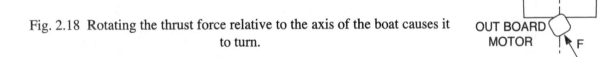

Fig. 2.18 Rotating the thrust force relative to the axis of the boat causes it to turn.

A second approach to steering a boat is possible if it is equipped with twin propellers. In this case, the boat is driven in a straight line when both propellers are turning at the same angular velocity. However, if one propeller is slower than the other, the thrust forces developed differ as shown in Fig. 2.19. If $F_2 > F_1$, as shown, the boat will turn to starboard.

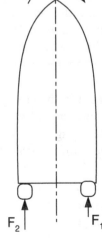

Fig. 2.19 Using different angular velocities on twin propellers to turn a boat.

It is also possible to steer by causing the skirt on one side of the hovercraft to touch the surface momentarily. When contact with the surface is made, a drag force is created that will cause the hovercraft to turn. This procedure is difficult to control as balance of the hovercraft must be reestablished as soon as the turn is complete. Otherwise, the hovercraft will continue to rotate until the gap between the skirt and the surface is reestablished. The choice of a method of steering is yours, but note that the commercial designs introduced earlier in this chapter employ twin fans at the stern of the hovercraft to steer the boat using the same principle indicated in Fig. 2.19.

If the hovercraft is operating on a solid surface, braking is easy. Turn off the plenum fan and the plenum pressure quickly decreases. The hovercraft will drop to the surface and skid to a stop. On water it is much more difficult to brake. It is possible to turn the fan off, but hovercraft designed for water applications have skirts that provide buoyancy for the vehicle. When the plenum pressure decays, the hovercraft settles into the water and the drag forces increase; however, the vehicle will continue to move. If you are using fans for propulsion, your can turn the hovercraft through 180° and use the fans to counteract the vehicles momentum.

Balance

Balance of the hovercraft is essential particularly if it is to operate on a solid surface. If the hovercraft is not in balance, it will tilt to one side, causing the skirt to drag on the surface and steering to become very difficult if not impossible. To insure balance, the center of pressure and the center of gravity must coincide. The center of pressure is the centroid of the area over which the pressure acts. For regular shapes, the center of pressure is obvious as indicated by the black dots presented in Fig. 2.20. For areas with two axes of symmetry, the center of pressure occurs at the intersection of these two axes. For an area with only one axis of symmetry, such as the triangle shown in Fig. 2.20, it is necessary to compute[2] the location of the center of pressure.

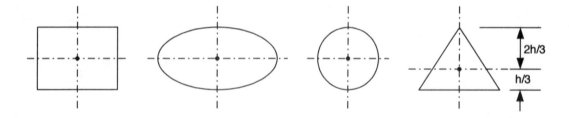

Fig. 2.20 The dot locates the center of pressure for common geometrical shapes.

The center of gravity pertains to the distribution of weights over the deck area of the hovercraft. Suppose the hovercraft is designed with the layout of components illustrated in Fig. 2.21. The components include two axial fans, a centrifugal fan, the NXT microcontroller, sensors, relays and an auxiliary battery pack. The figure also includes the centerlines for the rectangular deck with the center of pressure located with a black dot. The axial fans used for propulsion and steering are placed at the aft end of the hovercraft, but they are symmetrical with respect to the longitudinal axis. The sensors are mounted in the bow and are also symmetrical with respect to the longitudinal axis. The NXT and the relay cluster are located in the aft portion of the craft on the longitudinal centerline. The outlet nozzle of the centrifugal fan was placed at the intersection of the centerlines. However centrifugal fans usually are not symmetrical and as a consequence its center of gravity may be shifted relative to the longitudinal axis.

---

[2] Locating centroids and centers of pressure will be covered in a course on Statics (ENES 102) at the University of Maryland.

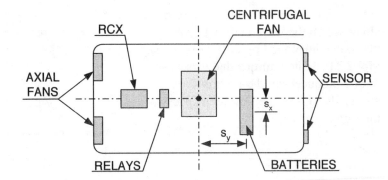

Fig. 2.21 Layout of components on
model hovercraft.

As shown above, the hovercraft is not in balance. The components to the left of the transverse centerline will cause the aft end to tilt downward. Also the weight of the centrifugal fan is not symmetrically distributed relative to the longitudinal centerline. To balance the hovercraft model, the battery pack is positioned with s_x and s_y adjusted so that the moments about the longitudinal and transverse centerlines are in balance.

In practice, balance can be achieved by suspending the model with a piece of string or wire attached at its center of pressure, as shown in Fig. 2.22. All of the components are mounted on the model except the battery pack. The battery pack is then placed on the deck as shown in Fig. 2.22 and shifted in both the x and y directions until the deck is level. The distances s_x and s_y are measured to establish the exact location of the battery pack that provides the necessary balance to prevent tipping when operating the hovercraft.

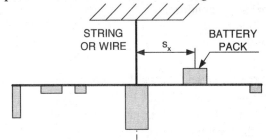

Fig. 2.22 Placement of the battery pack to achieve
balance of the hovercraft.

Unbalance Due to Dynamic Forces

The procedure illustrated in Fig. 2.22 is effective for establishing a static balance of the vertical forces acting on the hovercraft. However, when the hovercraft is in motion dynamic forces in the horizontal direction act to propel and steer the vehicle. These forces produce an unbalanced condition and if they are sufficiently large, the hovercraft will tilt and its skirt will contact the surface of the course at one or more points. A friction force F_f develops at each contact point in the direction that opposes the motion. Let's consider contact at one point with the development of a friction force F_f, as illustrated in Fig. 2.23.

This force generates a torque that causes the hovercraft to turn left in an uncontrolled manner.

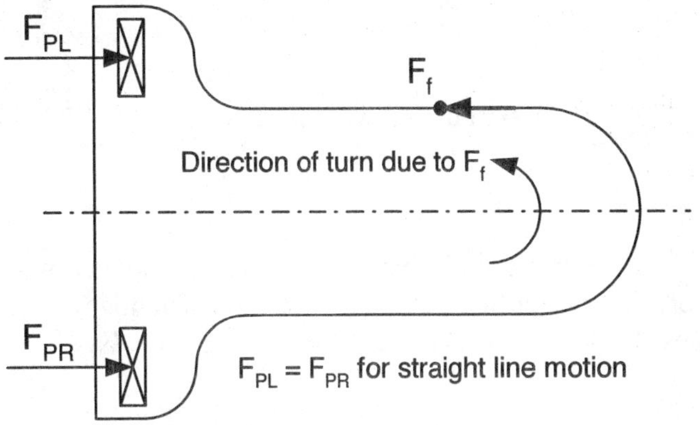

F_{PL}

F_f

Direction of turn due to F_f

Fig. 2.23 Friction force applied at contact point at the skirt surface interface causes the hovercraft to execute an unscheduled turn and lose control.

F_{PR}

$F_{PL} = F_{PR}$ for straight line motion

To restore control, a sensor must be employed to detect the deviation from the specified track along which the hovercraft is to follow. The sensor's output is a feedback signal that is used to adjust the thrust developed by the two propulsion fans. The adjusted propulsion forces F_{PL} and F_{PR}, shown in Fig. 2.24, create a torque that offsets the torque produced by the friction force F_f. and tends to turn the hovercraft back to the right.

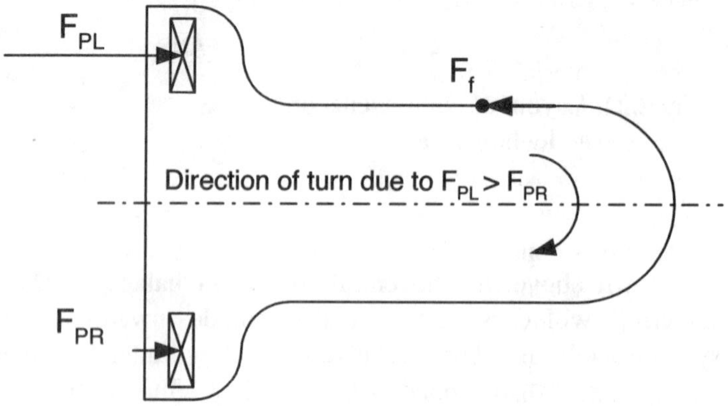

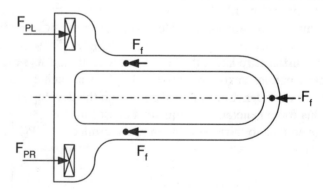

Fig. 2.24 Propulsion fans used to control the hovercraft's natural instabilities

Again sensors are necessary to detect when the hovercraft has realigned with the target track. At this instant, the output from the sensors is employed as feedback signals to adjust the propulsion forces so that $F_{PL} = F_{PR}$, which is necessary for straight line tracking when the hovercraft regains balanced condition.

A Strategy for Dealing with an Unbalanced Hovercraft

Experience has shown that it is extremely difficult to maintain perfect balance as the hovercraft moves along a specified track even on a very smooth surface. Moreover, unexpected friction forces that occur at contact points at various locations along the skirt disrupt motion and steering strategies. An approach to this problem is to develop either a control strategy or a hovercraft design that minimizes the effects of the unpredictable friction forces.

Consider first the design of a hovercraft with a skirt that is configured so that three contact points extend below the plane formed by the base of the skirt, as illustrated in Fig. 2.25. If the lift force F_L becomes slightly less than the weight W of the hovercraft, the skirt will contact the surface of the track at these three specified points. Friction forces will develop at these contacts, but the magnitude of these friction forces can be minimized and the turning torques they generate will tend to cancel out.

Fig. 2.25 A skirt design with three contact points.

Better stability in steering can be achieved with this design; however, the penalty is the drag force F_D produced by the sum of the three friction forces. If we assume the friction forces are equal, it is possible to write the equation for the drag force F_D as:

$$F_D = 3F_f = \mu(W - F_L) \tag{2.2}$$

where μ is the coefficient of friction between the contact points and the surface of the track.

Clearly, there are two ways to minimize the drag force. The first way is to adjust the lift force so that the quantity $(W - F_L)$ is nearly zero by adjusting the air flow rate and the gap size. The second way is to select a material with a low coefficient of friction μ for the contacts on the skirt. Materials such as polyethylene or Teflon exhibit low friction coefficients in the range of 0.10 to 0.15 when sliding over smooth hard surfaces. By minimizing the quantity $(W - F_L)$ and using contact supports fabricated from low friction material, the drag force can be minimized. Sensors providing feedback signals will still be required with this design approach in order to adjust the propulsion forces F_{PL} and F_{PR} to maintain the hovercraft's controlled motion along the track.

A second strategy for dealing with the unbalanced condition that causes deviation from straight line tracking involves controlling the speed of the hovercraft. If the speed of the hovercraft is relatively high when the unbalance condition occurs, the vehicle turns abruptly and the deviation from the track is significant. When the deviation from the track is excessive, the sensors used to maintain control are no longer in a position to measure the tracking target and systematic control is impossible. To avoid these excessive deviations, the speed of the hovercraft should be limited so that the turns, which occur when the vehicle becomes unbalanced, are less abrupt. More time is available for corrective actions (sensing, feedback and adjusting the power to the propulsion fans) when the speed is limited. With additional time, adjustments to the propulsion forces can be made bringing the hovercraft back on to the target track before the sensors are rotated out of their tracking positions.

Autonomous Control

Autonomous control means that the entire mission is automatic. You interact with the hovercraft only once—to start the process. Autonomous control may be achieved in a number of different ways, but they all involve a microprocessor (or microcontroller). In this textbook, the LEGO NXT unit has been selected as the controller to be employed in achieving autonomous control. A photograph of the NXT is presented in Fig. 2.26 and a thorough description of its features is presented in Chapter 6.

The NXT controller uses a 32-bit, 48MHz ARM7 processor with 256KB flash memory and 64KB of RAM. The controller is housed inside a battery-powered brick that is equipped with ports for sensors and motors. The NXT case contains a tray for six AA batteries or a rechargeable lithium battery. It has four input ports to accommodate sensors, three output ports to power motors and lights and a standard USB 2.0 port. A monochrome LCD monitor with a graphical user interface controlled by four buttons lets the user select and run stored programs.

If the NXT controller is procured with a Mindstorms kit, three servo-motors are provided to power your inventions. These motors can also be used to sense rotation to within one degree. The kit also includes four sensors: ultrasonic for range-finding and motion detection; sound for measuring noise level; touch, which responds to contact by activating a momentary switch; and light, which measure the intensity of light reflected from a surface. The motors and sensors are connected to the NXT controller's ports by wires with RJ12 connectors that resemble (but don't match) Ethernet plugs.

Programming the NXT may be accomplished using several different codes. The most popular methods for programming are based on LabVIEW software from National Instruments. Recently National Instruments has released a toolkit to allow standard LabVIEW users to create and download NXT programs. The ROBOLAB 2.9 software, which was used with the older RCX controller, has been modified to allow programming of both the RCX and the NXT microcontrollers. Both of these software tools employ an icon oriented programming environment where icons are dragged onto a working diagram to create a sequence of conmmands. These commands include loops, forks and jump and lands. The details of programming in ROBOLAB 2.9 will be described in Chapter 6. Programs to control the sensors and the motors are downloaded from a computer to the NXT controller by USB cable or wirelessly through Bluetooth communications.

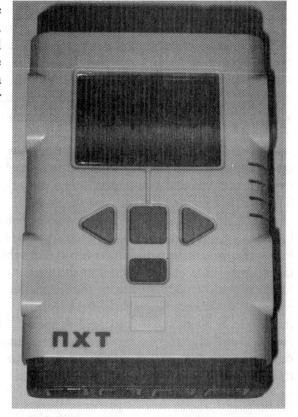

Fig. 2.26 The NXT controller.

Sensors

Autonomous operation of the hovercraft requires either timing each maneuver necessary to complete the mission or controlling actuators based on feedback signals from sensors. Timing a maneuver is the easiest method to program because the NXT controller contains a clock that is accurate to 0.1 seconds. With time of operation used to control forward motion or turns, sensors are not required. However, the different times, which are programmed to control the mission, must be known precisely for each operation and each operation must be reproducible.

Sensors that indicate angle of rotation, contact, temperature, light intensity, proximity, etc. are available to provide feedback signals. These feedback signals are interpreted by the NXT and, depending on the programming, actuators are employed to adjust the course of the hovercraft so as to complete the assigned mission. Sensor feedback control provides a reliable means of control when each operation is not easily reproducible.

Auxiliary Power Supplies

The NXT unit is powered by either six AA alkaline or NiMH batteries that provide adequate power for the microprocessor and for operation of small motors and actuators found in the MindStorm kit marketed by the LEGO Educational Division. The six alkaline batteries provide a 9 V supply voltage and the six NiMH batteries provide a 7.2 V supply voltage. However, the missions described previously require fans that will draw more power than the NXT can safely supply. The output ports on the NXT can deliver about 500 mA at about 9 volts into a low resistance circuit before entering a thermal shutdown mode to prevent damage to internal components in the NXT. Because the current draw by the levitation fan is several amps, it is clear that an auxiliary power supply will be required. The auxiliary power supply (i.e., battery) is sized to provide the necessary current at the voltage required by the fans (probably 12 Vdc). The 500 mA output from the NXT controller is more than sufficient to activate a transistor or relay (switch) that connects the battery supply to one or more fans.

2.7 HOVERCRAFT SKIRTS

Skirts are added to the plenum chamber to provide a deeper cushion of air upon which the hovercraft rides, as indicated in Fig. 2.27. The purpose of the skirt is to enable the hovercraft to pass over rough surfaces while maintaining its normal speed. The skirt is often fabricated from a flexible material so that it can deform easily when an obstacle is encountered. It is usually semi-circular in shape so that retains its form as the plenum pressure is increased. The semi-circular shape is conducive to inward deformation when obstacles are encountered yet resistive to the deformation due to the plenum pressure.

If the surface over which the hovercraft travels is very smooth, skirts fabricated from rigid materials have the advantage of more precise geometry and simplicity of construction. The edge of the skirt can form a near perfect plane that enables the gap size to be uniform about its periphery.

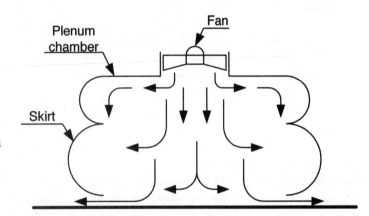

Fig. 2.27 Adding a skirt to the plenum chamber to enable the hovercraft to move over obstacles.

2.8 HOVERCRAFT STRUCTURES

The hovercraft structure must be light and strong. Weight is important because the plenum pressure—which is limited by the fan your team selects—must be sufficiently high to lift the weight of the structure, the hovercraft's components, and cargo. Aircraft modeling methods and materials are often employed to fabricate very light structures. Materials used for airplane models include balsa wood, foam, and fabrics that are light and strong. However, these airplane-like structures are not robust and may not withstand impacts that are inevitable during the hovercraft's mission.

More suitable materials include thin gage plywood, foam board, polypropylene and styrene sheets. Styrene sheets can be cold formed or vacuum formed when heated. Polypropylene sheets can easily be cut with a hobby knife and cold formed. Pieces of both of these plastics can be joined with suitable adhesives. Epoxy sheets reinforced with either glass or carbon fibers are also available. The reinforced epoxies exhibit a very high strength to weight ratio, but they are expensive when compared to other plastic sheet materials.

Rubber tubes and fabric are often used to fabricate soft skirts because they are flexible and will deflect when the hovercraft encounters an obstacle. Attachments are made with staples, cord lacing and/or adhesives.

2.8.1 Types of Hovercraft Structures

Raft Construction

There are several methods used to construct the hovercraft structure. The raft construction method, illustrated in Fig. 2.28, incorporates a flat raft-like piece of foam or foam board under the main deck. The raft is attached rigidly to the main deck with stringers located at intervals about its periphery. It is also used to support the edges of the flexible skirt. The attachment to the flexible skirt can be made with lacings or with a continuous piece of fabric or thin plastic sheet. If fabric or a thin plastic sheet is used, holes must be provided in the sheet/fabric to permit the air flow from the lift fan to enter the plenum chamber under the raft. The raft construction method is often used when the hovercraft is expected to traverse a body of water. The volume of the foam used in constructing the raft is sufficient to provide the buoyancy needed to prevent the hovercraft from sinking in the event of a loss of power to the lift fan. The raft also contributes to the stiffness of the main deck

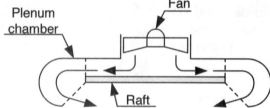

Fig. 2.28 Cross section showing raft construction details.

Box Construction

The box construction method, illustrated in Fig. 2.29, incorporates a box-like structure that is inset into the main deck. The box supports the lift fan and extends both forward and aft to provide space for transporting cargo. It is tied into the main deck about its periphery and contributes to the stiffness of the deck. Holes are provided in the sides of the box to enable the air flow from the lift fan to inflate the bag skirt. It is also used to support the edges of the flexible (bag) skirt. The attachment to the skirt can be made with lacings or with a continuous piece of fabric or thin plastic sheet, as described above.

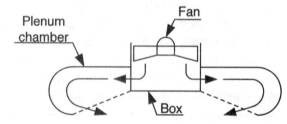

Fig. 2.29 Cross section showing box construction details.

Shell Construction

The box construction method, illustrated in Fig. 2.30, incorporates a rigid shell that serves as the main deck and the closure for the plenum chamber. A flexible skirt can be added if the course for the hovercraft mission includes obstacles. However, if the surface for the course is smooth, a skirt is not required. From a construction viewpoint, the shell design is the simplest because it does not involve building of an internal structure or a flexible skirt. However, fabricating the structure involves molding the shell, which requires considerable skill. An alternative to molding is to fabricate the rigid skirt from a plastic pipe cut longitudinally to form a C-section or to use foam pipe insulation. The C-section can be formed into a circular ring and attached to the deck with adhesive to produce the shell like structure.

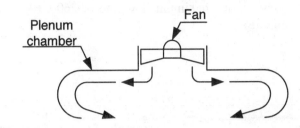

Fig. 2.30 Cross section showing shell construction details.

Double Deck Construction

The double deck construction method, illustrated in Fig. 2.31, is similar to the raft construction method because it incorporates a flat deck-like board under the main deck. The two decks are spaced apart by a predetermined dimension and fastened together with bolts. One edge of a bag skirt is attached to the upper deck and the other edge is attached to the lower deck. Some of the air from the lift fan is deflected by the lower deck and inflates the bag skirt. A hole cut in the lower deck permits the remaining sir to flow into the plenum chamber to lift the hovercraft. The spacing between the two decks and the size of the hole in the lower deck control the pressure in the skirt and the plenum pressure. The double deck hovercraft is relatively easy to build and the dimensions of the hole or the between deck spacing are easy to modify if the skirt and plenum pressures require adjustment.

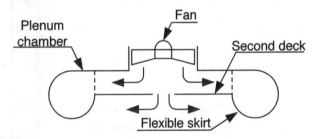

Fig. 2.31 Cross section showing double deck construction details.

2.9 HOVERCRAFT COMPONENTS

The layout of the components shown in Fig. 2.21 indicates many of the primary components that will be used in the design and fabrication of your team's prototype hovercraft. These components include fans, the NXT, relays, batteries and sensors. The selection of these components is a significant part of the design process. For this reason, a brief review of each of the primary components is presented in the paragraphs below:

Fans

Centrifugal Fans or Blowers

There are two basic types of fans—axial flow and centrifugal flow. The centrifugal fan, often called a blower, is presented in Fig. 2.32. It has an inlet port, a small motor that drives a rotor, and an output port. The outlet is tangent to the rotor and perpendicular to the axis of the motor. The centrifugal fan generates a higher pressure than the axial fan and is often used for developing the lift force. It has the disadvantage that it is relatively large, heavy and requires significant power to generate high flow rates.

The characteristic curves showing pressure, RPM (angular velocity) and current draw as a function of flow rate for this blower are presented in Fig. 2.33. The blower operates at about 4,000 RPM and draws from 4 to 11 amperes depending on flow rate. The pressure developed by the fan is conventionally measured in inches of water and decreases from 2.3 in. of H_2O at zero flow to zero inches at its maximum flow rate of 260 CFM (cubic feet per minute).

Fig. 2.32 Photograph of a centrifugal fan (blower).

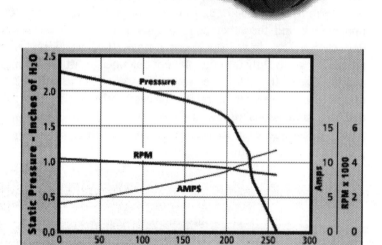

Fig. 2.33 Characteristic curves that describe important blower parameters.

The blower, shown in Fig. 2.32, has the advantage of a high flow rate together with a relatively high pressure. Its disadvantage is that it draws large currents (12 A) and it is heavy with a weight of nearly 5 lb. Perhaps your team will accept the disadvantages because this blower delivers 200 CFM of air at a pressure of 1.6 in. of H_2O. Or your team may look for a different blower that requires less current and does not weigh as much. The decision will largely depend on the weight estimate of the overall vehicle and on the size and geometry the vehicle must be to meet the mission requirements.

A large number of fan characteristic curves can be found by performing an internet search for fan vendors. The authors particularly partial to EBM Papst webpage at http://ebmpapst.us/..

Axial Fans

An axial flow fan, depicted in Fig. 2.34, does not use centrifugal forces to develop air pressure. Instead it has a motor embedded in its center hub that drives fan blades that propel air down the axis of the fan. These fans are lighter than the centrifugal fans for the same flow rates; however, the pressures developed are significantly lower.

The characteristic curve for a 12 Vdc axial fan 120 × 120 × 38 mm in size is presented in Fig. 2.35. Note at the nominal 12 V, the maximum flow rate is about 105 CFM; however, when the flow rate is zero the static pressure is only 0.275 in. of H_2O. The weight of this fan is only 0.62 pounds and the rated current is 0.58 A. Axial flow fans are used for a variety of applications that require large volumes of low pressure air, such as for cooling of electronic equipment.

Fig. 2.34 A small axial fan with direct motor drive.

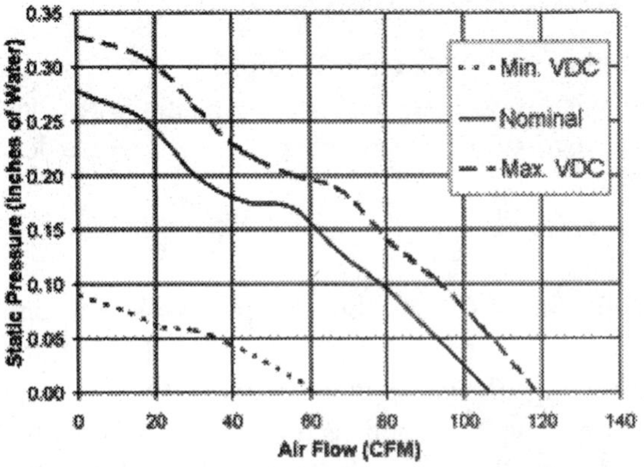

Fig. 2.35 The pressure as a function of flow rate for an axial fan.

When your team selects a fan, be certain to study the characteristic curves as they will markedly affect the performance of the hovercraft. The analysis of hovercraft performance that uses flow rate and pressure from the fan's characteristic curve will be described in Chapter 3.

Relays

Relays are electromechanical switches that employ electromagnets to close one or more sets of contacts. In general, the purpose of a relay is to use a small amount of power to energize an electromagnet which moves an armature that is capable of switching much larger power levels into a circuit. For example, you might want activate the electromagnet using 6 V and 40 mA (240 mW), while the armature can support 12 Vdc at 12 amps (144 W).

As indicated in Fig. 2.36, relays are relatively simple devices that consist of only four parts:

- Electromagnet
- Armature
- Spring
- Set of electrical contacts

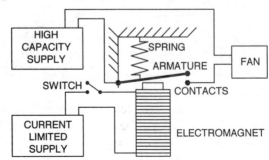

Fig. 2.36 Schematic diagram illustrating the use of a relay.

In this figure, it is evident that a relay consists of two separate and completely independent **circuits**. The first circuit, shown at the bottom of Fig. 2.36, drives the electromagnet. In this circuit, a switch controls current flow to the electromagnet. When the switch is closed, the electromagnet is activated. The armature acts as another switch in the second circuit. When the electromagnet is activated, the armature closes on the contacts completing the second circuit that turns on the fan. When the current to the electromagnet is interrupted, the spring pulls the armature up, separating the contacts and opening the fan circuit.

When procuring[3] relays, it is necessary to check on several design features:

- The voltage and current needed to activate the electromagnet.
- The maximum current capacity of the contacts.
- The number of armatures (generally one or two)
- The number of contacts (poles).
- If the contact or contacts are normally open (NO) or normally closed (NC).

Another switching method employs transistors that can be activated with small currents at low voltages. Transistors are capable of switching large current flows on and off at an incredibly fast pace. Transistors for this and other applications are described in Chapter 4.

Batteries

While the NXT has three output ports that can supply up to 500 mA, this amount of current is not sufficient to drive the fans required to lift and propel your team's hovercraft. An auxiliary power supply consisting of one or more batteries will be required. There are a large number of different types of batteries available and each type comes in many different sizes. Which type is the most suitable for the mission? What size battery will minimize weight while providing energy for the duration of the mission? Let's explore the type of batteries currently available and some of their characteristics.

Table 2.2 lists several different types of batteries. It is possible to compare the voltages, capacities and relative energy output normalized by the weight of a cell, because the data listed in Table 2.2 is for AA size batteries. The lithium battery exhibits the highest performance with a capacity relative to its weight of 300 W-h/kg. These batteries are relatively expensive; they are usually employed with digital camera or watches where long life with low current drain is the primary design requirement.

Table 2.2
Comparison of the characteristics of several types of batteries

Size	Chemistry	Cell Voltage	Capacity A-h	Capacity W-h	W-h/kg	Cost $
AA	Lithium	1.5	2.9	4.35	300	3.70
AA	Alkaline	1.5	2.85	4.28	185	0.60
AA	NiMH*	1.2	2.0	2.4	83	2.48
AA	CZn	0.8	1.1	0.88	62	0.45
AA	NiCad*	1.2	1.0	1.2	52	2.21

* Indicates these batteries are rechargeable.

---

[3] An excellent source of many electronic components including electromagnetic relays and transistors is Digi-Key (www.digikey.com).

The Nickel-Metal-Hydride (NiMH) and the Nickel-Cadmium (NiCad) batteries, also relatively high in cost, are both rechargeable. They are used in applications where the current and power demands are relatively high and the battery life is extended by recharging. These batteries can be recharged several hundred times before they fail. Thus their relatively high cost is mitigated.

The lowest cost batteries are the common carbon-zinc and the alkaline types; however, these batteries cannot be recharged. They are commonly used in flashlights, smoke detectors, door openers, toothbrushes, etc. The superiority of the alkaline battery is evident as its capacity approaches that of the lithium battery at a much more modest cost.

Another important type of battery is the sealed lead acid (SLA) that is used as a power source for automobiles. The battery is available in a box like package, with six individual cells connected in series to produce an output of 12 Vdc. The cell size is varied to produce batteries with many different capacities. The smallest SLA commonly available has a capacity of 1.3 A-h, while a much larger SLA battery on your automobile can deliver 25 A for more than two hours. High current drain is possible with this type of battery, with cold cranking amperes quoted in excess of 500 A. It is also rechargeable with a relatively long lifespan (4 to 6 years). With a capacity relative to its weight of 52 W-h/kg, it is comparable to a Ni-Cad battery.

When your team selects a battery, consider voltage, capacity, current drain, weight, cost and the advantages of recharging. Making this selection will provide you with an opportunity to perform a design trade-off study. Additional information on the electrical characteristics of batteries is presented in Chapter 4.

Sensors

Sensors are required in most autonomous missions to provide feedback signals that enable course corrections to be made. Sensors represent significant business opportunities in today's market. They are commercially available for measurements of force, torque, distance, strain, temperature, light intensity, air quality, water quality, proximity, etc. Several different types of sensors are described in detail in Chapter 5. Also included in Chapter 5 is a discussion of the methods of interfacing sensors with the NXT. The NXT contains a microprocessor, four sensor input ports, and three power output ports that allow you to create an autonomous hovercraft, which can be programmed to perform specified tasks. The NXT can be programmed with a wide variety of software programs such as LabVIEW, NXC (Not Exactly C) and le JOS NXJ (Java), but in this textbook, we will describe ROBOLAB as the programming approach. Two way communication with the NXT is also possible with Bluetooth signals from transmitters and receivers located in the NXT and your computer.

Several sensors have been developed by the LEGO Educational Division to interface with the input ports on the NXT. These sensors include touch (a contact switch), light (a LED and photodiode), temperature (a thermistor), distance (a sonar), sound (a microphone) and rotation. Other sensors are available from third party suppliers. Unless time is used to control the action of the propulsion fans, one or more sensors will be required to provide the feedback signals necessary for maneuvering the hovercraft over the track associated with the assigned mission.

Mechanical Components

LEGO, over the last twenty years, has evolved from a hands-on children's toy to a surprisingly complete prototyping tool for relatively complex engineering systems. In addition to structural elements, such as plates, beams, bars and columns, its accessories include motors, sensors, gears, wheels, belts, toggles, controllers and infrared transmitters and receivers. Compared to other prototyping alternatives, LEGO components are relatively inexpensive. The availability of components and the familiarity that most students have with the LEGO concept allows design teams to rapidly construct models without significant laboratory support or other college infrastructure. Also rapid construction permits the student teams to try several different design approaches enabling significant improvements to be implemented in the limited time available during a semester.

You should be aware that we do not consider the LEGO components as toys. They provide a design space sufficient to teach both mechanical and electrical concepts of power, motion, communication and control. Have fun, but learn about gears, power transmission, sensors, motors, microprocessors, IR communication, programming, feedback loops, electrical circuits and control in the process.

2.10 HOVERCRAFT DEVELOPMENT

When designing the hovercraft, we usually divide the development process into four major phases, which include:

1. The design of a system consisting of the many subsystems needed to meet the mission requirements.
2. The preparation of an assembly kit and the construction of the hovercraft prototype.
3. The generation of a control algorithm capable of autonomously controlling the hovercraft's motion.
4. The verification of the hovercraft prototype's performance through testing.

We will facilitate this process by providing you with descriptions of the hovercraft to be designed, the mission that the system is to perform and the time constraints on both the development of the system and its ability to complete the mission. The instructor will provide the test environment; however, it will be necessary for you to schedule testing times in advance because many design teams share the same facilities. Your instructor will schedule the time of the final trials of your hovercraft model. It will be the responsibility of each design team to have their hovercraft operational and ready for testing at the designated time.

We suggest that you make full use of the library and the Internet to search for literature and patent files. There is absolutely nothing to be gained by reinventing the wheel. Copying a design is common practice—it is called reverse engineering. Reverse engineering involves a complete dissection of a competitor's product to gain a detailed understanding of all aspects of each component. Each component is then redesigned to improve its performance, the appearance of the new product and to reduce its cost. If you are able to find a description of a previous hovercraft in the literature and decide to reverse engineer it, be certain to understand the function of each component in the system. In your design, make a serious attempt to improve performance by making design changes to optimize your vehicle's performance for the particular product specifications you have been given.

You can understand the design requirements for the hovercraft by reviewing the product specification provided at the start of the course [3]. The product specification, written relatively early in the development process, is usually prepared in tabular format. It tersely states design requirements and design limits or constraints. Remember, we always design with constraints and/or limits. The specification also includes the performance criteria and design targets for weight, size, power, cost, etc.

We will place explicit design constraints that limit your flexibility in designing the various subsystems needed in the assembly of the hovercraft. First, there is a maximum limit[4] of $25.00 per team member for the cost of materials, supplies and components that may be purchased for use on the hovercraft. The idea is to keep the out-of-pocket expenses for team members at an affordable level,

---

[4] Your instructor may choose to increase or decrease this arbitrary limit on student expenditures depending on the system design requirements.

while giving the team adequate funds to build high-quality, high-performance systems for each mission. We envision that three sub-teams of two to four members each will work to develop the three subsystems that are integrated to develop a complete hovercraft. With 9 members, the maximum amount that can be expended on a hovercraft model is $225.00. Other design constraints on size, weight and power requirements may be imposed by your instructor.

2.10.1 Design Strategy

After each member of the development team understands the design specification and design limitations, you are ready to generate a design strategy. There are two interrelated aspects to the design strategy. The first entails the mechanical features such as lift fans, propulsion fans, relays, auxiliary power supplies, as well as structural details. The second involves sensors, control logic and the programming to execute that logic. Design strategies are simply ideas for performing each function involved in the operation of a system that performs a specified task within the constraints imposed upon the system. These design strategies enable your team to meet the product specifications. To illustrate design strategies, let's examine the functions (systems and processes) involved in designing a hovercraft.

1. The structural system supports all of the components, such as fans, the NXT, batteries, sensors, etc. It also contains the cushion of air upon which the hovercraft floats.
2. The lift system includes the lift fan and the controls for turning that fan on and off.
3. The propulsion system includes a means of providing thrust and the control of the magnitude and direction of the thrust vector.
4. The power system includes auxiliary power supplies and control of these supplies.
5. The sensing system includes the sensors required to generate feedback signals enabling control of the hovercraft's speed and direction.
6. The master control system enables autonomous control to successfully complete the mission.

We anticipate that Team A will take responsibility for functions 1 and 2, Team B will be responsible for functions 3 and 4, and Team C will be responsible for functions 5 and 6. It is essential that the three subsystem teams communicate continuously during the design phase, because all the subsystems developed must be successfully integrated to produce an efficient hovercraft that can complete the mission without mishaps.

2.10.2 Preparing the Assembly Kit

When your design is complete, prepare a parts list that indicates the name of each part, size and quantity that will be used in constructing the hovercraft. To demonstrate the technique used in preparing a parts list, consider the example presented in Fig. 2.37.[5]

After the parts list is complete, the parts can be procured from the appropriate suppliers[6] and arranged in an assembly kit. When the procurement activity is concluded and the assembly kit is

---

[5] This parts list is an example and is not complete. Your team's parts list is to contain all of the parts required to build a complete model as well as the source and model number of each part.

[6] Information describing the characteristics of many of the components required to design the hovercraft can be found on the web. These components can be ordered on-line or purchased at local retail outlets such as Home Depot, Lowe's, or Radio Shack.

complete, the team can begin to assemble the hovercraft. Assembly will be rapid because the design is finalized and all the parts have been identified and placed in a convenient container.

PARTS LIST FOR A HOVERCRAFT

NUMBER	NAME OF ITEM	DESCRIPTION OR DRAWING NO.	QUANTITY	PRICE
1	LIFT FAN	SUPPLIER NAME AND MODEL NO	1	$ 38.00
2	PROPULSION FAN	SUPPLIER NAME AND MODEL NO	2	$ 8.75
3	DECK	1/4 X 4 X 4 FOAM BOARD --DWG. NO. 100-01	1	$ 1.15
4	SKIRT	4 in. DIA. DRAINAGE PIPE	8 FT	$ 6.25
5	RELAYS	SUPPLIER NAME AND MODEL NO	3	$ 2.85
6	MAIN BATTERY	SUPPLIER NAME AND MODEL NO	1	$ 12.30
7	SECOND BATTERY	SUPPLIER NAME AND MODEL NO	1	$ 2.50
8	PC BOARD	SUPPLIER NAME AND MODEL NO	1	$ 7.45
9	TERMINALS	RADIO SHACK --- ASSORTED SIZES	1	$ 1.75
10	WIRE	RADIO SHACK --- STRANDED #16	25 FT	$ 1.75
11	WIRE	RADIO SHACK --- STRANDED #22	25 FT	$ 1.10
12	EYE SCREWS	HOME DEPOT ---- SMALL WITH 1/2 in. EYE	12	$ 0.60
13	SENSOR	SUPPLIER NAME AND MODEL NO	1	$ 32.00
14	SEALANT	HOME DEPOT ---- LARGE TUBE	1	$ 3.85
15	RCX BRICK	LEGO EDUCATIONAL DIVISION	1	$125.00

Fig. 2.37 Parts list for a model hovercraft.

2.10.3 Programming for System Control

The NXT controls the actions of the hovercraft that your team designs. To exercise this control, the NXT is programmed (given instructions to stop or start output fan motor ports in response to sensor signals, etc.) The program, written on a computer, is transmitted by cable or Bluetooth signals to the microcontroller. The microcontroller is essentially the brains of the NXT. Programming may be accomplished in several languages such a C, not exactly C (NEC) or with a graphical language. Three different graphical languages are commonly employed including: the MindStorms' code NXT-G, ROBOLAB 2.9 and a LabVIEW Tool Kit for the NXT. ROBOLAB is based on LABVIEW, a popular program available from National Instruments that is often used in industry to process signals from sensors and to drive controllers. We have selected ROBOLAB as the programming environment for this course because it has different levels of complexity and is based on LabVIEW yet customized for programming the NXT. These different levels enable the beginner to start with the easiest method of programming and then to assemble more challenging programs as he or she becomes more proficient.

The three levels or modes of programming in ROBOLAB are **Pilot**, **Inventor** and **Investigator**. **Pilot** is the easiest programming mode because it involves only serial programming thus insuring that the program will always compile and execute. **Inventor** represents the second level of complexity because it is possible to create a program containing all the typical programming elements such as constants, variables, loops and functions. **Investigator** is the most complex level of programming. **Investigator** contains all of the features in **Inventor** together with the ability to log data and to perform data analysis. For all three levels, programming is accomplished by selecting icons from drop down palettes and dragging them into the correct position on a block diagram. For higher level programming, the icons are wired together to connect input and output ports enabling jumps, forks and loops.

Methods of programming in **Inventor** and **Investigator** will be described in detail in Chapter 6.

2.10.4 Verifying System Performance

When your hovercraft is complete and you have written the program to control its actions, it is time to verify its performance. Ideally, verification experiments should be conducted several times during the design process. Subsystems can be operated to check their performance prior to assembly of the entire system. Preliminary checks at several stages in the design process permit the team to detect design flaws early and to remedy them. Finding errors early in the process is important. If you ship a product with serious flaws and a company must issue a recall, it is a very expensive proposition. Even more serious is an accident due to a faulty product. Litigation is expensive even if the company wins the case. If the company loses, punitive damages are often many millions of dollars. Testing products released to the public to insure their safety is an essential part of doing business.

Verifying system performance in the restrictive design environment created in this course is much easier. You will know in advance the design requirements and the performance targets that the system is to achieve. When the system design and assembly is complete, operate the system in a simulated environment. Measure the time required for the hovercraft to meet the specified mission requirements. Observe the response of the system. Does it operate at the correct speed? Is the program efficient? Are the sensors providing the necessary feedback signals? Is the deck sufficiently rigid? Is the skirt contacting the surface uniformly? Is the flow rate from the lift fan sufficient? Is the pressure developed in the plenum chamber adequate? Is the hovercraft in balance, and other questions of similar nature? The time involved to verify the performance of a prototype is usually relatively low, thus allowing sufficient time to redesign and introduce improvements.

2.11 DESIGN CRITERIA

Your design experience this semester is constrained to some degree by the number of components that you can locate, study, analyze, discuss and procure in a relatively short period of time. Also, you cannot ask a company to develop a product specifically for your application. For this reason, you will not have to deal with several design goals encountered when designing in a much larger design space. An unconstrained design space permits you to design with any standard component or with specialized components that can be manufactured according to your specifications. How many standard components are there in the marketplace? The answer probably depends on the definition of a standard part, but it is reasonable to consider that about 200,000 mechanical parts are available from a well-stocked supply house and probably another 200,000 components are available from an electronic supply outlet. If your company has manufacturing facilities, more specialized components can be produced to meet your design requirements. Clearly, a more challenging design space exists in the real world.

To prepare you for this world, you should be aware of design criteria with which engineers are concerned with as they develop products. These include:

1. We design products that meet the needs of a specific set of customers.
2. We always seek a quality product. Quality means different things to different people. In this instance, we define a quality product as one that is reliable, meets the specified performance requirements, and is durable. Your team may wish to add to this definition.
3. We often attempt to achieve high-performance in a product with additional power, installing superior controls, reducing weight or by using new higher strength materials.
4. We strive to design a product that can be produced at low cost. However, balance between cost, performance and quality must be achieved. Low cost that is achieved by sacrificing reliability and/or performance will result in eventual failure of the product in the marketplace.

5. Our products should be reliable. They should be capable of meeting the performance specifications without the need for repair. For higher priced products, a limited amount of periodic maintenance may be justified if it prolongs the product's life significantly.

6. Products should be designed for ease of manufacturing and assembly. Manufacturing and assembly are separate operations and could be listed separately. They are both very important as manufacturing costs and assembly time markedly affect the total cost of the product.

7. We attempt to design products for easy maintenance. In many situations, spare parts and skilled technicians are in short supply or non-existent. Easy maintenance is an important element in customer satisfaction.

8. Product weight is important in most designs. Usually lighter is better because using less material results in lower costs for raw materials and shipping.

9. Product size is also important with small typically being better than large. This is particularly important in electronic products where performance increases while cost decreases as feature sizes are reduced.

10. Last, but certainly not least, the product must be safe to operate. Hazards associated with products are covered in detail later in this text. Before beginning the design of any of your systems, we **strongly recommend** that you read Section 15.6 and understand the **many different hazards** associated with the design and operation of products.

Your team should develop many different ideas to support the strategy upon which the design of your hovercraft is based. Each idea or concept should be discussed in considerable detail prior to incorporating it in the model's final design. The more completely you understand the design concepts, the better prepared you will be to conduct design trade-off analyses. In a design trade-off analysis, you compare several design ideas, select the best one and discard the others.

2.12 DESIGN TRADE-OFF ANALYSIS

To perform a design trade-off analysis, we consider each design concept individually and list its strengths and weaknesses. Factors usually considered in this analysis include the design criteria listed above such as: size, weight, cost, ease of manufacturing, performance, appearance, etc. After the strengths and weaknesses of each design idea have been listed, we can compare and evaluate the design concepts and select the most suitable concept for our prototype development.

Design concepts are generated for each function of the system. You will have several functions to consider in the design of the overall hovercraft system with many design decisions involved with each function. For example, the hovercraft involves the six functions listed below. There are many design features that must be considered in designing the components (parts) necessary to meet each functional requirement. We will list some of the key design decisions associated with each of the six major functions.

1. Structural System – Supports all of the components, such as fans, NXT, batteries, sensors, etc. It also contains the cushion of air upon which the hovercraft floats.

   - Type of structure
   - Deck material
   - Size and shape of the deck
   - Size and location of openings in the deck
   - Attachment of components to the deck
   - Fabrication of the deck
   - Skirt material
   - Dimensions of the skirt

- Attachment of the skirt to the deck
- Sealing to minimize air leakage
- Locating the main battery supply to balance the deck

2. Levitation System – Includes the centrifugal fan and controls for that fan.

   - Development of a weight estimate
   - Analysis of required plenum pressure
   - Analysis of leakage flow
   - Selection of a lift fan or impeller to provide adequate pressure and flow rate.
   - Estimate of power requirement for the lift fan or impeller
   - Determine control method for the lift fan or impeller
   - Procure lift fan or impeller

3. Propulsion System – Includes a means of providing thrust and the control of the magnitude and direction of the thrust vector.

   - Develop strategy for steering
   - Analysis of forces required for propulsion
   - Selection of fans or other propulsion devices
   - Design mounting methods for fans or propellers
   - Determine power requirements
   - Specify control method for the propulsion fans or propellers
   - Procure propulsion fans or propellers

4. Power System – Includes external auxiliary power supplies and the control of these supplies.

   - Determine mission duration
   - Calculate power requirements (voltage, current and wattage)
   - Determine maximum current drain for each battery
   - Select method of controlling auxiliary power supplies
   - Perform design trade-off study to select battery chemistry and size
   - Procure the batteries required
   - Design mounting system for batteries and relays or transistors.

5. Sensing System – Provides feedback signals enabling the control of the hovercrafts speed and direction.

   - Analyze the mission requirements to establish a control strategy
   - Develop a data file that provides specification for various sensors
   - Conduct a design trade-off analysis to establish the most suitable sensing system
   - Procure the sensor or sensors used to provide feedback signals to the NXT
   - Calibration of sensors for use with NXT
   - Specify any auxiliary power requirements if necessary
   - Design mounting system for the NXT and sensors

6. Master Control System – Enables autonomous control to successfully complete the mission.

   - Procure a NXT unit to use as the microcontroller.
   - Procure a copy of the ROBOLAB software to program the NXT
   - Plan the strategy for executing the mission

- Review sensor data to insure that the sensors will provide the necessary feedback signals
- Write the program that will lift the hovercraft, propel it along the mission's course, steer it through the required turns, and stop it at its final destination.

The procedure in design is to generate ideas for performing each function, expanding these ideas with more detail until they are understood by all of the team members and finally to perform a design trade-off analysis where the merits and faults of each concept are evaluated. When the design proposals are selected for each of the different functions involved in the selection process, the winning concepts must then be integrated into a seamless system that meets the mission requirements.

We often refer to the part of a product that performs some function as a subsystem. The collection of all of the subsystems constitutes a complete system, which when manufactured and assembled, becomes a prototype. The integration of the various subsystems is often difficult, because one subsystem influences the design of the others. The manner in which the subsystems interact is defined as the interface between subsystems. In a mechanical application, the fit of a shaft in a bearing and/or the fit of a seal against a shaft are examples of interfaces. In an electronic application, the timing of the communication signals between subsystems is an example of an interface. Frequently, interfaces between the subsystems are troublesome and difficult to manage. It is vitally important that your team control the many interfaces created by the six different functions to enable seamless integration of all of the subsystems without loss of effectiveness and efficiency of the hovercraft system.

2.13 PREPARING DESIGN DOCUMENTATION

The final step in designing a hovercraft involves preparing the design documentation package. The package is an extended engineering document and contains:

- Engineering sketches or photographs (preferably digital) illustrating the design concepts.
- Engineering assembly drawings describing the arrangement of various components that fit together to form the complete system.
- Engineering drawings of all of the component parts in sufficient detail to permit the component to be manufactured by anyone capable of reading an engineering drawing or to be procured by a purchasing agent.

You are to define the names of all the unique parts to enable your team members to properly identify them. Assembly drawings are important because we must be able to convey the construction sequence for both subassemblies and the complete model. However, with the advent of digital photography it is possible to replace the assembly drawings with assembly photographs. Sketching will always be important as a quick means of conveying your ideas. We encourage you to use sketches in discussions with team members to convey your ideas more clearly in a graphical format.

We recognize that many of you may need instruction in graphics, and have included three chapters in this text to help you learn the techniques required to prepare three-view and pictorial drawings and to construct tables and graphs.

The design documentation package also contains a parts list that identifies every unique component employed in the assembly of the model. A typical example of a parts list for the hovercraft was presented previously in Fig. 2.37. This figure shows the quantity of each part required for the assembly of a single prototype.

An engineering report is also included as part of the design package. This report supports the design by describing the strategy and the key features incorporated in the hovercraft's development. Your report should treat each function involved in the design and describe the concepts that your team considered. The rationale for each design concept that was adopted, based on systematic design trade-off studies, is an essential element of the report. The report should also present a flow diagram of the program written for the controller. If the program is complex, a description of the programming strategy should also be included. Additionally, design analyses that predict the performance of your hovercraft are included. These analyses are completed before actually testing the model and the results from the analyses and the tests are useful in assessing the merits of your design. More information on the preparation of an engineering report is included in Chapter 12.

2.14 FINAL DESIGN PRESENTATION

The design phase of the product development process is concluded with a final design review. This is a formal review of the design of the hovercraft system and is presented to the class (peers), to the instructor and to any others in the audience. It is the team's responsibility to describe all of the unique features of the design and to predict the performance of the product. It is the responsibility of the class (peer group) and the instructor to question the feasibility of the design and the accuracy of the predicted performance. If you note a shortcoming or a flaw in the design during the design briefing, identify the problem to the team presenting their work. As a peer reviewer of another team's design, be tactful in your critique. Criticism, when provided for a colleague, is always a difficult and delicate task. Offer constructive suggestions in good faith and in good taste. If you are on the receiving end of criticism, do not become defensive. The individual offering the critique is not attacking your capabilities; he or she is actually trying to give you a suggestion that might help you achieve a better design.

The purpose of the design review is to locate design deficiencies and errors. It is much better to correct errors in the paper stage of the design process, not later in the hardware phase when it is considerably more difficult and costly to fix problems due to a poor design.

2.15 HOVERCRAFT ASSEMBLY

When the assembly kit is complete and checked against the parts list, you begin the final design phase— model evaluation. The first part of the evaluation is to assemble or build the model. This step is often called "first article build," because it is the first time that we have attempted to fit together all of the components used in the assembly. The "first article build" is successful if:

- All of the parts are available
- The parts all fit together properly
- The tolerances on each part and each feature are correct
- The surface finish on all of the parts is acceptable

Product design is often judged against the four F's—form, fit, finish and function. The "first article build" permits us to assess how well the team performed with regard to the first three of the Fs—form, fit and finish. Form, fit and finish of the various parts that your team manufacturers are dependent on your team's tool room skills and patience. Most of the components required for the hovercraft may be procured directly from suppliers. Only a few parts will require your skill with hand tools. Work smart and carefully to develop a model that looks professional. Function depends on the merits of your design. If your team develops functional subsystems and if they are properly integrated, the hovercraft will perform well and the mission will be conducted successfully.

We anticipate that each team will make a few errors in the preparation of the design drawings and changes will be required. The natural tendency is to remove and replace the offending parts to correct (modify) the design and move on with the "first article build." This behavior is acceptable only if the team revises the drawing of this part to reflect the modification made to correct the design deficiency. In the real world, the integrity of the drawing package is much more important than the prototype. The second and all subsequent assemblies will be fabricated from the details shown in your engineering drawings and not by examining the prototype. The prototype is often scrapped after it has been built and tested.

2.16 MODEL EVALUATION

After the assembly of the hovercraft is complete, it should be carefully inspected to insure that it is safe. Even simple products can be **dangerous.** For example, a large button that can be pulled from a doll is a potential danger if a child swallows it. In this course, the dangers are relatively small because we are dealing with low voltages and small, light-weight models. However, be certain that you have avoided introducing pinch points. Pinch points are openings that close due to motion of one or more parts when a product is operated. Fingers, hair and/or clothing caught in pinch points clearly represent a hazard). **Also, carefully shield any fan openings to avoid the possibility of a person inserting his or her finger into a rotating fan blade.**

The final step in this development process is to test your hovercraft to determine if it is functional and efficient. Typically the instructor prepares a test facility and schedules a time interval for evaluating each team's model. Your hovercraft will be judged based on the criteria announced by your instructor. He or she usually will select criteria from the following listing:

1. Performance: Ability to achieve the specified mission.
2. Control: Ability of system to perform quickly and effectively.
3. Cost: Minimize part count and avoid using expensive components.
4. Design innovation: Novelty and creativeness.
5. Quality of assembly: craftsmanship.
6. Appearance: Style

2.17 TEAMWORK

In this course, you are required to participate on a large development team. There are three reasons for this requirement. First, the projects are too ambitious for an individual to complete in the time available. You will need the collective efforts of the entire team to complete the design, construction and testing of a hovercraft model during the semester. Your instructor will press each development team to complete each phase of the design on schedule.

Second, it is a course objective that you to begin to learn teamwork skills. Experience shows that most students entering the engineering program have not fully developed these skills yes. From k-12, the educational process has focused on teaching you to work as an individual, often in a setting where you competed against other students in your class. We will insist that you begin functioning as a team member where cooperation, following, and listening are at least as important as individual effort. Leadership is important in a team setting, but cooperation and following the lead of others are also critical elements for successful team performance.

The final reason is to better prepare you for the real world that you will enter upon graduation with a B. S. degree in engineering. You will probably be assigned to a development team very early in your career if you take a position in industry. A recent study [4] by the American Society Of Mechanical Engineers (ASME), the results of which are shown in Table 2.3, ranked teamwork as the most important skill to develop in an engineering program. Teamwork was also the first skill, in a list of 20, considered important by managers in industry. We believe that this course will be instrumental in introducing you to the team working skills so necessary for a successful career in engineering.

As you work as a team member in the development of a hovercraft system, you will be introduced to several of the 20 topics listed in Table 2.3. As you begin to understand the design process, we trust that you will develop an appreciation for these skills.

Table 2.3
Important Skills for New Mechanical Engineers with Bachelor Degrees
Priority Ranking by Industrial Representatives

1. Teams and Teamwork	11. Sketching and Drawing
2. Communication	12. Design for Cost
3. Design for Manufacture	13. Application of Statistics
4. CAD Systems	14. Reliability
5. Professional Ethics	15. Geometric Tolerancing
6. Creative Thinking	16. Value Engineering
7. Design for Performance	17. Design Reviews
8. Design for Reliability	18. Manufacturing Processes
9. Design for Safety	19. Systems Perspective
10. Concurrent Engineering	20. Design for Assembly

2.18 OTHER COURSE OBJECTIVES

While your experiences in designing a model hovercraft with autonomous control are the primary objectives of this course, there are several other important educational objectives. You will quickly recognize a critical need for graphics as you attempt to describe your design concepts to fellow team members. We have included three chapters on graphics to help you learn the basic skills required at this entry level. These chapters include materials on three-view and pictorial drawings, graphs and tables.

As you proceed with the design of your hovercraft, we frequently will require you to communicate both orally and in writing. Design reviews before the class, provide you the opportunity to learn presentation skills such as style, timing and the preparation of visual aids. Preparing the design report gives you experience in developing complete, high-quality engineering drawings and in writing technical reports that contain text, figures, tables and graphs.

Another objective of the course is to introduce you to graphical programming. A chapter has been included to describe the usage of ROBOLAB to control your hovercraft model. Material is also included in this chapter that describes flow charts in facilitating the development of control strategies.

The final objective is to begin the development of your design analysis skills. We understand that you are beginning your studies. For this reason, we will present key equations that mathematically model the electrical and mechanical components used in many of the subsystems that you employ in the design of your hovercraft. We do not expect you to completely understand all of the theoretical aspects of the relatively complicated subsystems involved. Most of you will take several courses later in the curriculum dealing with these subjects in much greater detail. However, at this stage of your career, we want you to begin to appreciate the relationship between analysis and design. To help you with the analysis, we have included a chapter on the fluid mechanics and design analysis, a chapter on sensors and another chapter on electrical circuits. While the coverage is very brief and introductory, it will provide you with a better understanding of both mechanical and electrical systems.

REFERENCES

1. Valenti, M. "Teaching Tomorrow's Engineers," Special Report, Mechanical Engineering, Vol. 118, No. 7, July 1996.
2. Ulrich, K. T., S. D. Eppinger, 2nd Edition, Product Design and Development, McGraw-Hill, New York, NY, 2000.
3. Cross, N., Engineering Design Methods, 3rd Edition, Wiley, New York, NY, 1994.
4. Anon, "Integrating the Product Realization Process into the Undergraduate Curriculum," ASME report to the National Science Foundation, 1995.
5. Macaulay, D., The New Way Things Work, Houghton Mifflin Co., Boston, 1998.
6. Horowitz, P. and W. Hill, The Art of Electronics, 2nd Edition, Cambridge University Press, New York, 1989.
7. Dally, J. W., Riley, W. F. and K. G. McConnell, Instrumentation for Engineering Measurements, 2nd Edition, Wiley, New York, 1993.
8. Martin, Fred g., Robotics Explorations: A Hands-On Introduction to Engineering, Prentice-Hall, Upper Saddle River, NJ, 2001.
9. Capozzoli, P. and C. B. Rogers, "LEGOs and Aeronautics in Kindergarten Through College," AIAA, Vol. 96, 1996, p. 2245.
10. Cyr, M., V. Miragila, T. Nocera and C. Rogers, "A Low Cost Innovative Methodology for Teaching Engineering Through Experimentation," Journal of Engineering Education, Vol. 86, No. 2, 1997, pp. 167-171.
11. Portsmore, M., M. Cyr and C. Rogers, "Integrating the Internet, LabVIEW, and Lego Bricks into Modular Data Acquisition and Analysis Software for K-College," Proceedings of the ASEE 2000 Annual Meeting, Session 2320, St. Louis, June 19-23, 2000.

LABORATORY EXERCISES

2.1   Measure the static head developed by a lift fan or by a suitably confined impeller.
2.2   Measure the thrust developed by a small axial fan.
2.3   Calibrate a proximity sensor over operational range of the sensor.

EXERCISES

2.1   List ten products that you have used today and the companies that manufacture and market them.
2.2   Suppose that you do not want to work on product development, manufacturing or sales. What opportunities remain for you in engineering?
2.3   Write an engineering brief describing the characteristics of a winning product.
2.4   Write the most important equation in engineering and business.
2.5   Why is the Ford Motor Company reluctant to develop a completely new model of an automobile?
2.6   What is a design concept? How many concepts will your team generate in developing a propulsion system for the hovercraft model?
2.7   What are the main subsystems in the development of a hovercraft model?
2.8   Why is it important to prepare a parts list during a development program?
2.9   What is an assembly kit and why do we prepare one in the development of the first model?
2.10   Why do engineering executives invest scarce company funds in assembling a prototype?

2.11 What are the safety considerations with which you must be concerned in testing your hovercraft models?

2.12 Why do you believe industry representatives ranked teams and teamwork as the most important skill for new engineers?

2.13 Prepare a milestone schedule showing the date when each major design phase will be completed to ensure overall project success.

2.14 Prepare a strategy to guide the programming of the hovercraft to win the land race mission.

2.15 Prepare a strategy to guide the programming of the hovercraft that will successfully negotiate the specified path on the navigation mission.

2.16 Prepare a strategy to guide the programming of the hovercraft to quickly traverse the course on the land and water mission.

2.17 Prepare a strategy to guide programming the hovercraft to win the demolition mission.

2.18 Prepare a strategy to guide programming the hovercraft to win the water race mission.

2.19 Prepare a strategy to guide the programming the hovercraft to win the cargo mission.

Notes

CHAPTER 3

FLUID MECHANICS AND DESIGN ANALYSIS

3.1 INTRODUCTION

Developing new products requires a close relationship between design and analysis. When designing, we select components, layout the structure, place components, test subsystems, etc. At the same time engineers perform analyses to insure that all aspects of the design will function correctly and that the system will meet specifications. The hovercraft development provides an opportunity for your team to design a unique vehicle and to perform analyses. For example, before you can select a lift fan, an analysis must be performed to establish the pressure required to develop sufficient force to lift the hovercraft. You must also determine if the flow rate of the fan is adequate to sustain the leakage rate. The selection of the propulsion fans also requires prior analysis to insure that they will develop sufficient thrust to propel the hovercraft.

In design you often find that several different solutions are possible. Again analysis aids you in selecting the solution that is the most suitable while considering the constraints on your design. As we introduce analysis methods, we will describe solution space. Solution space covers a very wide range of analytical possibilities. At its center are the parameters that insure a successful design. At its extremes are the parameters that will lead to a design that often fails to meet its objective.

In this chapter we will introduce the theories necessary to perform several different design analyses associated with the development of a hovercraft. We will then perform analyses to insure that the functions (systems and processes) involved in designing a hovercraft operate satisfactorily. These functions are listed below:

1. The structural system supports all of the components, such as fans, impellers, propellers, the RCX, batteries, sensors, etc. The structural system also contains the cushion of air upon which the hovercraft floats.
2. The lift system includes the lift fan or impeller and the controls for that fan.
3. The propulsion system includes a means of providing thrust and the control of the magnitude and direction of the thrust vector.
4. The power system includes external auxiliary power supplies and control of these supplies.
5. The sensing system provides feedback signals enabling control of the hovercraft's speed and direction.
6. The master control system enables autonomous control to successfully complete the mission.

3.2 PRESSURE-MASS DENSITY-HEIGHT RELATIONSHIPS

Let's consider a body of water that is at rest as shown in Fig. 3.1. From the surface, we place a reference axis y that points downward into the depth of the water. Next examine a small element that is Δy thick with a cross sectional area of ΔA, located on the left side of this figure. A pressure p acts downward on its top surface and a pressure $p + \Delta p$ acts upward on its bottom surface. Note the pressure always act perpendicular to the surface upon which it acts and is increasing with the depth of the water.

In addition to these pressures, a downward force ΔW acts on the element. This force is due to the weight of the water contained within the element.

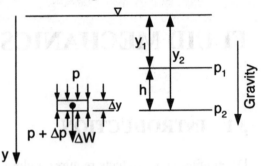

Fig. 3.1 Pressures created at depth in a static pool of water.

Let's write the equilibrium equation for the elemental volume of water depicted in Fig. 3.1. The equilibrium equation implies that the sum of the forces upward equal the sum of the forces downward. Thus, we write

$$(p + \Delta p)\Delta A \text{ (upward)} = p\, \Delta A + \Delta W \text{ (downward)}$$

This equation reduces to:

$$\Delta p \Delta A = \Delta W \tag{3.1}$$

The weight of the water in the elemental volume is given by:

$$\Delta W = \gamma\, \Delta A\, \Delta y \tag{a}$$

where $\gamma = \rho g$ is the specific weight, g is the gravitational constant and ρ is the mass density (mass/volume) of water.

Substituting Eq. (a) into Eq. (3.1) yields:

$$\Delta p = \gamma\, \Delta y = \rho g \Delta y \tag{3.2}$$

Summing both sides of Eq. (3.2) gives:

$$\sum_{p_1}^{p_2} \Delta p = \gamma \sum_{y_1}^{y_2} \Delta y \tag{b}$$

this reduces to:

$$p_2 - p_1 = \gamma\,(y_2 - y_1) = \gamma\, h = \rho g h \tag{3.3}$$

Equation (3.3) indicates that the pressure at some point in a liquid with a specific weight γ is dependent only on the height of the liquid above that point. This fact allows the vertical height or "head" of a specified liquid to be used as a measurement of pressure.

3.3 MANOMETERS

There are many transducers that can be employed to measure pressure. However, if the pressure is relatively low and constant, a manometer is a simple and effective measuring device. To show this simplicity, consider the U-tube manometer illustrated in Fig. 3.2. The U-tube is filled with a colored liquid and then inserted into a chamber with an unknown pressure p_x.

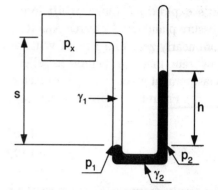

Fig. 3.2 A U-tube manometer measuring the unknown
pressure p_x.

The chamber pressure forces the liquid down on one side of the U tube and up on the other side. It is evident that $p_1 = p_2$ because they are at the same datum plane (level). From Eq. (3.3) we can write:

$$p_1 = p_x + \gamma_1 s \qquad\qquad \text{and } p_2 = 0 + \gamma_2 h \qquad\qquad (a)$$

Equating p_1 and p_2 leads to:

$$p_x = \gamma_2 h - \gamma_1 s \qquad\qquad (3.4)$$

If the liquid in the manometer is colored water and the fluid in the pressure chamber is air, then it can be assumed that the specific weight $\gamma_1 << \gamma_2$ and Eq. (3.4) reduces to:

$$p_x = \gamma_2 h \qquad\qquad (3.5)$$

It is clear from Eq. (3.5) that a simple U-tube manometer can be used to measure the air pressure in the plenum chamber under steady state conditions.

3.4 LIFT FORCES

Let's next consider the lift forces that are generated when we introduce pressure into the plenum of the hovercraft. The hovercraft will be in equilibrium in the vertical direction when the force developed by the pressure equals the total weight W of the hovercraft, as indicated in Fig 3.3.

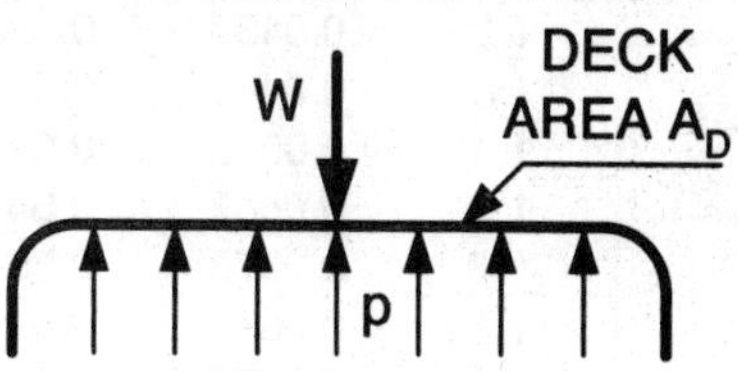

Fig. 3.3 Equilibrium diagram for the hovercraft (vertical direction).

For the hovercraft to be in equilibrium and floating on a cushion of air, the lift force due to the pressure must equal the downward force due to the weight W of the hovercraft structure and all of the components mounted on its deck. The lift force F_L is given by:

$$F_L = pA_D \qquad\qquad (3.6)$$

where A_D is the area of the deck of the hovercraft exposed to the pressure p.

The floating condition for the hovercraft based on equilibrium is:

$$F_L = W \qquad (3.7)$$

Let's explore the range of lift forces that can be generated by various combinations of the plenum pressure p and the deck area A_D. Reference to Fig. 2.33, Fig. 2.35 and Fig. 2.37 indicates that pressures from nearly zero to 2.3 in. of water are possible depending on the choice of the fan or impeller and the flow rate that the fan or impeller generates. To begin this analysis, let's convert the pressure measurement given in inches of water to pounds per square inch (psi).

From Eq. (3.5), we write:

$$p = \gamma h = \left[\frac{62.4 \, \text{lb/ft}^3}{1728 \, \text{in.}^3/\text{ft}^3} \right] h = \left(0.03611 \ \text{lb/in.}^3 \right) h \qquad (3.8)$$

The results from Eq. (3.8) are shown in the second column of the spreadsheet presented in Table 3.1 for p varying in steps of 0.1 from zero to 2.5 inches of water. Using Eq. (3.6) the lift force F_L was calculated for hovercraft deck areas ranging from 10 to 1000 in.[2] The results show that relatively high lift forces can be generated (in excess of 90 lb), but these larger lift forces require large deck areas and high plenum pressures.

Table 3.1
Lift force as a function of plenum pressure and deck area.

DETERMINATION OF LIFT FORCE AS A FUNCTION OF PRESSURE AND AREA								
		A=10 in^2	A=20 in^2	A=50 in^2	A=100 in^2	A=200 in^2	A=500 in^2	A=1000 in^2
p	p	Lift Force	Lift Force	Lift Force	Lift Force	Lift Force	Lift Force	Lift Force
in.of H2O	psi	lb	lb	lb	lb	lb	lb	lb
0.0	0	0	0	0	0	0	0	0
0.1	0.0036	0.04	0.07	0.18	0.36	0.72	1.81	3.61
0.2	0.0072	0.07	0.14	0.36	0.72	1.44	3.61	7.22
0.3	0.0108	0.11	0.22	0.54	1.08	2.17	5.42	10.83
0.4	0.0144	0.14	0.29	0.72	1.44	2.89	7.22	14.44
0.5	0.0181	0.18	0.36	0.90	1.81	3.61	9.03	18.06
0.6	0.0217	0.22	0.43	1.08	2.17	4.33	10.83	21.67
0.7	0.0253	0.25	0.51	1.26	2.53	5.06	12.64	25.28
0.8	0.0289	0.29	0.58	1.44	2.89	5.78	14.44	28.89
0.9	0.0325	0.32	0.65	1.62	3.25	6.50	16.25	32.50
1.0	0.0361	0.36	0.72	1.81	3.61	7.22	18.06	36.11
1.2	0.0433	0.43	0.87	2.17	4.33	8.67	21.67	43.33
1.3	0.0469	0.47	0.94	2.35	4.69	9.39	23.47	46.94
1.4	0.0506	0.51	1.01	2.53	5.06	10.11	25.28	50.56
1.5	0.0542	0.54	1.08	2.71	5.42	10.83	27.08	54.17
1.6	0.0578	0.58	1.16	2.89	5.78	11.56	28.89	57.78
1.7	0.0614	0.61	1.23	3.07	6.14	12.28	30.69	61.39
1.8	0.0650	0.65	1.30	3.25	6.50	13.00	32.50	65.00
1.9	0.0686	0.69	1.37	3.43	6.86	13.72	34.31	68.61
2.0	0.0722	0.72	1.44	3.61	7.22	14.44	36.11	72.22
2.1	0.0758	0.76	1.52	3.79	7.58	15.17	37.92	75.83
2.2	0.0794	0.79	1.59	3.97	7.94	15.89	39.72	79.44
2.3	0.0831	0.83	1.66	4.15	8.31	16.61	41.53	83.06
2.4	0.0867	0.87	1.73	4.33	8.67	17.33	43.33	86.67
2.5	0.0903	0.90	1.81	4.51	9.03	18.06	45.14	90.28

Before it is possible to proceed with either the design or the analysis of the design, it is necessary to estimate the weight of the hovercraft structure and all of its components. A preliminary weight estimate is shown in Table 3.2. It includes a selection of the components that were described previously in Chapter 4. The results listed show an estimated weight of 6.70 lbs for this hovercraft. It may prove necessary to modify this listing and add or change components if the design analysis of shows that the flow rate is not sufficient to provide a large enough gap between the skirt of the hovercraft and the surface over which the hovercraft's mission is to be executed.

Table 3.2
Estimated weight of hovercraft structure and components

WEIGHT TABLE			
Component	Quantity	Weight lb	Total Weight lb
Lift Fan (Impeller)	1	1.5	1.5
Propulsion Fan	2	0.6	1.2
Relays	3	0.1	0.3
NXT	1	0.7	0.7
Sensors	2	0.1	0.2
Main Battery (SLA)	1	1.0	1.0
Auxiliary Battery	1	0.1	0.1
Deck	1	0.75	0.75
Skirt	1	0.5	0.5
Fasteners	20	0.005	0.1
Wiring	4 ft	0.025	0.1
Tie Downs	10	0.01	0.1
Mounting Hardware	3	0.05	0.15
Total Weight (lb)			6.70

An examination of the results of this estimate indicates that most the weight is due to the three fans, the main battery and the NXT (65.2%). The estimates of these heavy components are accurate because the weight of each component was available from the suppliers. The weight of several other components, such as the relays, sensors, fasteners, tie downs, etc., was estimated using conservative numbers. The question is what should be the design target for the total weight. Let's select 7.5 lbs, which provides a safety factor of 1.12 in achieving lift-off.

3.5 SIZING THE STRUCTURE

At this point in the analysis, we have nearly enough information to size the deck of the hovercraft. The information needed to complete this phase of the analysis is related to the characteristic curve of the lift fan or impeller. Let's assume that the impeller shown in Fig. 2.36 has been selected with its characteristic operating curve, as shown in Fig. 3.4. Examination of this curve shows that the pressure from the fan drops steadily as the flow rate increases. To avoid the lower pressures at the higher flow rates, let's assume[1] the impeller operates at a minimum pressure of 1.0 in. of H_2O and a maximum flow rate of 72 CFM.

Examination of Table 3.1 shows a number of different pressure area combinations that yield lift forces that exceed 7.5 lbs. If we study the results for a plenum pressure of 1.0 in. of H_2O, it is clear that

---

[1] The operating point for the impeller or fan depends on the impedance of the flow channel. This impedance is largely controlled by the size of the gap between the hovercraft's skirt and the surface. As this gap goes toward zero, the pressure generated by the fan or impeller will increase toward its static limit.

the deck area of the hovercraft will have to be slightly larger than 200 in². From Eq. (3.6), it is possible to determine the required deck area A_D of the hovercraft for a pressure of 1.0 in. of H_2O as:

$$A_D = \frac{F_L}{p} = \frac{7.5}{0.0361} = 208 \text{ in}^2 \tag{a}$$

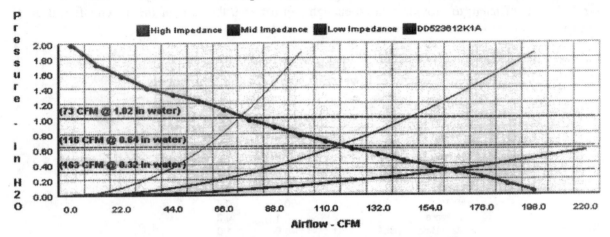

Fig. 3.4 Characteristic operating curve for an impeller.
Comair Rotron Model DD523612K1A Part No. 039832.

It is clear from the data listed in Table 3.1 and Table 3.2 that the impeller selected[2] is sufficient to lift the hovercraft if the deck area is 208 in² or larger if a plenum pressure of 1.0 inches of H_2O is developed. The next question to consider in this design analysis is if the flow rate is sufficient to maintain a reasonable size gap between the skirt on the hovercraft and the surface upon which it must travel.

3.6 DETERMINING THE LEAKAGE RATE

Continuity Equation

To determine the leakage rate through the air gap around the perimeter of the hovercraft, it is necessary to introduce two very important equations used in fluid mechanics—the conservation of mass relation and Bernoulli's equation. Let's consider the flow of a fluid through a stream tube as illustrated in Fig. 3.5. Under steady state conditions, the mass[3] of fluid entering the stream tube is the same as that leaving it.

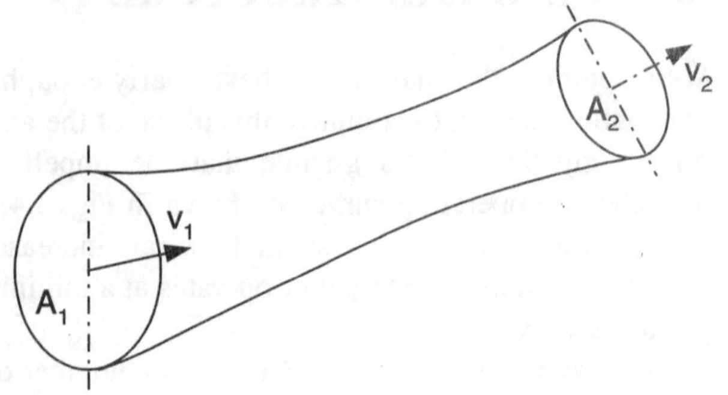

Fig. 3.5 Flow into and out of a stream tube.

---

[2] Many other impellers are available with higher and lower flow rates. Design trade-off studies can be conducted to balance the benefits of increased flow rates with the increased weight and power requirements of the larger capacity impellers.
[3] This statement is based on the principle of conservation of mass.

Because no fluid is lost or stored as it passes through the steam tube, we can write the conservation of mass condition as:

$$A_1 v_1 \rho_1 = A_2 v_2 \rho_2 \tag{3.9}$$

where v is the velocity of the flow.

If we assume the flow is incompressible,[4] $\rho_1 = \rho_2$ and Eq. (3.9) reduces to:

$$A_1 v_1 = A_2 v_2 = Q \tag{3.10}$$

where Q is the flow rate.

It is evident from Eq. (3.10) that the product of the velocity and the area is a constant everywhere along a stream tube.

Bernoulli's Equation

Bernoulli's equation is based on the principle of conservation of energy. To illustrate this concept, consider frictionless flow though the piping system shown in Fig. 3.6. At station 1 the fluid, flowing with a velocity v_1, is at a pressure p_1 and a specific weight γ_1. As the fluid travels from station 1 to station 2, the pipe changes elevation and the cross sectional area of the pipe decreases. The fluid at station 2 is flowing with a velocity v_2 and it exhibits a pressure p_2 with a specific weight γ_2. If energy is not added or lost as the fluid moves from station 1 to station 2, then the principle of conservation of energy leads to Bernoulli's equation that is written as:

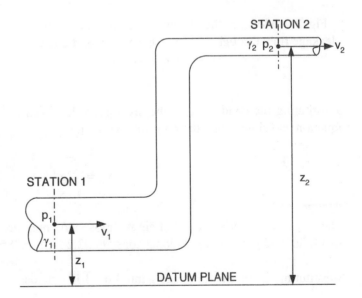

Fig. 3.6 Flow conditions in a piping system without energy added or lost.

---

[4] In this analysis we assume that air is incompressible. Of course air is compressible, but we can make this assumption because the changes in pressure encountered when pressurizing the plenum chamber are very small. In other words, the low pressure air generated by the impeller can be treated as an incompressible fluid without introducing serious errors.

$$\frac{p_1}{\gamma_1} + \frac{v_1^2}{2g} + z_1 = \frac{p_2}{\gamma_2} + \frac{v_2^2}{2g} + z_2 \qquad (3.11)$$

The first term $\frac{p}{\gamma}$ may be considered as the work that can be performed by the fluid with a pressure p.

The second term $\frac{v^2}{2g}$ is related to the fluid's kinetic energy. Finally the third term z is the fluid's potential energy relative to a defined datum plane. Each of these terms can also be considered as heads—the pressure head, the velocity head and the potential head. It is evident from Eq. (3.11) that the sum of the three terms involving pressure, velocity and height above a datum plane are constant at every point along a stream tube.

Leakage Flow Rate

We will use Bernoulli's equation to derive a simple equation that can be employed to estimate the leakage flow rate. First, we model the plenum chamber of the hovercraft as illustrated in Fig. 3.7. Assume the air pressure is sufficient to lift the hovercraft forming a gap between the surface and the edge of the skirt. The flow exits this gap at a velocity v_2. As the flow vents to the atmosphere, the pressure at the exit is given by $p_2 = 0$. Inside the plenum chamber, the lift impeller is maintaining a pressure p_1. We also assume that the plenum chamber is large relative to the size of the impeller duct so that the velocity of the air within the plenum can be set at $v_1 = 0$. The change in the elevation between station 1 and 2 is negligible.

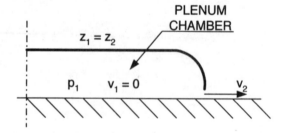

Fig. 3.7 Model of the plenum chamber showing the leakage flow through the gap between the skirt and the surface.

Substituting the conditions on the pressure, velocity and elevation at stations 1 and 2 into Bernoulli's equation and simplifying the resulting expression gives:

$$v_2 = \sqrt{2g\frac{p_1}{\gamma_1}} \qquad (3.12)$$

where γ_1 is the specific weight of air at atmospheric pressure (14.7 psi), which is 0.0763 lb/ft^3 or 4.416×10^{-5} lb/in^3 and g is the gravitational constant 32.2 ft/s^2 or 386 in./s^2.

Substituting the results for γ_1 and g into Eq. (3.12) yields:

$$v_2 = \sqrt{2g\frac{p_1}{\gamma_1}} = \sqrt{\frac{(2)(386)10^5}{4.415}p_1} = 4.182 \times 10^3\sqrt{p_1} \qquad (3.13)$$

where the units for v_2 and p_1 are in./s and psi, respectively.

The results in Eq. (3.13) indicate that the velocity of the air leaking from the gap between the skirt of the hovercraft and the surface increases as the square root of the plenum pressure. Recall from Section 3.5 that we assumed the impeller generated a minimum pressure of 1.0 in. of H_2O and a flow rate of 72 CFM. A plenum pressure of 1.0 in. of H_2O corresponds to a pressure $p_1 = 0.0361$ psi. Substituting this value into Eq. (3.13) yields:

$$v_2 = 794.6 \text{ in./s} = 3{,}973 \text{ ft/min} \tag{a}$$

Next recall Eq. (3.10) and note that $Q_{Impeller} = 72$ CFM at an operating pressure of 1.0 in. of H_2O. Then from this equation, it is possible to write:

$$A_g = \frac{Q_{Impeller}}{v_2} \tag{3.14}$$

Substituting numerical values for v_2 and Q_{Fan} into Eq. (3.14) yields:

$$A_g = \frac{Q_{Impeller}}{v_2} = \frac{72}{3{,}973} = 0.0181 \text{ ft}^2 = 2.610 \text{ in.}^2 \tag{b}$$

The result in Eq. (b) indicates that the fan operating at 72 CFM can sustain a total gap area of 2.610 in². Next, let's establish the height of the gap between the skirt and the surface. This dimension is important, because the gap must be sufficiently large to permit the hovercraft to travel over slightly rough surfaces.

In Section 3.5 the deck area A_D was calculated as 208 in.² if the lift force was to be at least 7.5 lb. However, we did not establish the shape of the deck. Is it to be square, rectangular or circular? To minimize leakage, it is important to minimize the perimeter of the deck. It is well known that the optimum shape to maximize area while minimizing perimeter is a circle. Hence, the design of the hovercraft structure incorporates a circular plenum chamber. The diameter D of a circle with an area A_D is given by:

$$D = \sqrt{\frac{4A_D}{\pi}} = \sqrt{\frac{4(208)}{\pi}} = 16.27 \text{ in.} \tag{c}$$

The circumference C of this circle is given by:

$$C = \pi D = \pi (16.27) = 51.13 \text{ in.} \tag{d}$$

Finally, the dimension g of the gap is given by:

$$g = \frac{A_g}{C} \tag{3.15}$$

Substituting the results from Eq. (b) and Eq. (d) into Eq. (3.15) gives:

$$g = \frac{A_g}{C} = \frac{2.610}{51.13} = 0.0511 \text{ in.} \tag{e}$$

For a lift impeller operating at $Q_{Impeller}$ = 72 CFM at a pressure of 1.0 in. of H_2O, a gap of 0.0511 in. can be maintained in a steady state. If the surface over which the hovercraft travel is rough, there is a possibility that the skirt will snag and possibly cause the hovercraft to turn or to stop. Is it possible to increase the gap by changing the operating point so that the fan generates higher flow at lower pressure? Is it possible to add a second impeller and to double the flow rate thereby increasing the clearance between the hovercraft skirt and the surface of the course? Another option is to select a large impeller with a higher flow rate. The larger impellers weigh more, require more deck area for mounting and draw higher currents.

The impeller selected for the analysis presented above is one of many commercially available. It may not be the most suitable model for the hovercraft your team is designing. However, the analysis method demonstrated above enables your team to explore solution space and to select a more suitable lift fan or impeller.

3.7 DETERMINING THE PROPULSION FORCE

Let's next determine the propulsion force or forces necessary to drive and steer the hovercraft. Recall that the hovercraft is floating on a cushion of air so it can be moved with relatively small forces. Frictional and drag forces that are usually a significant factor in propulsion are not a consideration in the design of a hovercraft that is to travel at relatively low velocities over short distances while floating over a relatively smooth surface.

For a hovercraft, propulsion forces are generated by momentum transferred due to air flow. This air flow can come from a number of sources—the most common of which are aft mounted fans as indicated in Fig. 2.4. Opening an orifice and bleeding air from the plenum chamber would also provide a thrust force due to the momentum transferred by that air flow. Finally, tilting the hovercraft forward so that the gap about the circumference was not equal would generate non symmetrical air flow that in turn would generate forces on the hovercraft to drive it forward.

Let's consider an axial fan similar to the one shown in Fig. 2.38 to provide a propulsion force F_P. According to the impulse- momentum principle, we can write:

$$\mathbf{F} \, \Delta t = \Delta(\mathbf{M}) = \Delta \mathbf{mv} \tag{3.16}$$

where $\mathbf{F} \, \Delta t$ is the impulse due to the propulsion force F_P.

$\Delta \mathbf{M} = \Delta \mathbf{mv}$ is the change in momentum of the air as it passes through the axial fan.

Using Eq. (3.10) it is possible to express the change in momentum as:

$$\Delta mv = (A_0 \rho_0 v_0) \Delta t \, v_0 - (A_i \rho_i v_i) \Delta t \, v_i \tag{3.17}$$

where the subscripts i and o refer to the inlet and outlet conditions of the flow at the fan. Note that the incompressibility assumption indicates that mass density $\rho_0 = \rho_i = \rho$. We also assume that the velocity v_i of the hovercraft is low compared to the exit velocity v_0 of the axial fan; hence it is evident that $v_0 = v_i$. Consequently, Eq. (3.17) reduces to:

$$\Delta mv = (A_0 \, \rho \, v_0) \Delta t \, v_0 = Q \, \rho \, v_0 \, \Delta t \tag{3.18}$$

Substituting Eq. (3.18) into Eq. (3.16) yields[5]:

$$F_P = Q\rho\, v_0 \qquad (3.19)$$

Let's employ Eq. (3.19) to determine the propulsion force from a single axial fan. To begin refer to the characteristic curve of an axial fan, which is illustrated in Fig. 3.8.

Reference to Fig. 3.8 indicates that flow rates from the fan of about 80 to 100 CFM at low pressures are possible. The fan will be exhausting into the atmosphere so small pressures will be required to achieve these flow rates. Let's use 80 CFM as a conservative estimate of the flow rate from this axial fan. The output velocity v_0 corresponding to this flow rate is given by Eq. (3.10) as:

$$v_0 = \frac{Q}{A_0} = \frac{80}{0.0833} = 960 \text{ ft / min} \qquad (a)$$

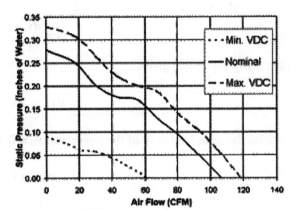

Fig. 3.8 Pressure as a function of air flow for a 120 × 120 × 38 mm axial fan.

where the fan area excludes the hub where the motor obstructs the flow.

The mass density for air at standard temperature and pressure is given by:

$$\rho = \frac{\gamma}{g} = \frac{0.0765}{32.2} = 2.376 \times 10^{-3}\, \frac{\text{lb} - \text{s}^2}{\text{ft}^4} = 0.66 \times 10^{-6}\, \frac{\text{lb} - \text{min}^2}{\text{ft}^4} \qquad (b)$$

Finally, we substitute these results into Eq. (3.19) to obtain the propulsion force F_P as:

$$F_P = Q\rho\, v_0 = (80)(0.66 \times 10^{-6})(960) = 0.0507 \text{ lbs} \qquad (c)$$

The force produced by this fan is relatively small—slightly less than one ounce. Will this be sufficient to drive the hovercraft and if so what velocities can be achieved?

---

[5] In this relation A and ρ are scalar quantities and F_P and v_0 are vectors. Because the flow is uniaxial, the directions of both F_P and v_0 are collinear and coincide with the axis of the fan. Because of this fact we can treat them as scalar quantities.

3.8 DETERMINING THE SPEED OF THE HOVERCRAFT

Let's assume that we have two propulsion fans mounted on the aft end of the hovercraft producing a total thrust force of 0.1 lb. This force accelerates the hovercraft from its berth according to Newton's second law that states:

$$F = ma \tag{3.20}$$

where m is the mass of the hovercraft that is equal to $W/g = 7.5/32.2 = 0.2329$ lb-s^2/ft.

The acceleration a of the hovercraft is given by:

$$a = \frac{F}{m} = \frac{0.1}{0.2329} = 0.4293 \ \frac{ft}{s^2} \tag{a}$$

Recall the definition of acceleration, which is:

$$a = \frac{dv}{dt} \tag{3.21}$$

Integrating Eq. (3.21) gives:

$$v(t) = a\,t + v_0 \tag{3.22a}$$

Because the initial velocity of the hovercraft $v_0 = 0$ at $t = 0$, Eq. (3.22a) reduces to:

$$v(t) = a\,t \tag{3.22b}$$

Recall the definition of velocity, which is:

$$v = \frac{ds}{dt} \tag{3.23}$$

where s is the distance traveled.

Substituting Eq. (3.22b) into Eq. (3.23) and integrating gives:

$$s = \tfrac{1}{2}\,at^2 + s_0 \tag{3.24a}$$

This relation reduces to:

$$s = \tfrac{1}{2}\,at^2 \tag{3.24b}$$

if $s_0 = 0$ at $t = 0$.

Under a constant acceleration of 0.4293 ft/s^2, the hovercraft will slowly gain speed with time of operation. The speed and distance traveled as a function of time t is presented in Table 3.3.

Table 3.3
Velocity and distance traveled of a hovercraft at a constant acceleration of 0.4293 ft/s².

Time (s)	Velocity v (ft/s)	Distance s (ft)
1	0.429	0.21
2	0.859	0.86
5	2.15	5.37
10	4.29	21.5
20	8.59	86.9
30	12.9	193
40	17.2	343
50	21.5	537
60	25.8	773

These results indicate that the hovercraft slowly acquires velocity but after 60 seconds it has reached a speed of 25.8 ft/s (17.6 MPH). These speeds are much too high to perform the maneuvers required on certain missions; however, lower speeds can be achieved by cycling the propulsion fans on and off, by selecting much smaller fans or by stopping and restarting the lift fan periodically to limit the hovercraft's velocity.

To steer the hovercraft, one of the fans is turned off. The thrust from the other fan is off axis and its force creates a moment that turns the hovercraft. To minimize this turning radius, the fans should be separated as far as possible. The results in Table 3.3 also indicate that a hovercraft operating in a straight line at full throttle will travel about 773 ft in 60 seconds. The speeds possible with small fans, which generate propulsion forces measured in fractions of an ounce, are sufficient to complete any of the missions described previously within a few minutes, if the hovercraft does not snag on an obstacle or the surface of the course.

3.9 SIZING THE AUXILIARY POWER SUPPLY

It is clear that the power required to drive the lift and propulsion fans must be supplied from an auxiliary power source. What is the required capacity of this source? If we select a source that has excess capacity, we incur a weight penalty. If we select one that is too small, we may exhaust the battery before the mission is complete. The current drawn by the fan or impeller motor varies with the flow rate. At a flow rate of 72 CFM, the 12 Vdc motor driving the impeller shown in Fig. 2.37 draws about 2.85 A. The propulsion fans are rated at 6 W at 12 Vdc, which implies a maximum current draw of 0.5 A for each motor. Consequently, we will need an auxiliary power supply that can deliver about 4 A for sufficient time to complete a mission.

Let's assume that your team has a strategy for completing a mission in 6 minutes (1/10 h). The capacity C_B of the battery is given as the product of the amperes multiplied by the time that this level of current is delivered.

$$C_B = I\,t \qquad\qquad (3.25)$$

where C_B is the capacity measured in (Ah) and I is the current draw in A and t is the time in h.

For the 6 minute mission, the capacity of the auxiliary supply is $C_B = 4 \times 1/10 = 0.4$ Ah. This value is small compared to the capacity of the battery in your automobile, but you certainly do not want to deal with the weight of an auto battery.

There are three types of batteries that provide this capacity and current drain—Nickel Metal Hydride, Nickel Cadmium and Sealed Lead Acid (SLA). All three of these types are rechargeable that is also a requirement to avoid battery replacement after each mission. Go online and explore the battery sizes available, weight of the battery packs and the cost of the installation before making your selection. We found a SLA battery that delivers 0.8 Ah at 12 V that weighs 1.0 lb at a cost of about $10.00. Can your team find a better battery for this mission?

3.10 DESIGN CONSIDERATIONS FOR THE DECK AND SKIRT

The two main structural elements of the hovercraft are the deck and the skirt. The deck should be sufficiently strong and rigid to carry the weight of the fan or impeller, the batteries and all of the other lighter components. Reference to the preliminary weight listings in Table 3.2 indicates that the weight of the deck estimated at 0.75 lb and as such it is a not significant factor in the total weight target of 7.5 lb. It is probably advisable to use thin plywood or glass reinforced plastic sheets as deck material and attempt to achieve a deck weight of less than 0.75 lb.

The skirt design is dependent on the mission. If the mission includes traveling over water, the skirt must provide a buoyancy force that is sufficient to float the hovercraft. In this case, the skirt is usually made from an inflatable tube. Another option would be to fabricate the skirt from foam plastic, which has the advantage of easier fastening to the deck.

If the mission is over a solid surface, there are more options available for skirt materials. An inflatable tube is still an option, but it is not easy to fasten the tube to the deck in an air tight manner. Rings of closed cell plastic foam may be employed but they are more rigid and prone to snagging. Relatively thick plastic sheet (4 to 6 mils) has been used by some model makers[6] to produce a relatively large flat tube as illustrated in Fig. 3.9. The air from the lift impeller inflates this plastic tube and then is vented into the plenum chamber. The plenum pressure then lifts the cushions and the hovercraft together to form a gap along which the escaping air flows.

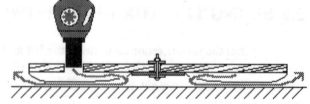

Fig. 3.9 Plastic sheeting used to form a cushion like skirt. Design and drawing by William J. Beaty.

The arrangement shown in Fig. 3.10, utilizes a ¾ segment of corrugated tubing for the skirt. This tubing is made from a relatively thin plastic material that is easy to cut and to staple. The tube is stapled to the top of the deck and then the reentrant corner is filled with a sealant to form an air tight plenum chamber. Slicing the tube to form the ¾ segment provides a structure that is semi-rigid, yet sufficiently flexible to adjust for slight surface irregularities that might be encountered.

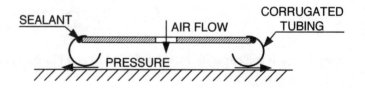

Fig. 3.10 A segment of corrugated tubing fastened and sealed to the deck serves as a skirt.

---

[6] This design is described by William J. Beaty a research engineer at the University of Washington. He spends his free time running SCIENCE HOBBYIST, a large website for amateur science and science education. You can locate an interesting article on hovercraft construction by Mr. Beaty on www.amasci.com/me.html.

Another approach is to use either a container[7] or the lid to a container found in a kitchenware department of a retail outlet. These containers are unusually made from relatively thick polyethylene and they come in a large number of different shapes and sizes. The walls of say a dishpan can be employed to form the skirt. The bottom of the dishpan may not be sufficiently rigid for the deck, but it can be reinforced with a thin sheet of plywood. Attachment of the plywood sheet to the plastic pan is easily accomplished with either staples, tacks or small screws.

It is clear that there are a number of different approaches to designing and constructing a hovercraft structure that has a strong and rigid deck, an air tight plenum chamber and skirt with sufficient flexibility to accommodate modest surface roughness.

3.11 SENSOR SELECTION

Sensor selection depends on the mission requirements. If the mission calls for the hovercraft to move from point a to point b, a line following strategy is often employed. In this case, a line is made delineating the course between points a and b. This line is usually made with black tape on a white background to provide the maximum contrast for detecting reflected light. The light reflected from the tape is measured with a light sensor to provide a feedback signal. Other missions may require either contacting obstacles or avoiding them. In these cases either touch sensors or proximity sensors are used to provide feedback signals. Regardless of the mission, the sensors used must interface with the NXT controller. Information pertaining to the design of sensors that interface with the NXT is provided in Chapter 5.

Another approach is to procure sensors from the LEGO Educational Division because they are designed to interface with the NXT. Sensors available from the LEGO Educational Division include:

- Touch sensors—are normally open limit switches housed in peg mounted package.
- Temperature sensors for the RCX controller—are temperature-sensitive resistors (thermistors). They are housed in a 2×3 brick with a small probe extending out about 3 L from the short side of the brick. They measure temperatures from $-20°$ C to $50°$ C. Conversion cables are available to adapt this sensor to the NXT controller. The NXT converts the resistance measurement to a temperature, which is displayed on its LCD.
- Light sensors—are active devices (requiring power) that include a light emitting diode (LED) and a phototransistor built into a peg mounted package. The sensor can be used either to detect incident light or to detect reflected light from the LED.
- Rotation sensors—are incorporated in the LEGO servo motors include an optically coupled encoder with a resolution of 1.0 degrees.
- Ultrasonic sensors—are distance measuring instruments. They send out an ultrasonic signal and measure the time of transit of the signal. From the speed of sound and this time measurement, the distance to the sound reflecting object is determined.

3.12 MASTER CONTROL SYSTEM

The NXT serves as the master controller. Its three output ports A, B and C can control three relays or transistors that provide 12 Vdc with sufficient current to operate the lift and propulsion fans. The sensors are connected to the input ports 1, 2, 3 or four on the NXT. The type of sensors used will depend on the mission and the strategy for accomplishing it. The signals that they provide will serve as the basis for controlling the power to the fan and/or impeller motors.

---

[7] Such as dishpans or storage containers of all sizes.

The program for controlling the operation of the hovercraft's motors so as to successfully complete the assigned mission is to be written in ROBOLAB. A detailed description of programming in ROBOLAB is presented in Chapter 6.

3.13 SUMMARY

In this chapter we introduced the theory necessary to perform the design analyses associated with developing a hovercraft. We performed several analyses to insure that the functions (systems and processes) involved in designing a hovercraft would operate satisfactorily.

Because the pressures developed by fans are often expressed in terms of inches of water, we derived the pressure height relation based on the equilibrium equation. Then showed how a manometer measures pressure with the height of a column of water. Next we determined the lift force that could be created with a plenum chamber at a pressure p and a deck area of A_D. The results were tabulated for a large number of pressures and areas to provide a "solution space" for lift requirements. An example of the first estimate of the weight of the hovercraft was presented. The operating characteristics of a centrifugal fan and/or impeller were used to select an operating flow rate and plenum pressure. These values were then used to determine the deck area or the hovercraft.

The continuity relation was derived based on the principle of conservation of mass and Bernoulli's equation was derived based on conservation of energy. These two equations were then used to determine the leakage flow rate. By equating the leakage flow rate with the flow rate generated by the fan the gap area was determined. A gap dimension of 0.0511 in. was determined for a plenum chamber with a circular base for the impeller selected and the assumptions made in the analysis.

The propulsion force developed by a small axial flow fan was calculated base on the momentum-impulse principle. Using this result we then determined the velocity and distance traveled by a hovercraft as a function of time of operation.

Guidance was provided for sizing the auxiliary power supply and the selection of the battery chemistry. Several design concepts for the hovercraft's skirt were introduced and materials and methods of fabrication were briefly described. Finally a section describing sensor selection was included. The LEGO sensors which interface easily with the RCX were described. Also we discussed a proximity sensor, which issues a signal when objects about 240 mm away are encountered. This sensor, which has been adapted to the RCX, may be useful for several of the missions.

Finally, the RCX was identified as the master controller. It will be necessary to develop a strategy for completing the specified mission and to program the RCX in ROBOLAB to execute this strategy.

LABORATORY EXERCISES

3.1 Construct a manometer capable of measuring a head of 200 mm of water. Could you use this manometer to measure the outlet pressure from a tank of propane used as a fuel for a barbeque grill?

3.2 Construct a plenum chamber and pressurize it with your team's lift fan or impeller. Measure the plenum pressure with a manometer as you vary the gap between the plenum chamber and a smooth flat surface. Vary the gap from zero to 3.0 mm in steps of 0.5 mm.

3.3 Using the plenum chamber from Laboratory Exercise 3.2, measure the flow rate of your lift fan as a function plenum pressure. This can be accomplished by cutting a series of circular holes in the side of the plenum chamber to serve as orifices. The flow rate can be determined for each orifice diameter if the plenum pressure is measured.

EXERCISES

3.1   Determine the pressure in units of pounds per square inch (psi) of the water in a swimming pool at a depth of 8 feet.

3.2   What is the pressure of air in Pa if a mercury barometer is reading 76 mm of Hg?

3.3   Reproduce Table 3.1 using SI units instead of U. S. Customary units.

3.4   Derive an expression similar to Eq. (3.8) for the pressure in Pa if the head is measured in mm of H_2O.

3.5   After you have discussed design concepts and have searched the Internet for components, generate a table similar to that shown in Table 3.2 that estimates the weight of your team's hovercraft.

3.6   Establish the target weight of your hovercraft, compute the safety factor for this weight and prepare a brief statement justifying the estimated and target weights.

3.7   After your team has selected its lift fan and/or impeller, assume a flow rate and pressure of its operation. Describe the rationale for this assumption.

3.8   Based on the results of Exercises 3.6 and 3.7, determine the area required for the deck of your team's hovercraft.

3.9   Based on the result of Exercise 3.7, calculate the velocity of the air leaking through the gap between the skirt and the surface.

3.10   Based on the results of Exercises 3.7 and 3.9, compute the diameter of the plenum chamber and the height of the gap.

3.11   Select an axial flow fan and, from its characteristic curve, choose an air flow rate that the fan can maintain when used as a source of propulsion. Determine the velocity of the air flow from this fan.

3.12   Using the results of Exercise 3.11, calculate the propulsion force generated by this fan.

3.13   If you use two of the fans selected in Exercise 3.11 for propulsion, determine the velocity achieved by the hovercraft after 32s. Also determine the distance traveled.

3.14   Perform a design trade-off study and select the chemistry of the batteries used in the auxiliary power supply. Write an engineering brief discussing your findings.

3.15   Determine the capacity of the auxiliary power supply if it is to supply 5 A for a mission with a duration of 12 minutes.

3.16   Prepare a drawing of the hovercraft including its skirt. Discuss the materials selected and describe the manufacturing methods your team intends to employ in fabricating the model.

3.17   Based on the assigned mission prepare a strategy for control of the hovercraft to complete the mission in a minimum time.

3.18   Select the sensor or sensors necessary to implement the strategy prepared in Exercise 3.17.

3.19   Write a program in ROBOLAB that will activate the lift fan or impeller.

3.20   Write a program in ROBOLAB that will activate both propulsion fans and drive the hovercraft forward for 13s.

3.21   Write a program in ROBOLAB that will steer the hovercraft to port until it turns through 90°.

3.22   Write a program in ROBOLAB that will enable the hovercraft to follow a black line formed by 2 in. wide tape on a white background.

3.23   Prepare a plan for procuring a proximity sensor to use on the navigation mission.

Notes

CHAPTER 4

BASIC ELECTRIC CIRCUITS

4.1 INTRODUCTION

This chapter introduces you to the basic quantities, concepts, and physical devices used in electric circuits. Let's begin with definitions of charge, current, and voltage, the basic quantities of electricity and other components—resistance, capacitance, and inductance—that characterize circuit elements. Only the simplest concepts governing circuits will be described in this chapter. The basic passive components—those that do not require an external power source—will be described. Two active components, a diode and a transistor, will also be described.

4.2 BASIC ELECTRICAL QUANTITIES

Electricity involves charge carriers—either electrons, which are negative charge carriers, or holes which are positive charge carriers. Since we cannot see, smell or touch an electron, electricity has always been somewhat mysterious to most of us. Holes, which are atoms with an electron missing from its outer ring, are even more obscure. Nevertheless, we will try to introduce many of the basic electrical concepts in a simple, straightforward manner and investigate the mystery.

Electricity can be understood most simply by drawing an analogy between charge flowing through a circuit and water flowing through a pipe. The analogy is not perfect, but it provides a great deal of insight into the basic operation of electricity.

4.2.1 Charge

Charge (q) is the fundamental quantity of electricity. It is the substance of which current is made, just as a volume of water is the substance of which water current is made. Charge is measured in units of coulombs (C). One electron has a charge of 1.602×10^{-19} C. We can think of charge as a substance (free electrons) that is contained within a conductor such as a wire. This is analogous to a hose full of water. Charge does not really **do** anything until it moves. Excess charge can create an electrostatic force, but nothing happens until the charge moves.

4.2.2 Current

Current (I) is charge in motion. The amount of current flow is measured in amperes (A), which are coulombs per second. Just as we speak of the rate at which a volume of water moves through a pipe, we can think of current as a flow rate. The higher the current, the higher the rate at which charge is moving by a point in a conductor. This concept is expressed in equation form as:

$$I = dq/dt \qquad (4.1)$$

Th e current I is the rate of change in or movement of charge per second.

4.2.3 Voltage

Voltage is another fundamental electrical quantity. The unit of voltage is volts (V). It is the electrical equivalent of pressure. Just as a pump creates pressure to force water through a pipe, a voltage source creates pressure to push charge through a conductor.

A voltage source is a circuit element that delivers a specified voltage (electrical pressure) regardless of the current flowing. The most basic voltage source is a battery. A flashlight battery pushes charge through a conductor that comprises the light bulb filament, creating friction within the filament, and causing the bulb to glow. A 6-V battery pushes four times as hard as a 1.5-V battery. The symbol for a constant voltage source (called a **direct current**, or dc, source) such as a battery is shown in Fig. 4.1a. A more general voltage source is shown in Fig. 4.1b.

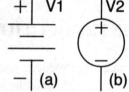

Fig. 4.1 (a) dc voltage source symbol
(b) general voltage source symbol.

4.2.4 Resistance

Resistance measures the difficulty of pushing current from one place to another. The unit of resistance is ohms (Ω). The higher the resistance of a device, the more difficult it is to push current through it. Pumping water through a small-diameter pipe requires more pressure to achieve the same flow rate as pumping through a larger-diameter pipe. A high-resistance electrical device is analogous to a small-diameter pipe, resisting more strongly the flow of charge. Likewise, a low-resistance device is analogous to a large-diameter pipe that offers little resistance to flow.

A resistor is a circuit element that has a fixed resistance regardless of the input voltage. A resistor in a circuit loop with a voltage source is shown in Fig. 4.2. The connecting lines between the voltage source and the resistor are wires, which can be thought of as resistors with zero resistance. In our water analogy, wires are extremely large pipes that offer little resistance to water flow.

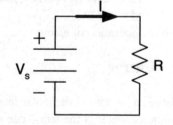

Fig. 4.2 A simple circuit loop containing a dc voltage source and a resistor.

The relation between the applied voltage V_s, the resistance R, and the current flow I is given by:

$$V_s = I\,R \qquad (4.2a)$$

$$I = V_s / R \qquad (4.2b)$$

Equations (4.2a) and (4.2b) are both expressions of Ohm's law. When V_s is expressed in volts and I in amperes, R is given in ohms (Ω).

The resistance R of a conductor is given by:

$$R = \rho L / A \qquad (4.3)$$

where ρ is the specific resistance of the metal conductor (Ω-cm).
   L is the length of the conductor (cm).
   A is the cross-sectional area of the conductor (cm^2).

We can fabricate resistors by winding very-small-diameter, high-resistance wire on coils or by evaporating thin films of metal on ceramic substrates. The relation in Eq. (4.3) allows us to select the metal and to size the conductor to provide a specified resistance.
   A resistor is a dissipative component. When current flows through a resistor, power is lost and heat is generated. The power dissipated by the resistor is given by:

$$P = V_d\,I \tag{4.4}$$

where P is the power expressed in watts (W) and V_d is the voltage drop across the resistor.

Substituting Eqs. (4.2a) and (4.2b) into Eq. (4.4), we obtain:

$$P = I^2 R = V_d^{\,2} / R \tag{4.5}$$

We have used subscripts with the voltage in Eqs. (4.2a) and (4.2b) and the subscript d with the voltage in Eq. (4.5). For the simple circuit shown in Fig 4.4, the supply voltage V_s is equal to the voltage drop across the resistor V_d. In this case, the subscripts can be interchanged. Clearly, it is easy for us to determine the power lost if we know the current flowing through a resistor or the voltage drop across the resistor.

4.2.5 Capacitance

Capacitance is a bit more difficult to understand than the previous quantities. However, it also has a water-related analogy. It is used to characterize devices that have a degree of "springiness" – devices in which charge can be pumped in but which tend to push charge back when the external pressure is removed. The ease with which charge can be pushed into the device measures its capacity. The unit of capacitance is farads (F).
   A capacitor is a passive electrical component that has a fixed capacitance. The circuit symbol for a capacitor is shown in Fig. 4.3. The capacitor ordinarily consists of two flat, parallel-plate electrodes separated by a dielectric, which also serves as an electrical insulator. The capacitance C of a flat-plate capacitor is given in units of picofarad and is determined by:

$$C = k\,\varepsilon\,A/h \tag{4.6}$$

where ε is the dielectric constant; A is the area of the plates h is the distance between the two plates as shown in Fig. 4.3, and k is a proportionality constant; 0.00885 for dimensions in millimeters.

Fig. 4.3 A flat parallel plate capacitor.

When a voltage V is applied across the capacitor, it stores a charge q that is given by:

$$q = C V \qquad (4.7)$$

where C is the capacitance given in farads denoted with the symbol F.
       q is the charge on the capacitor given in coulombs.

When the capacitor is charged, it becomes a storage device. The electrical energy stored (w) is determined from:

$$w = C V^2 /2 \qquad (4.8)$$

where w is the stored energy in joules when C is expressed in farads.

When a voltage is first applied across the terminals of a capacitor, current flows and it becomes charged. However, when the capacitor is fully charged and maintained at a constant voltage, no current flows through it. The capacitor conducts current only when the voltage across it changes with respect to time.
The water analogy for a capacitor is a balloon or membrane stretched across the cross-section of a pipe to block the flow of water. As pressure is applied to attempt to pump water through the pipe, the balloon begins to stretch, which allows a small amount of water to flow. As the balloon stretches, it resists more and more until the backpressure of the balloon resisting the flow equals the forward pressure of the pump. At that point, the water ceases to move.

4.2.6 Inductance

Another passive component is the inductor, whose symbol is shown in Fig. 4.4. We produce an inductor by winding a coil of small-diameter, insulated wire about a core. The inductance L is given by:

$$L = 4\pi N^2 \, A\mu/ \, \ell \text{ x } 10^{-9} \qquad (4.9)$$

where L is the inductance in Henrys
       N is the number of turns of wire in the coil.
       μ is the permeability of the media within the coil.
       A is the cross sectional area of the coil.
       ℓ is the length of the coil.

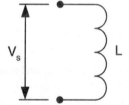

Fig. 4.4 An inductor coil across an alternating voltage source.

When a step voltage is first applied across an inductor, it acts like an open circuit and does not permit current to flow through it. Gradually, current flow through the inductor begins to increase until the inductor offers no resistance at all. In a sense, the inductor has an internal inertia with respect to current. The inertia must be overcome for current to flow. When current begins to flow, the inertia of the inductor tends to keep the current flowing even when the external voltage source is removed. Inductance L is a measure of the electrical inertia of the inductor.
A solenoid, which is a coil with an inductance, is commonly used to develop a magnetic field. The solenoid produces a magnetic force on a metal plunger that extends into the core of the coil. A solenoid engages a relay that switches your battery across the starter motor when you start the engine on your auto.
In dc circuits, an inductor coil acts like a small resistor and it has little effect on the current flowing in the circuit. However, when the coil is exposed to a current that fluctuates with respect to time (an alternating current) there is a significant voltage drop across the coil that is given by:

$$V_d = L\,(dI/dt) \qquad\qquad (4.10)$$

where (dI/dt) is the rate of change of the current with respect to time.

The water analogy for an inductor is a water wheel in a channel through which water flows. At first, the inertia of the stationary wheel prevents the water from moving through the channel and past the wheel. As the water pushes on the wheel, the wheel begins to turn and speed up, allowing more and more water to pass. Eventually, the wheel is turning at the rate the water would flow without the presence of the wheel. If the water pressure on the up stream of the wheel is turned off, the rotational inertia of the wheel keeps it turning. It pulls water from the upstream side and pushes it downstream. With time, the wheel begins to slow since there is no external pressure, until the wheel and the water flow both cease.

4.2.7 Diode

The diode is another important circuit element. While no particular electrical quantity is associated with diodes, they act in some respects like a current switch or a check valve. A diode allows current to flow freely in one direction (ideally, just like a wire) but does not permit current to flow (ideally) in the opposite direction. In a sense, a diode is the electrical equivalent of a check valve in a pipe carrying water. The symbol for a diode is shown in Fig. 4.5. The direction of the arrow indicates the direction in which the diode allows current to flow. We will discuss a common use of diodes later in the chapter.

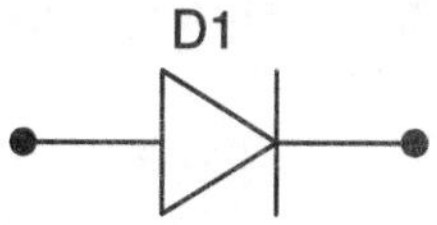

Fig. 4.5 Circuit symbol of a diode

4.3 KIRCHHOFF'S LAWS

4.3.1 Kirchhoff's Voltage Law

We introduced Ohm's law with a simple circuit loop containing a voltage source and a resistor as shown in Fig. 4.2. This is a good beginning; however, what happens when you begin to insert additional components in the circuit? You will need to develop a few more tools useful in analyzing slightly more complex circuits. Let's begin by adding one additional resistor to the simple circuit loop as shown in Fig. 4.6a. Then determine the relation for the current flowing in this slightly more complex circuit.

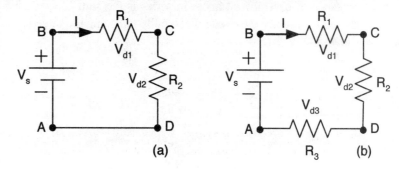

Fig. 4.6 Circuit loops containing two or three resistors.

To approach this problem, we need to use **Kirchhoff's voltage law** that states:

The sum of the voltages changes about a circuit loop is zero.

If we refer to the drawing of the circuit shown in Fig. 4.6a, we can write Kirchhoff's law in an equation format as:

$$+ V_{A-B} - V_{B-C} - V_{C-D} - V_{D-A} = 0 \qquad (4.11)$$

In the interpretation of Eq. (4.11), we move in the direction of the current flow (clockwise in this case). Moving from node A to B, we pass through the battery that increases the voltage at B relative to A by an amount V_s. This is the reason for the + sign shown on the term V_{A-B}. Continuing around the circuit loop, we proceed from node B to C and pass through resistor R_1. There is a voltage drop V_{d1} as the current passes through this resistor equal to $I R_1$. The fact that the voltage drops across R_1 is the reason for the minus sign on term V_{B-C}. Moving onto the next leg of the circuit between node C and D, we encounter a second voltage drop $V_{d2} = I R_2$. Because the voltage decreases as we move from node C to D, a minus sign is assigned to the term V_{C-D}. Finally, we consider the leg of the circuit between nodes D - A. We have a perfect conductor between these two points with no loss or gain of voltage. For this reason, $V_{D-A} = 0$. Substituting these four voltage changes into Eq. (4.11) gives:

$$V_s - I R_1 - IR_2 - 0 = 0$$

$$I = V_s / (R_1 + R_2) \qquad (4.12)$$

If we apply the same analysis techniques to the circuit with the three resistors shown in Fig. 4.6b, it is easy to add another term to the relation shown above and write:

$$V_s - I R_1 - IR_2 - IR_3 = 0$$

$$I = V_s / (R_1 + R_2 + R_3) \qquad (4.13)$$

Examine Eqs. (4.12) and (4.13), and observe that the resistors in a series arrangement in the circuit add together to act like a single resistor with an equivalent resistance R_e given by:

$$R_e = R_1 + R_2 + R_3 = \sum R \qquad (4.14)$$

Equation (4.14) leads to a well-known resistor rule for resistors connected in a series arrangement. The effective resistance of several resistances in series is the sum of the individual resistances. The equivalent resistance is used to simplify the circuit diagram by replacing the three-resistor circuit with a single equivalent resistor as shown in Fig. 4.7.

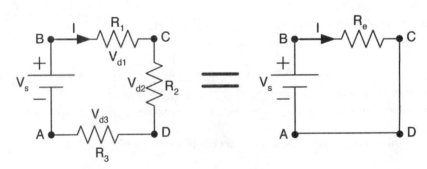

Fig. 4.7 An equivalent circuit with an equivalent resistor replaces a more complex circuit.

4.3.2 Kirchhoff's Current Law

What is the effect if the resistors are not all connected in a series arrangement? What if we have a parallel arrangement of the resistors such as shown in Figs. 4.8a and 4.8b? To analyze this type of a circuit, we need to introduce **Kirchhoff's current law**. This simple law states:

The sum of the currents flowing into a node must be equal to
the sum of the currents flowing from that node.

Fig. 4.8 Circuits with parallel arrangements of resistors.

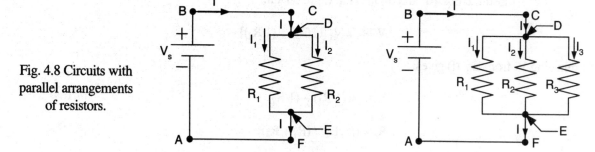

Let's refer to node X, presented in Fig. 4.9, with currents flowing into and out it. Kirchhoff's current law permits us to write:

$$I_1 + I_2 - I_3 - I_4 = 0 \qquad (4.15)$$

To deal with the resistors in parallel, as shown in Fig. 4.8a, we apply Eq. (4.15) to the current flow at node D and write:

$$I - I_1 - I_2 = 0 \qquad (a)$$

Fig. 4.9 Node X with current flows in and out.

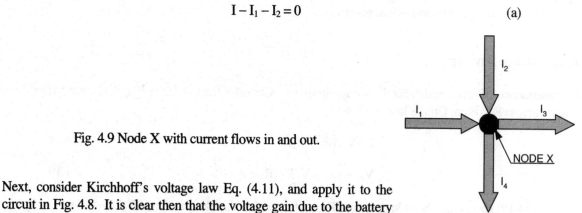

Next, consider Kirchhoff's voltage law Eq. (4.11), and apply it to the circuit in Fig. 4.8. It is clear then that the voltage gain due to the battery is offset by the voltage drop across the two parallel resistors between points D - E. We write this fact in an equation:

$$V_s - V_{D-E} = 0 \qquad (b)$$

We know that the current divides at node D with I_1 flowing through R_1 and I_2 through R_2 to produce the voltage drop V_{D-E}. This knowledge permits us to write:

$$V_{D-E} = I_1 R_1 = I_2 R_2 = V_s \qquad (c)$$

Equation (c) gives relations for both I_1 and I_2:

$$I_1 = V_s / R_1 \quad \text{and} \quad I_2 = V_s / R_2 \tag{d}$$

Following the form of Eqs. (d), we write:

$$I = V_s / R_e \tag{e}$$

where R_e is the equivalent resistor that replaces the two parallel resistors in Fig. 4.8a to give the same equivalent circuit as shown in Fig. 4.7.

Let's substitute Eqs. (d) and (e) into Eq. (a) to obtain:

$$V_s/R_e = V_s \left[(1/R_1) + (1/R_2) \right] \tag{f}$$

Eliminating V_s from Eq. (f) gives:

$$1/ R_e = (1/R_1) + (1/R_2) \tag{4.16a}$$

or

$$R_e = (R_1 R_2) / (R_1 + R_2) \tag{4.16b}$$

If we find three resistors in parallel, we follow the same procedure in the analysis of the circuit and show that the equivalent resistance for the three parallel resistors is:

$$1/R_e = (1/R_1) + (1/R_2) + (1/R_3) \tag{4.17}$$

While resistors connected in series were summed to give the equivalent resistance, resistors in parallel follow a different rule. For parallel resistors, the reciprocals are summed and set equal to the reciprocal of the equivalent resistance. This fact implies that the equivalent resistance for a parallel arrangement of resistors is less than that of the smallest individual resistance.

4.3.3 Voltage Divider

A consequence of series resistance is voltage division. Consider the circuit in Fig. 4.10 and determine the voltage across R_2. From Ohm's law, we have:

$$I = V_s /(R_1 + R_2) \tag{4.18}$$

$$V_{d2} = IR_2 = V_s R_2/(R_1 + R_2) \tag{4.19}$$

Equation (4.19) shows that the voltage across R_2 is a fraction of V_s. The fraction is $R_2/(R_1 + R_2)$.

Fig. 4.10 Voltage division circuit.

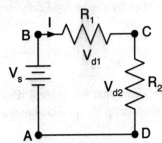

Another consequence of Kirchhoff's laws is that voltage sources in series add just as resistors do as long as the orientation of the source is taken into account. That is why two 1.5 V flashlight batteries are used in an ordinary flashlight; the voltage of the two batteries adds to provide a total voltage source of 3 V. On the other hand, if one of the batteries is inserted upside down, it will act **against** the voltage of the first battery and give a net zero voltage. In that case, the flashlight will not work.

At this point, you should recognize that Ohm's law and Kirchhoff's two laws can be employed to analyze circuits. Also, you should be able to take a complex circuit and use equivalent resistances to reduce the circuit to its most elementary form.

4.4 ALTERNATING AND DIRECT CURRENT

4.4.1 Definitions

Most circuit systems are classified as either direct current (dc) or alternating current (ac). A typical dc system is one that is powered by a battery, such as a flashlight or an automotive electrical system. Typical ac systems are those that are powered from a wall outlet in the house, such as a blow dryer or television. Direct current simply means that the current is constant in the circuit; it does not oscillate in amplitude with time. Alternating current oscillates with time usually in the form of a sine wave.

Paradoxically, the terms ac and dc are used to classify voltage sources as well as current, since constant voltage sources ordinary produce constant current and alternating voltage sources ordinary produce alternating current. The rate at which a periodic signal cycles is called the **frequency**, which is measured in cycles per second, or **hertz** (Hz). The **amplitude** is the peak value of the periodic waveform. The **RMS** (root-mean-square) value is the square root of the average squared value. For a sine wave, the RMS is equal to amplitude divided by $\sqrt{2}$. The 110 volts ac available at a wall outlet is actually 110 volts RMS. The peak value of the voltage is about 156 V. In the United States, ac supply is a sine wave that oscillates at 60 Hz. A graph of voltage as a function of time in a household power supply (110 V ac) is shown in Fig. 4.11.

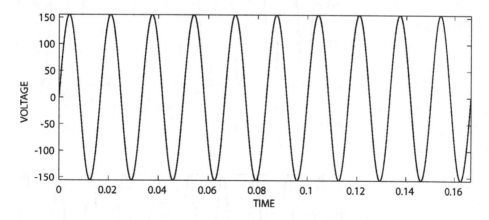

Fig. 4.11 Voltage as a function of time in wall outlet

4.4.2 Converting ac to dc

In many cases, the power is supplied from an ac outlet, even though the internal circuitry of a product like a PC requires a dc voltage source. In these cases, a circuit must be used to convert ac voltage to dc. Two types of circuits that accomplish this are the **half-wave rectifier** and the **bridge rectifier**. Both rectifiers take advantage of the diode, which allows current to flow one way but not the other.

Half-Wave Rectifier

A half-wave rectifier is shown in Fig. 4.12. It consists simply of a diode in series with an ac power supply and the load (the device to be powered—a resistor in this case). When V1 is positive, the diode acts like a wire and current passes through unimpeded. However, when V1 is negative, no current is allowed to flow backward, so the voltage drop across R1 is zero. The input and output voltages for this circuit are shown in Fig. 4.13.

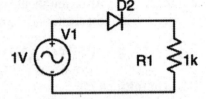

Fig. 4.12 Half-wave rectifier (diode) in series with load R1.

Note that in the graph showing output voltage the negative voltage values are truncated by the action of the diode. The dotted line in each figure shows the average value of the voltage on each graph. The input voltage oscillates equally on either side of zero, so the average value is zero. On the other hand, the output voltage is always nonnegative, so the average output voltage is positive. This output voltage waveform actually has both a dc component and an ac component. For some applications, this output can be used as a dc supply; the ac component (the fluctuations around the average value) can be ignored. In other cases, further filtering is required to smooth out the fluctuations to provide a more constant voltage.

Fig. 4.13a Input voltage.

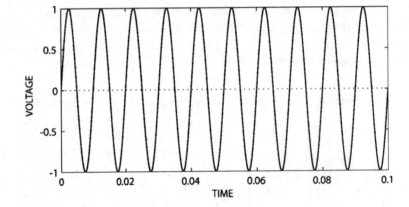

Fig. 4.13b Output voltage.
(Dotted lines show average voltage.)

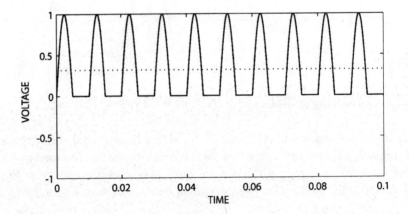

Bridge Rectifier

A bridge rectifier, shown in Fig. 4.14, operates on the same principle as the half-wave rectifier. It is a more complex circuit, but the performance is better. The network of four diodes allows the current to flow only in one direction regardless of the sign of the input voltage. Hence, its output voltage waveform is modified as shown in Fig. 4.15.

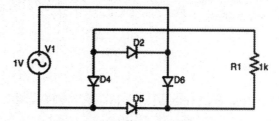

Fig. 4.14 Bridge rectifier in series with a load R1.

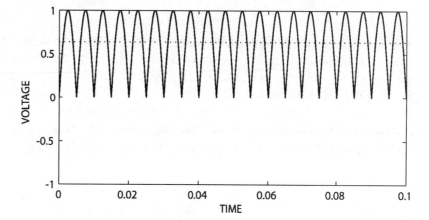

Fig. 4.15 Output voltage for bridge rectifier with the input defined in Fig. 4.13. (Dotted lines show the average voltage.)

It is clear that the bridge rectifier converts the negative-going voltage lobes to positive voltage lobes. Thus, the average value (the dc component) is twice as high as for a half-wave rectifier. This doubling of the average value comes at the expense of using four diodes instead of one.

Capacitor Filter

It is possible to smooth out the bumps in either the half-wave-rectified signal or the bridge-rectified signal by inserting a capacitor in series with the load. A capacitor-filtered half-wave rectifier is shown in Fig. 4.16. The capacitor begins to accumulate charge when the source voltage is positive. When the source is negative, the capacitor begins to discharge and acts like a temporary battery to maintain the voltage across R.

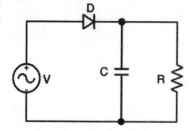

Fig. 4.16 Capacitor-filtered half-wave rectifier.

A graph of voltage as a function of time from a simulated capacitor-filtered half-wave rectifier circuit is shown in Fig. 4.17. The solid trace is the input voltage, and the dashed trace is the output. Notice that the output does not reach the same peak value as the input. This difference is because the capacitor does not charge instantaneously; it is still charging when the peak input begins to drop. However, when the input drops below the output voltage, the capacitor slowly discharges, sending current through the resistor and keeping the output voltage higher than it would be otherwise. Thus, the average value (the dc component) is increased.

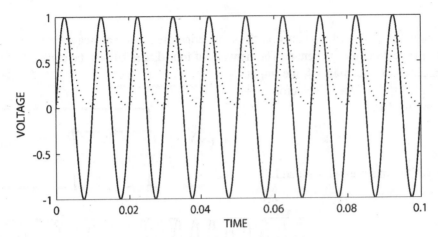

Fig. 4.17 Simulated input (solid) and output (dashed) voltages of a capacitor-filtered half-wave rectifier.

4.5 TRANSISTORS

Transistors are semiconductor devices that are used either as amplifiers or as high-speed electronic switches. The most widely used amplifier is the bipolar junction transistor illustrated in Fig. 4.18. The devices are planar, therefore, they can be fabricated using lithographic methods in P and N doped silicon. The devices are extremely small with areas of 10^{-9} m^2.

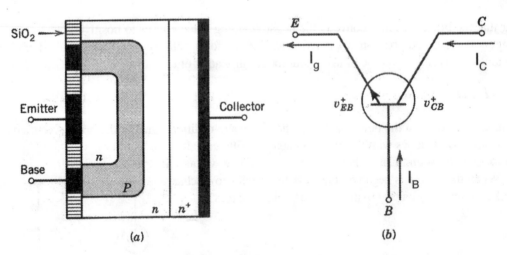

Fig. 4.18 Representation of a NPN bipolar transistor.
(a) Planar structure in silicon.
(b) Circuit symbol.

The transistors are three-terminal devices with the base represented by B, the collector by C, and the emitter by E. The theory of operation of the bipolar transistor is beyond the scope of this book; however, the transistor will act as a current amplifier because relatively small base currents I_B produce large collector currents I_C. For example, when a NPN transistor is connected in a common emitter configuration with voltage sources and a resistive load R_L, as shown in Fig. 4.19, the transistor acts to amplify the input signal I_i. The signal current I_i causes a variation in the base current I_B that in turn produces a variation in the collector current I_C along the load line as shown in Fig. 4.19b. The time-varying part of the collector current represents the amplified output current that is drawn from the source V_{CC} and flow through the load resistance R_L. The gain G is given by:

$$G = \frac{I_0}{I_i} \tag{4.20}$$

where I_0 is the sinusoidal component of I_C.

The gain G depends on the base and collector characteristics of the transistor and V_{CC} and R_L. Signal gains for a single transistor are in the range from 10 to 100.

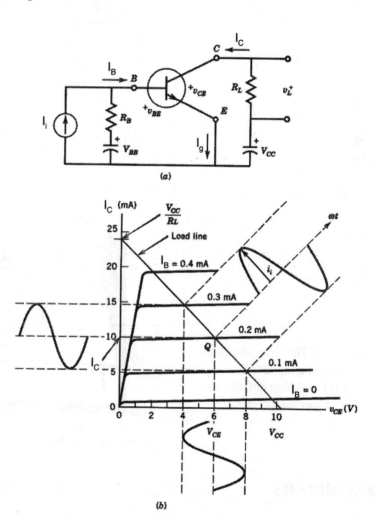

Fig. 4.19 A basic NPN transistor
     current amplifier.
      (a) Circuit diagram.
    (b) Operating characteristics.

     Transistors are also employed as very-high-speed electronic switches that are either open or closed depending upon the voltage applied to the base. When operated as a switch, the transistor is connected into the simple circuit shown in Fig. 4.20a. Since both P/N junctions are reversed biased, practically no collector current flows and the transistor is operated in the cutoff region of Fig. 4.20b at point 1 when the input voltage (current) to the base is zero. At point 1 the collector current is small (5 µA) with an applied voltage $V_{CC} = 5$ V. This condition corresponds to a cut off resistance of 1 MΩ and the switch, whose contacts are the collector and emitter terminals, is open.

When a positive voltage is applied at the input, the base current increases (say to 0.3 mA) and the operation of the transistor moves along the load line of Fig. 4.20 to point 2. At this point, the transistor is operating in a saturated condition and the voltage drop V_{CC} across the transistor is very small. The collector current is about 30 mA at a saturation voltage of 0.3 V which yields a switch resistance of about 10Ω. In this state, the transistor is considered as a closed switch.

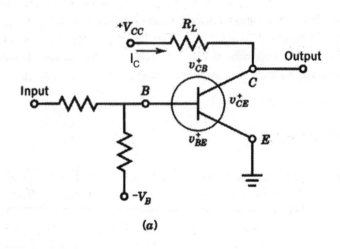

(a)

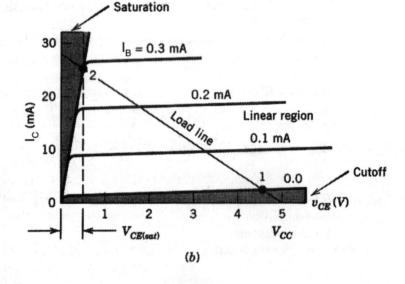

Fig. 4.20 The transistor in a
switching application.
(a) Switching circuit.
(b) Operating regions.

(b)

4.6 SUMMARY

We have described the basic electrical quantities of a typical circuit. In addition, we have defined Ohm's law, resistance, and the power dissipated by current flow through a resistor. The other common passive components—the capacitor and inductor—have also been introduced.

Circuit analysis is vitally important for all engineers regardless of their discipline. For this reason, we have introduced Kirchhoff's two laws, and given two examples of circuit analysis leading to the equations for the equivalent resistance of circuits with series and parallel-connected resistors.

We have also discussed the differences between ac and dc systems. Circuits that convert from ac to dc with different levels of efficiency, depending on the number and type of components used, have been presented. Finally, the diode and transistor have been introduced. The diode is used to control the direction of flow of electrical current and the transistor can be employed as an amplifier or as a switch.

REFERENCES

1. Irwin, J. D., Basic Engineering Circuit Analysis, 8th ed., John Wiley & Sons, New York, 2004.
2. Irwin, J. D. and Kerns, D. V., Jr., Introduction to Electrical Engineering, Prentice-Hall, 1996.
3. Horowitz, P. and Hill, W., The Art of Electronics, 2nd ed., Cambridge University Press, New York, 1989.
4. Dorf, R. C. and Svoboda, J. A., Introduction to Electric Circuits, 6th ed., John Wiley & Sons, New York, 2003.

EXERCISES

6.1   Draw the symbols used on drawings to represent a resistor, a capacitor and an inductor. Also, write the relations for the voltage across these components in terms of the current flow.

6.2   If you design a circuit containing an equivalent resistance of 2,000 Ω with a 9-V battery power supply, find the current drained from the battery. If the battery has a capacity of 2A-h (ampere-hours), determine the life of the battery. Assume that it cannot be recharged. What is the power dissipated by the resistor?

6.3   A size D, 1.5 V, alkaline battery has a capacity of 1200 hours when the battery is discharged through a 100 Ω resistor. Determine the current flow through this resistor, and state the capacity in terms of ampere-hours (Ah).

6.4   A 9 V alkaline battery has a capacity of 6 hours when the battery is discharged through a 100 Ω resistor. Determine the current flow through this resistor, and state the capacity in terms of ampere-hours (Ah).

6.5   Write Kirchhoff's two laws. Use circuit diagrams to illustrate these two laws.

6.6   Draw a circuit with four series-connected resistors and give the expression for the equivalent resistance. If these resistors have values of 1000, 400, 1500 and 10,000 Ω, find the equivalent resistance. Draw a circuit diagram that contains the equivalent resistor, which is also equivalent to your initial circuit diagram.

6.7   Draw a circuit with four parallel-connected resistors, and give the expression for the equivalent resistance. If these resistors have values of 1000, 400, 1500 and 10,000 Ω, find the equivalent resistance. Draw a circuit diagram that contains the equivalent resistor, which is also equivalent to your initial circuit diagram.

6.8   Trace the current path through a bridge rectifier when the source voltage is positive. Trace the current path through a bridge rectifier when the source voltage is negative.

Notes:

CHAPTER 5

SENSORS

5.1   INTRODUCTION

Transducers are electromechanical devices that convert a mechanical change such as displacement or force into an electrical signal that can be monitored as a voltage after some degree of signal conditioning.   A wide variety of transducers are commercially available for use in measuring mechanical quantities such as force, torque, strain, velocity, acceleration, etc. Transducer characteristics include: range, linearity, sensitivity and operating temperatures. The sensor that is incorporated into a transducer to produce an electrical output determines its characteristics. For example, a photodiode that detects reflected light intensity provides a transducer that senses the color or reflectivity of a nearby surface. A simple limit switch can also serve as a sensor to detect contact. The touch sensor marketed by LEGO for the NXT utilizes a short-throw, normally-open, contact switch as its sensor. When the switch is closed, the circuit resistance across its terminals goes from infinity to nearly zero providing an abrupt voltage change that is monitored by a circuit contained within the NXT. The sound sensor for the NXT incorporates a very small microphone that converts the pressure associated with a sound wave into an electrical signal. This signal is processed by circuits contained within the NXT and then can be used as a feedback signal for controlling one or more of the output ports on the NXT.

Sensors used in transducer design include switches, potentiometers, differential transformers, strain gages, capacitors, piezoelectric and piezoresistive crystals, thermistors, etc. Important features of a few of these sensors are described in this chapter.

5.2   POTENTIOMETERS

The simplest type of potentiometer, shown schematically in Fig. 5.1, is the slide-wire resistor. This sensor consists of a length L of resistance wire attached across a voltage source v_s. The relationship between the output voltage v_o and the position x of a wiper, as it moves along the length of the wire, can be expressed as:

$$v_o = (x/L)v_s \qquad\qquad x = (v_o/v_s)L \qquad\qquad (5.1)$$

Clearly, the slide-wire potentiometer can be used to measure a displacement x if v_0 is measured and v_s and L are known.

Resistors fabricated from a short length of straight wire are not feasible for most applications, because the wire's resistance is too low. A sensor with a low resistance imposes excessive power drain on the voltage source. To alleviate this difficulty, high-resistance, wire-wound potentiometers are obtained by winding the wire around an insulating core, as shown in Fig. 5.2. The potentiometer illustrated in Fig. 5.2a is used for linear displacement measurements.   Cylindrically shaped potentiometers, similar to the one illustrated in Fig. 5.2b, are used for angular displacement measurements. The resistance of a wire-wound potentiometer can range between 10 and 10^6 Ω depending upon the wire material, its diameter and the length of the coil.

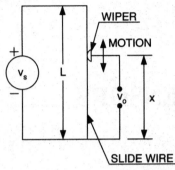

Fig. 5.1 Slide-wire resistance potentiometer.

The resistance of a wire-wound potentiometer increases in a stepwise manner as the wiper moves from one turn to the adjacent turn. This step change in resistance limits the resolution of the potentiometer to L/n, where n is the number of turns in the length L of the coil. Resolutions ranging from 0.05 to 1% are common, with the lower limit obtained by using many turns of very small diameter wire.

The active length L of the coil controls the range of the potentiometer. Linear potentiometers are available in many lengths up to about 1 m. Arranging the coil into a helix extends the range of the angular-displacement potentiometer. Helical potentiometers are commercially available with as many as 20 turns; therefore, angular displacements as large as 7,200° can be measured with relative ease.

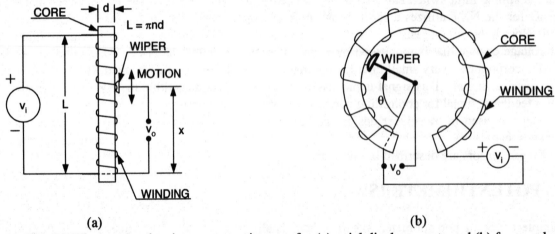

(a)   (b)

Fig. 5.2 Wire-wrapped resistance potentiometer for (a) axial displacements and (b) for angular displacements.

In order to improve resolution, potentiometers have been introduced that utilize thin films of conductive plastic with controlled resistivity instead of wire-wound coils. The film resistance on an insulating substrate exhibits very high resolution together with lower noise and longer life. For example, a resistance of 50 to 100 Ω/mm can be obtained with conductive plastic films that are used for commercially available potentiometers with a resolution of 1 μm or less. The frictional force that must be overcome to move the wiper is depended on the construction details of the particular potentiometer; however, friction forces from 0.5 to 1.0 N are common. The life expectancy for potentiometers fabricated with conductive plastics is commonly specified at 20×10^6 strokes. An example of a linear motion potentiometer with a single end shaft is presented in Fig. 5.3. Note the three terminals used for electrical connections.

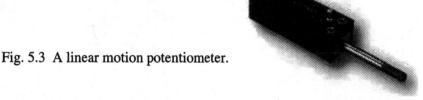

Fig. 5.3 A linear motion potentiometer.

The dynamic response of both the linear and the angular potentiometer is severely limited by the inertia of the shaft and wiper assembly. Because of its large inertia, the potentiometer is used only for static or quasi-static measurements where a high frequency response (bandwidth) is not required.

Potentiometers are used primarily to measure large displacements—10 mm or more for linear motion and 15° or more for angular motion. Potentiometers are relatively inexpensive yet accurate; however, their main advantage is simplicity of operation, because only a voltage source and a digital voltmeter (DVM) are required for a complete instrumentation system. The primary disadvantages of potentiometers are a limited frequency response that precludes their use for dynamic measurements and the force required to overcome friction of the wiper.

5.3   PHOTOELECTRIC SENSORS

In many applications where direct contact cannot be made with the object being examined, a photoelectric sensor can be employed to make physical measurements by monitoring changes in the intensity of light reflected off the object of interest. When light impinges on a photoelectric sensor, it either creates or modulates an electrical signal. Most photoelectric devices employ semiconductor materials and operate by either generating a current or by changing the semiconductor's resistivity. These devices are photodetectors that respond quickly to changes in light intensity.

5.3.1   Photoconducting Sensors

Photoconductive cells, illustrated in Fig. 5.4, are fabricated from semiconductor materials, such as cadmium sulfide (CdS) or cadmium selenide (CdSe), which exhibit a strong photoconductive response. When a photon with sufficient energy strikes a molecule of, say CdS, an electron is driven from the valence band to the conduction band and a hole remains in the valence band. This hole and the electron both serve as charge carriers, and with continuous exposure to light, the concentration of charge carriers increases and the resistivity decreases. A circuit used to detect the resistance change ΔR of the photoconductor is also shown in Fig. 5.4. The resistance of a typical photoconducting sensor changes over about 3 orders of magnitude as the incident radiation varies from very dark to very bright.

When a photoconductor is placed in a dark environment, its resistance is high and only a small current flows. If the sensor is exposed to light, the resistance decreases significantly (the ratio of maximum to minimum resistance for R_d in Fig. 5.4 ranges from 100 to 10,000 in common commercial sensors); therefore, the output current can be quite large. The sensitivity depends on cell area, type of cell (CdS or CdSe) and the power limit for the cell. If the supply voltage (v_s) is set to supply the cell at its maximum power limit, then sensitivities of up to 0.2 mA/lx result for CdS type cells. A typical cadmium sulfide photoconductor exhibits a maximum dark resistance of about 1 to 2 MΩ. When subjected to a light intensity of about 2 foot-candles the resistance decreases to about 1 to 5 kΩ. The change of about three orders of magnitude is large, enabling many applications with very simple control circuits.

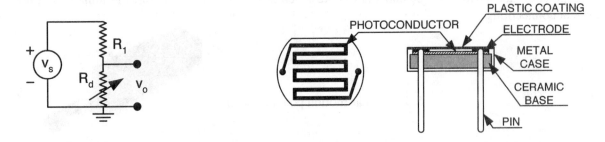

Fig. 5.4 When a photoconductor is used (modeled as R_d), the output voltage measured (shown as v_o) will vary with light intensity.

Photoconductors respond to radiation ranging from long thermal radiation through the infrared, visible and ultraviolet regions of the electromagnetic spectrum. The sensitivity, S, changes significantly with wavelength and drops sharply at both short and long wavelengths; consequently, photoconductive cells exhibit the same disadvantage as many other photodetectors. Their calibration depends on the wave length of the impinging light.

The photocurrent requires some time to develop after the excitation is applied and some time to decay after the excitation is removed.   The rise and fall times for commercially available photoconductors are usually about a second. Because of these relatively long delays, the CdS and CdSe photoconductors are not suitable for dynamic measurements. Instead, their simplicity, high-sensitivity and low-cost lend them to applications involving sensing, counting and switching based on a slowly varying light intensity.

5.3.2   Photodiode Sensors

Photodiodes are semiconductor devices that respond to high-energy particles and photons. Photodiodes operate by absorption of photons or charged particles and generate a current that flows in an external circuit, which is proportional to the incident power. Photodiodes can be used to detect the presence or absence of minute quantities of light and can be calibrated for extremely accurate measurements from intensities below 1 pW/cm^2 to intensities above 100 mW/cm^2.   Silicon is the most common semiconductor material used in fabricating planar diffused photodiodes. These photodiodes are employed in such diverse applications as spectroscopy, photography, analytical instrumentation, optical position sensors, beam alignment, surface characterization, laser range finders, optical communications and medical imaging instruments.

Planar diffused silicon photodiodes are P-N junction diodes. A P-N junction can be formed by diffusing either a P-type impurity (anode), such as boron, into a N-type bulk silicon wafer, or a N-type impurity, such as phosphorous, into a P-type bulk silicon wafer. The diffused area defines the photodiode active area. To form an ohmic contact it is necessary to diffuse another impurity into the backside of the wafer. The impurity is an N-type for a P-type active area and P-type for an N-type active area.   Contact pads are deposited on defined areas of the front active area and on its backside.   The active area is covered with an anti-reflection coating to reduce the reflection of the light for a specific predefined wavelength. The non-active area on the top is covered with a thick layer of silicon oxide. A schematic illustration of a planar diffused photodiode fabricated from N-type silicon is presented in Fig. 5.5.

By controlling the thickness of bulk substrate, the speed and responsivity of the photodiode can be controlled. When the photodiodes are biased, they are operated in the reverse bias mode, i.e. a negative voltage applied to anode and positive voltage to cathode.

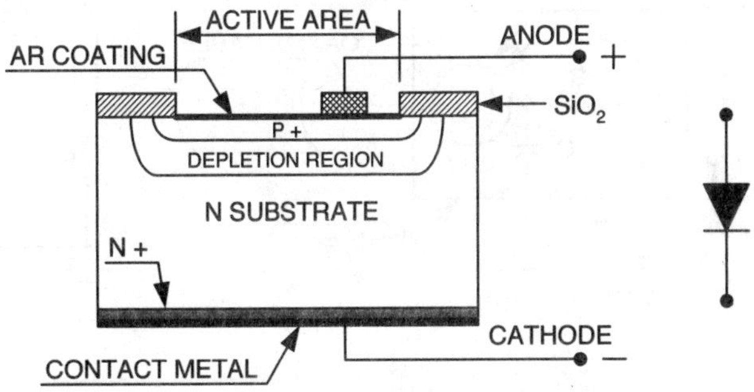

Fig. 5.5 Diagram of the diffusion areas used to create a P-N junction in a N-type silicon photodiode.

Electrical Characteristics of Photodiodes

A silicon photodiode can be represented by a current source in parallel with an ideal diode as shown in Fig. 5.6. The current source represents the current generated by the incident light, and the diode represents the P-N junction. In addition, a junction capacitance (C_J) and a shunt resistance (R_{SH}) are in parallel with the other components. A resistance (R_s) is in series in the circuit.

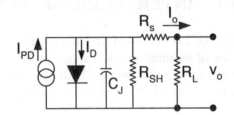

Fig. 5.6 Representative circuit for a photodiode.

Physically, the shunt resistance is the slope of the current-voltage curve of the photodiode at the origin, i.e. $v_0 = 0$. Although an ideal photodiode should have an infinite shunt resistance, actual values range from 10 MΩ to 1,000 MΩ. Experimentally the shunt resistance is determined by applying ± 10 mV across the diode, measuring the resulting current and calculating the resistance from Ohm's law. Shunt resistance is used to determine the noise current in the photodiode with no bias (photovoltaic mode). For superior photodiode performance, a very high shunt resistance is specified.

The series resistance of a photodiode arises from the resistance of the contacts and the resistance of the undepleted silicon shown in Fig. 5.5. It is used to determine the linearity of the photodiode in photovoltaic mode[1]. Although an ideal photodiode should have zero series resistance, typical values ranging from 10 Ω to as much as 1,000 Ω are measured.

The choice of using a semiconductor diode as a photovoltaic detector or as a photoconductive detector depends primarily on the frequency response required in the measurement. For light intensity fluctuations at the lower frequencies (less than 100 kHz), a photovoltaic circuit exhibits a lower noise voltage than a photoconduction circuit with reverse bias voltage. For higher frequencies required for measuring high-speed light pulses or high frequency modulation of a continuous light beam, the reverse bias serves to accelerate the electron/hole transition times, which improves the frequency response. Photoconductive diodes operate over a frequency range from DC (0 Hz) to 100 MHz and are capable of measuring the intensity of light pulses with rise times in the 3 to 12 ns range.

Semiconductor photodiodes are small, rugged and inexpensive. Because of these advantages, they have replaced the vacuum tube detectors in most applications. The photodiodes may be used in either mode with operational amplifiers to give a responsivity that is exceptionally high. Circuits showing photodiodes operating in the photoconduction mode and in the photovoltaic mode are presented in Fig. 5.7.

---

[1] Photodiodes act as photovoltaic devices when no bias voltage is applied.

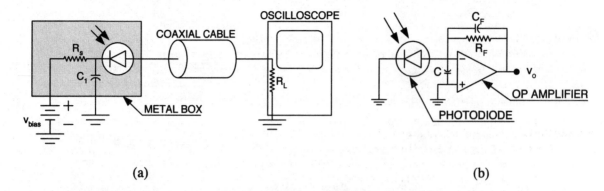

(a)   (b)

Fig. 5.7 Circuits used with photodiodes. (a) High intensity light at high frequency.
(b) Low intensity light with low frequency.

5.4   INTERFACING SENSORS WITH THE NXT CONTROLLER

The NXT is housed in a relatively small package with a monochrome LCD display and four buttons. In spite of its small size and simple interface, it is a powerful controller with two processors, four sensor ports, three motor ports, a USB port, and Bluetooth capability. A block diagram showing the primary components and their connections within the NXT controller is presented in Fig. 5.8.

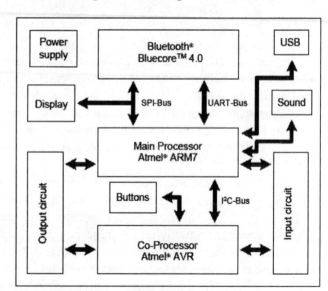

Fig. 5.8 Block diagram of the components
within the NXT controller.

5.4.1 Component Specifications for the NXT Controller

The NXT controller employs a number of electronic devices to provide its broad functionality. A description of the hardware specifications for the NXT controller is given below:

Main processor:
      Atmel® 32-bit ARM® processor, AT91SAM7S256
- 256 kB FLASH
- 64 kB RAM

- 48 MHz

Co-processor:
Atmel® 8-bit AVR coprocessor, ATmega48
- 4 KB FLASH
- 512 Byte RAM
- 8 MHz

Bluetooth wireless communication:
CSR BlueCoreTM 4 v2.0 +EDR System
- Supporting the Serial Port Profile (SPP)
- Internal 47 Kbytes RAM
- External 8 Mbits FLASH
- 26 MHz

USB 2.0 communication port:
Full speed port (12 Mbits/s)

Four input ports:
6-wire interface supporting both digital and analog interface
- A single high speed port, IEC 61158 Type 4/EN 50170 compliant

Three output ports:
6-wire interface supporting input from encoders

Display:
100 x 64 pixel LCD black & white graphical display
- View area: 26 × 40.6 mm

Sound output channel:
8-bit resolution
- Supporting a sample rate of 2-16 kHz

Four user-interface buttons:

Power supply:
6 AA batteries
- Alkaline batteries are recommended
- A rechargeable Lithium-Ion battery (1400 mAh) may be employed

Connectors:
6-wire industry-standard connectors
- RJ12 with right side adjustment

5.4.2 Input Ports on the NXT Controller

The NXT has four input ports that are used to connect sensors to the two processors with six wire leads. The six wire connector permits the acquisition of both analog and digital signals at each port, thus enabling the use of both analog and digital sensors with the NXT. A schematic diagram of the six pin connector is presented in Fig. 5.9. All of the ports have the same pin arrangement; however, the digital pins (5 and 6) are connected to a high speed RS485 controller to accommodate high frequency communications.

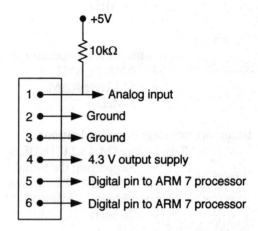

Fig. 5.9 Pin connections between the input ports and components within the NXT.

Pin #1, the analog input, is connected to a 10 bit A/D converter that is incorporated within the AVR coprocessor. This pin is also connected to a current generator which supplies power to the sensors connected to the NXT. The analog signals are sampled and converted to digital format with the same sampling rate for all of the analog sensors—333 Hz. The sensors require power for 3 ms before they are capable of making measurements. Pin #2 is a ground connection. These two pins have the same functionality as the input terminals on the RCX, which permits the use of legacy LEGO sensors with the NXT using appropriate lead wires.

Output power is provided from Pin # 4 for both the input and output ports. The maximum current that can be drawn from the NXT's power supply is 180 mA, which is divided among the seven ports. Hence, the maximum current capability of an individual port is about 20 mA. If the total current required exceeds the 180 mA limit, the output from pin #4 is automatically decreased without warning. If the power supply is accidentally shorted to ground, the NXT resets.

Pins #5 and #6 are used for digital communications using what is known as I²C protocol. The I²C is a two-wire communication interface developed by Phillips Semiconductor many years ago to connect a CPU to peripheral chips in a television set. Today this two wire interface is used in many applications. In the NXT, digital communication requires two lines—Pin #5 controls the timing signals from its clock and Pin #6 transmits information in both directions between the NXT and the sensors attached to it. The communication speed is 9600 bits/s for each port. Pins #5 and #6 are connected directly to two I/O terminals on the ARM7 processor. Firmware in the NXT creates four independent buses, one for each input port.

5.4.3 Output Ports on the NXT Controller

The NXT has three output ports that are used to connect actuators to the power supply within the NXT. A six wire digital interface is used with the output ports permitting the actuators to provide information back to the NXT without having to employ an input port. A schematic diagram of the six pin connector for the output ports is presented in Fig. 5.10. Note that all three ports have the same pin arrangement.

Pins #1 and #2 provide power to the motors and actuators. A second internal power supply within the NXT can provide a continuous current of 700 mA to each of the three output ports and a peak current of 1 A. The output from this power supply is pulse width modulated that can be controlled by software to brake or float. This internal power supply incorporates a built-in thermal protection system that automatically lowers the current output when overheating of the supply occurs.

Pin #3 is employed to provide a ground connection for the output power supply associated with Pin #4. Output power is provided from Pin # 4 for both the input and output ports. The maximum current that can be draw from the NXT's power supply is 180 mA, as described above.

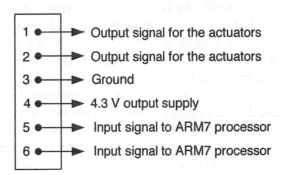

Fig. 5.10 Pin connections between the output ports
and components within the NXT.

Pin #5 and Pin #6 provide connections to input pins on the ARM7 processor through a Schmidt trigger. Tachometer pulses from the encoders in the LEGO servo motors are counted using this circuit and the firmware stored in the NXT. The signals from the servomotor are also used to establish the direction of rotation of the servo motors.

5.4.4 LEGO Sensors and the NXT Controller

Ports 1, 2 and 3 on the NXT are used to interface with both active and passive sensors. Recall that passive sensors do not require a separate power lead from the NXT to function. Moreover, the output from a passive sensor is a voltage drop across the sensor terminals due to a resistance or a voltage source placed across the terminals of the NXT. For example, the voltage source can be from an externally powered Wheatstone bridge, a solar cell or a thermocouple. The resistive loads may be from switches, potentiometers, thermistors or fixed value resistors.

Active sensors require power to function[2]. The light sensor from a LEGO Mindstorms Invention kit is an example of an active sensor because it requires power drawn from the NXT to operate the light emitting diode (LED) that illuminates the scene and activates the circuit containing the photodiode. The light reflected from the scene passes through a lens and impinges on a photodiode producing an output voltage that is monitored by the NXT and converted to a reading of light intensity. The supply voltage to the light sensor is provided through pin #1. A power generator connected to pin operates in a multiplexed mode. It is switched on to supply power to the light sensor for 3 ms and then off for 0.1s, which is sufficient to measure the output voltage from the photodiode. This design feature was added to the NXT to provide backwards compatibility with the RCX sensors.

The passive sensors that interface with the NXT controller include: the touch, the sound and the temperature sensors. These sensors do not need a separate power lead to function. The sound and temperature sensors generate a voltage that is related to the quantity they are measuring. The touch sensor generates a voltage within the NXT when its switch is closed. The output voltage from these sensors is sampled every 3 ms using the A/D converter in the AVR coprocessor.

The ultrasonic sensor included in the LEGO Mindstorm NXT kit is a digital sensor that requires an I^2C communication channel to function. The I^2C communication channel is available only on port #4 of the NXT. The communication is effected with two lines. The first line (Pin #5) provides the timing with clock signals and the second line (Pin #6) transmits information to and from the sensor (the slave). The communication speed is 9,600 bits/s.

---

[2] A more complete description of the sensors available from the LEGO Educational Division is provided in Chapter 6.

To begin, the NXT signals the sensor that it is about to transmit a message by sending a start signal. It then sends a 7 bit address to establish a connection to the specific sensor it is monitoring[3]. It then sends a command or data to the device (sensor) and waits for information to return. With the ultrasonic sensor, a high frequency short burst of sound (ping) is generated by a piezoelectric crystal transmitter located in the sensor. This sound wave propagates to some nearby object where it is reflected back to the sensor. The reflected sound wave (echo) is monitored by a receiver located in the sensor. By measuring the time of flight between the ping and echo, it is easy to determine the distance between the target and the sensor. Sound waves travel at about 333 m/s (depending on temperature and pressure); hence, propagation over a meter takes 3 ms. At the maximum range of 2.5 m for the LEGO sonar sensor, the transit maximum time to and from the target is $2 \times 2.5 \times 3 = 15$ ms.

5.4.5 Additional Features within the NXT Controller

Bluetooth® Communication within the NXT Controller

Wireless communication is possible with the NXT using Bluetooth® technology. Wireless communications between four NXTs are possible (one master unit and three slaves), but the master NXT can only communicate with one other NXT (slave) at a time. This capability is enabled by using Serial Port Profile (SPP), which is essentially a wireless serial port that may be used for incoming or outgoing communications. Wireless communications are possible between the NXT and personal computers or other devices equipped with Bluetooth technology. It is possible to send and receive messages between NXT controllers and between the computer and NXT as a program is executing. To conserve power used by Bluetooth, its range has been limited to about 10 m.

Communications to other Bluetooth units occur through four communication channels. Channel 0 is used by one or more of the slave NXTs sending information back to the master NXT. Channels 1, 2 and 3 are used by the master NXT for outgoing communications to the slave devices.

The Display on the NXT Controller

A monochrome liquid crystal display (LCD) with a resolution of 100×64 pixels is provided to enable readout and feedback to the operator. The viewing area on the display is 26×40.6 mm. There is an interface connecting the display to the ARM7 processor that operates at a frequency of 2 MHz. The display is updated in a continuous line by line replacement scheme that requires 17 ms for a total display update.

Power Management in the NXT Controller

Power is drawn from the six AA batteries housed in the NXT (either alkaline or lithium ion rechargeable). The power supply within the NXT provides three voltages—9 V from the batteries, 5 V and 3.3 volts for the ARM processor and the BlueCore® chip. The power supply is protected with a poly switch that is rated at 1.85 A for a continuous current and 3.3 A for the trip current. The current and power requirements placed on the power supplies depend upon whether the motors are loaded or not. These requirements are presented in Table 5.1.

---

[3] The fact that a 7 bit address is used to specify the sensor implies that up to 128 devices could be arranged in parallel and monitored in sequence with the NXT.

Table 5.1
Current and power requirements for the NXT

Supply Voltage (V)	Current		Power(9 Volt Battery)	
	Maximum (mA)	Normal (mA)	Maximum (mW)	Normal (mW)
No Load on Motors				
9	339	114	5,184	1,422
5	271	112	1,744	448
3.3	72	38	410	216
Load on Motors				
9	2,901	848	26,109	7,632
5	271	112	1,142	307
3.3	72	38	410	137

5.5   MEASURING VOLTAGE WITH THE NXT

The NXT can handle input voltages up to 5 V across Pin #1 and Pin #2 of its input ports. The circuit diagram of the input port on the NXT is presented in Fig. 5.11. An equivalent voltage measuring circuit in the NXT is shown in Fig. 5.12. The shaded box represents the NXT with its digital voltmeter (DVM), which is displayed on the LCD. A 10 kΩ resistor (R49 in Fig. 5.11) in series with a 5 V

voltage source (VCC5V in Fig. 5.11), internal to the NXT, are across the terminals of the DVM. External to the NXT is the sensor selected. In this case, the sensor is represented by an unknown voltage v_u in series with an external 10 kΩ resistor.

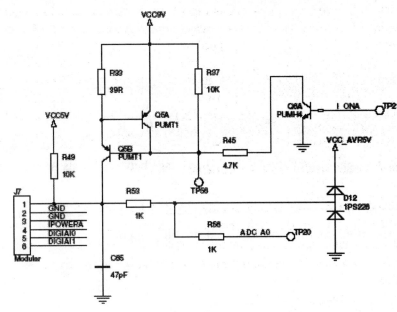

Fig 5.11 Circuitry in the NXT associated with the input ports

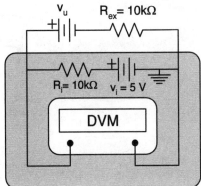

Fig. 5.12 Internal and external wiring for measuring voltage.

Let's perform a circuit analysis to determine the output voltage v_0 measured by the DVM. First, we will redraw the circuit shown in Fig. 5.12 to better identify the circuit loop that will be considered in the analysis. The new circuit diagram, presented in Fig. 5.13, shows loop 1 that includes two voltage sources and two resistors. We also show the current flow counterclockwise around this loop. Note the polarity of the two voltage sources, which implies that $v_u > 0$. Recall, Kirchhoff's voltage law, which states that the sum of the voltages around a loop must equal zero. Beginning at point A and applying this law permits us to write:

$$\Sigma V = + v_i - IR_i - v_u - IR_{ex} = 0 \qquad \text{(a)}$$

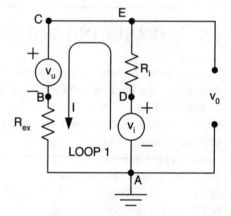

Fig. 5.13 Circuit diagram for the NXT showing the internal voltage source and resistor together with the external voltage source and resistor.

Solve Eq. (a) for the current flow I to obtain:

$$I = \frac{v_i - v_u}{R_i + R_{ex}} \qquad \text{(b)}$$

The voltage v_0 across the DMV is the voltage between points A and E, which is given by:

$$v_0 = v_i - IR_i \qquad \text{(c)}$$

Substituting Eq. (b) into Eq. (c) and simplifying yields:

$$v_0 = \frac{v_u R_i + v_i R_{ex}}{R_i + R_{ex}} \qquad \text{(d)}$$

If $R_i = R_{ex} = R$, then Eq. (d) reduces to:

$$v_0 = \tfrac{1}{2}(v_i + v_u) \qquad \text{(5.2)}$$

Set $v_i = 5$ V and Eq. (5.2) reduces to:

$$v_0 = \tfrac{1}{2}(5 + v_u) \qquad \text{(5.3)}$$

or

$$v_u = 2v_0 - 5 \qquad \text{(5.4)}$$

The DVM on the NXT does not look like a typical digital voltmeter, because it is incorporated in the display. The NXT contains a 10-bit analog to digital converter (ADC) that converts the voltage to a digital code that varies from 0 to 1023. The number displayed on the LCD may be in digital code (some number between 0 and 1023), binary (0 or 1), or proportional (0 to 100), depending upon the programming of the sensor port used in measuring the voltage.

The DVM can be used to monitor the four input (sensor) ports. Selection of the port to be monitored is made with the gray triangular buttons on the NXT following the procedure described in Chapter 6. The display shows a reading that is related to the sensor attached to this port. The relation between the voltage from the sensor and the reading will depend on the sensor selection and the programming of this port. Programming the ports for specific sensors is described later in Chapter 6.

The input and output ports on the NXT provide the input and output terminals where connections of sensors and actuators are made. On a casual inspection these ports look like standard terminals associated with telephone wiring. However, closer inspection shows six contacts at each port consistent with the 6-wire industry standard RJ12 connectors with right side adjustment. Connections are made to the NXT by pressure contact through these contacts. Leads with six wires are provided with connectors on both ends that are used to connect sensors and servo motors to the input and output ports in the NXT.

5.6 DESIGNING AND BUILDING SENSORS

There are a number of different sensors commercially available from LEGO—touch, sound, light, temperature and sonar (distance). In addition the angle of rotation of the servo motors can be monitored using the internal encoders in the servo motors. These are excellent sensors designed specifically to interface with the NXT. The touch, sound and temperature sensors are passive and do not require separate power leads to operate. However, the light sensor is active because the LED and circuit containing the photodiode both require power to function. In addition to the LEGO sensors available, a number of companies provide sensors that are compatible with the NXT to measure other quantities. Also, it is possible to design and build your own sensors using components that are available from local electronic supply stores. In fact, with Internet orders, it is possible to obtain almost any electronic device or component from regional supply houses in a few days.

Touch Sensor

Let's begin with the touch sensor because it is the simplest of all the sensors. It consists of a momentary switch, a LEGO support block to facilitate mounting in a LEGO design environment, and a connector that is compatible with the NXT. A schematic diagram of a momentary switch[4] that can be used for the touch sensor is shown in Fig. 5.14. As shown in Fig. 5.14, the momentary switch is in the normally open mode because a spring keeps the switch open until the push button is pressed, bringing the switch bar down on the contacts and closing the switch. When the button is released, the switch returns to the open position.

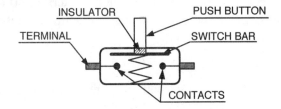

Fig. 5.14 A normally open momentary switch
     used in the design of a touch sensor.

---

[4] Momentary switches are available from a large number of companies through electronic suppliers. For this example, we selected a Panasonic detector switch part number ESE-11HS1 that was available from Digi-Key for $0.78. This switch, which is rated at 5 VDC and 10 mA, is actuated with a force of 35 g applied to its push button.

If the touch sensor is to be employed with the NXT, the electrical connections must be compatible with the ports on the NXT. The easiest way of connecting the switch to the NXT is to take a LEGO lead wire with port compatible connectors on both ends and cut it into two pieces. Take one of the two pieces and strip the insulation for about 6 to 8 mm from the ends of the two wires connected to pins #1 and #2. Then use a soldering iron to solder the two LEGO lead wires to the appropriate terminals on the contact switch. Next insert the connector into Port 1 on the NXT.

At this stage of the development of the touch sensor, check if the NXT responds when the push button on the switch is depressed. With the touch sensor connected to one of the input ports, depress the gray triangular buttons on the NXT until you have activated the View icon. Select the view icon and continue to press the gray triangular buttons until the touch sensor icon is displayed on the LCD. Press the square orange button to show the reading from the touch sensor. The display should read 0 indicating an open switch. Press the button closing the switch and note that the display reads 1. This reading indicates that the switch is functioning properly and will act as a touch sensor. For this simple sensor no calibration is required—it is either open with a zero reading on the NXT or closed with a reading of 1.

If the touch sensor is to be used in a LEGO design environment, it should be mounted on a LEGO plate or brick. The selection of either a plate or a brick will depend on the size of the switch and the method you choose to attach the switch to the brick or plate.

Angle Sensor

An angle sensor is also easy to design and build because it consists of a rotary potentiometer (see Fig. 5.2b), a LEGO support block to facilitate mounting in a LEGO design environment and a connector that is compatible with the NXT. Use the remaining end of the LEGO—NXT lead wire that was cut to provide a connector for the touch sensor to solve the connector problem. The most significant design problem is selecting a potentiometer. For this example, we have selected a single turn potentiometer that is encased in a very small rectangular case[5]. This single turn potentiometer is designed with an effective electrical angle of 270° ± 5°. Thus, it will serve as a sensor capable of measuring angles that vary from about 0 to 270°. The total resistance of the potentiometer is specified as 10 kΩ ± 10%. The potentiometer used in this experiment was checked and its resistance was measured as 10.8 kΩ. The resistive element over which the wiper moves is a conductive cermet.

This potentiometer has a threaded mounting shank that is ¼ in. in diameter with ¼ - 32 threads. It is fastened to a 2 × 2 LEGO angle as shown in Fig. 5.15. It was necessary to enlarge the hole in the LEGO angle to accommodate the threaded shank. We enlarged the hole by using several different drills with increasing diameter until a diameter of 1/4 in. was obtained The plastic used in fabricating the LEGO pieces is so soft that the drills can be turned into the preexisting hole by hand. We also trimmed off the four studs by holding the angle in a vise and removing the studs with a box cutter. This operation produced a flat surface so that the nut could be tightened to firmly fix the potentiometer to the LEGO angle. The potentiometer's shaft, which turns its wiper, is 1/8 in. (3.18 mm) in diameter. A 40 tooth LEGO gear was fitted onto this shaft to facilitate rotating the shaft of the potentiometer.

Fig. 5.15 The potentiometer mounted on a LEGO 2 × 2 angle brick.

---

[5] We have selected a 1 W - 10 KΩ ± 10% Bourns Series 50 precision potentiometer with a linear taper. The Digi-Key catalog number for this component is 51CAD-E24-A15-ND and its cost was $7.39.

To wire the potentiometer it is necessary to understand that the center terminal (lead) connects with the wiper. The outer terminals (leads) connect to the ends of the resistor element. If you measure the resistance across the outer terminals with an ohmmeter the reading should be 10 K ± 10% regardless of the angular position of the shaft. Next, connect the ohmmeter across the terminal on the left and the center terminal. Rotate the potentiometer in the clockwise direction as far as possible and observe that the resistance decreases as you rotate the shaft. The resistance across these two terminals approaches zero when the potentiometer wiper is in the full clockwise position.

Now that we understand the position of the wiper relative to the left hand terminal, solder lead wires to center and left hand terminals. The lead wire to the center terminal should connect to pin #1 and the lead wire to the left hand terminal should connect to pin #2 of the NXT input terminal.

Next prepare a program in ROBOLAB that will output the sensor reading of the sensor attached to port #1. Attach the sensor to port #1 and turn on the NXT and run the program. Observe the reading on the NXT display as you rotate the shaft. This reading varies from 1 to 532 as you rotate the shaft from the fully clockwise position to the fully counterclockwise position. Clearly, this reading is not directly related to the angle indicated by the wiper position. The reading is what is known as a raw value. The raw value varies from 0 for an input voltage of 0 V to 1023 for a voltage of + 5V across pins #1 and #2. The fact that we are recording raw values ranging from 1 to 532 implies that the voltage across the 10 kΩ potentiometer is varying from about 0 to about 2.5 V.

We are now ready to calibrate the sensor and to prepare a table showing the angle of rotation versus NXT readings. We will use the 40-tooth LEGO gear to indicate the angle of rotation by manually turning the gear one tooth per increment of rotation. We begin with the potentiometer near the fully clockwise position with a tooth in alignment with a suitable index mark and make the initial reading. We then advance the gear by one tooth, which is equivalent to an advance of 360°/40 = 9°, and then repeat the reading. This process is continued by rotating in increments in the counterclockwise direction until we reach the end of the travel of the potentiometer. Then we reversed the process and made the sequence of readings again while returning to the initial zero position. The raw value readings taken from the NXT display during the calibration process are presented in the Fig. 5.16.

An examination of the calibration curve in Fig. 5.16 indicates that the output from the angle sensor is not linear. Let's explore the reason for the non-linearity by performing a circuit analysis to determine the output voltage v_0 measured by the NXT. First, we prepare the circuit diagram, shown in Fig. 5.17, which includes the internal voltage source with v_i = 5.0 V, the internal resistor R_i = 10 MΩ and the sensor resistance R_s. Note the sensor's resistance is a variable that depends on the angular position of the wiper arm on the potentiometer. We also show the current flow counterclockwise around this loop. Beginning at point A and applying Kirchhoff's voltage law permits us to write:

$$\Sigma V = + v_i - IR_i - IR_s = 0 \qquad\qquad (a)$$

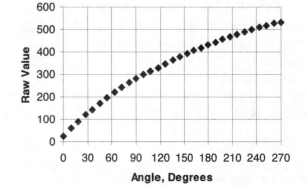

Fig. 5.16 Calibration curve for the 10 kΩ potentiometer used as an angle sensor.

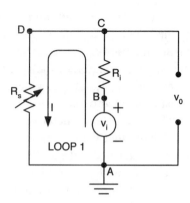

Fig. 5.17 Circuit diagram representing the angle sensor and the
internal components in the RCX.

Solve Eq. (a) for the current flow I to obtain:

$$I = \frac{v_i}{R_i + R_s} \qquad\qquad (b)$$

The voltage v_0 across the analog to digital converter in the NXT is the voltage between points A and C, in Fig. 5.17. This voltage is given by:

$$v_0 = v_i - IR_i \qquad\qquad (c)$$

Substituting Eq. (b) into Eq. (c) and simplifying yields:

$$v_0 = v_i - \frac{v_i R_i}{R_i + R_s} = v_i\left[1 - \frac{1}{1 + R_s/R_i}\right] = v_i\left[\frac{R_s/R_i}{1 + R_s/R_i}\right] \qquad (5.5)$$

Let's consider the maximum and minimum values for the sensor resistance R_s. If $R_s = 0$, its minimum value, Eq. (5.5) gives $v_0 = 0$. Recall the raw value displayed on the NXT was 1, which is nearly[6] 0. If $R_s = 10.8$ MΩ and $R_i = 10.0$ MΩ, then $R_s/R_i = 10.8/10.0 = 1.08$ and Eq. (5.5) gives $v_0 = 2.596$ V. The equivalent raw value is $(2.596/5.00)(1023) = 531$. It is encouraging to note that we recorded 532 for the raw value.

The reason for the non-linearity is the form of Eq. (5.5). The ratio R_s/R_i in the denominator causes the non-linearity. Non-linearity of sensors is not a significant problem today, because microprocessors incorporated with commercial instrumentation for sensors can be programmed to accommodate non-linear functions.

Light Sensor

A relatively complex light sensor is available from LEGO that interfaces with the NXT to measure light intensity on a scale of 0 to 100. This sensor incorporates a light emitting diode (LED) to illuminate the scene, a photodiode to measure either the direct light intensity or the reflected light, and circuitry to power both of these components. This is an active sensor that requires a separate power lead from the NXT to operate.

In many applications, less complex and lower cost light sensors fabricated from Cadmium Sulfide (CdS) photoconductors are sufficient. The characteristics of CdS photoconductors were

---

[6] The value of R_s is not quite zero, because resistance is introduced in the circuit due to lead wires and contact resistance at the potentiometer's wiper.

discussed previously in Section 5.4.1. They are variable resistors with the resistance changing from very high values when the sensor is dark to nearly zero resistance when the sensor is exposed to bright light and the CdS becomes a good conductor. We have discussed the design of angle of rotation sensor using a potentiometer that is a variable resistor. The development of a CdS light sensor is similar.

Only two parts are required to build this sensor: a CdS photoconductor (photocell) and a lead wire that is compatible with the NXT. We purchased a package of five photocells from Radio Shack. One of these photocells is illustrated in Fig. 5.18.

Fig. 5.18 A photocell and its two lead wires anchored with a dime.

After soldering the lead wires from the photocell to the wires for pins #1 and #2 on a NXT lead wire, we check to determine some of the characteristics of our light sensor. Using the same procedure developed in calibrating the angle sensor, we explore the readings from our light sensor that is connected to port 1. The raw value is displayed for this light sensor. In normal room light, the sensor gives a raw value that ranges from about 200 to 400 depending on its orientation. When exposed directly to a bright light source, the raw value decreases to about 15. If we cover the sensor with a black cloth, the raw value increases to about 900.

Typically, a low cost photocell is employed as a sensor to turn lights on or off and to detect significant changes in light intensity. To check if the photocell is effective for this application, write a suitable program for the NXT, connect the sensor to port 1, run the program and view the output from the light sensor on the LCD display. In normal room light, the reading is 1. Shield the sensor from the light and the reading goes to 0. The sensor is acting like a switch, turning on when exposed to light and turning off when shielded from light.

It is possible to use the CdS sensor to measure light intensity; however, the sensor is affected by light coming from many different directions. To improve its performance for measuring intensity, the sensing element should be placed in a short, small-diameter tube to shield it from stray light. Additional improvement in performance is achieved if a suitable lens is fitted to the front of this tube to focus the incoming light on the sensing element.

5.7 SUMMARY

Several basic sensors have been described in this chapter. These sensors can be used to measure an unknown quantity (such as the use of an electrical strain sensor to measure strain); however, in many other instances the sensors are one component of a more complex transducer such as a piezoelectric crystal in an accelerometer or a number of strain gages in a force transducer. Important characteristics of each sensor that must be considered in the selection process include:

1. **Size** - with smaller being better because of enhanced dynamic response and minimum interference with the process or event.
2. **Range**—with extended range being preferred to increase the latitude of operation.
3. **Sensitivity**—with the advantage to higher output signals that require less amplification.
4. **Accuracy**—with the advantage to devices exhibiting errors of 1% or less after considering zero shift, linearity and hysteresis.

5. **Frequency response**—with preference for wide-bandwidth sensors that permit application in both static and dynamic loading situations.

6. **Stability**—with very low drift in output over extended periods of time and with very small output signal changes with variations in temperature and humidity preferred.

7. **Temperature limits**—with the ability to operate from cryogenic to elevated temperatures preferred.

8. **Economy**—with reasonable costs preferred.

9. **Ease of application**—with reliability and simplicity always preferred.

The use of the NXT in measuring voltage was described. Kirchhoff's voltage law was employed to derive the relation for the voltage recorded by the DVM in the NXT and the voltage drop across a variable resistor. The methods employed to design and build a touch, angle of rotation, and light sensors were described in considerable detail. Examples were described and calibration methods were illustrated.

REFERENCES

1. Brindley, K.: <u>Sensors and Transducers</u>, Heinemann, London, 1988.
2. Sedra, A. S. and K. C. Smith: <u>Microelectronic Circuits</u>, 3rd ed., Holt, Rinehart, and Winston, New York, 1991.
3. Yang, E. S.: <u>Microelectronic Devices</u>, McGraw Hill, New York, 1988.
4. Wang, E., <u>Engineering with LEGO Bricks and ROBOLAB</u>, 2nd edition, College House Enterprises, Knoxville, 2005.
5. Baum, D, M. Gasperi, R. Hempel, and L. Villa, <u>Extreme Mindstorms: An Advanced Guide to LEGO Mindstorms</u>, Apress, Berkeley, 2000.
6. Erwin, B., <u>Creative Projects with LEGO Mindstorms</u>, Addison-Wesley, New York, 2001.
7. Dally, J. W., W. F. Riley and K. G. McConnell, <u>Instrumentation for Engineering Measurements</u>, 2nd edition, John Wiley, New York, NY, 1993.

EXERCISES

5.1 Describe the differences between a sensor and a transducer. Give an example of a transducer incorporating a displacement sensor. Give another example of a sensor that is also a transducer.

5.2 A slide-wire potentiometer having a length of 100 mm is fabricated by winding wire with a diameter of 0.10 mm around a cylindrical insulating core. Determine the resolution limit of this potentiometer.

5.3 If the potentiometer of Exercise 5.2 has a resistance of 2,000 Ω and can dissipate 2 W of power, determine the voltage required to maximize its sensitivity. What voltage change corresponds to the resolution limit?

5.4 A 20-turn potentiometer with a calibrated dial (100 divisions/turn) is used as a balance resistor in a Wheatstone bridge. If the potentiometer has a resistance of 20 kΩ and a resolution of 0.05%, what is the minimum incremental change in resistance ΔR that can be read from its calibrated dial?

5.5 Why are potentiometers limited to static or quasi-static applications?

5.6 List several advantages of the conductive-film type of potentiometer.

5.7 A new elevator must be tested to determine its performance characteristics. Design a displacement transducer that utilizes a 10-turn potentiometer to monitor the position of the elevator over its 100-m range of travel.

5.8 Design a circuit to turn on outside lights at your home as it begins to get dark. Use a photoconduction cell in the circuit and provide for an adjustment to control the intensity level for switch activation.

5.9 Sketch the circuit for a photovoltaic cell and write a paragraph explaining its operation. Write another paragraph stating the advantages and disadvantages of this light sensor.

5.10 Sketch the circuit for a photodiode used in the photoconduction mode and write a paragraph explaining its operation. Write another paragraph stating the advantages and disadvantages of this light sensor.

5.11 Write your own summary of three important topics in this chapter.

Notes

CHAPTER 6

PROGRAMMING IN ROBOLAB™

6.1   INTRODUCTION

The NXT incorporates a microprocessor capable of storing programs that provide instructions for controlling the fans and oth4er actuators used in your team's hovercraft. In this application, the control of the fans was complicated by the need to steer the hovercraft and to control its speed particularly around curves in the track. Moreover, the requirement is for autonomous control, which implies that the NXT must be programmed so that it provides command signals to the lift and propulsion fans that enable the hovercraft to negotiate the track in the minimum possible time.

We will be working with two types of control—sequential and decision-based. If sequential control is employed, the operator programs the controller to perform a sequence of tasks with the timing and power level associated with each task decided in advance. Programming for sequential control is relatively easy as it is simply a line of symbols (commands) without forks, jumps or loops. We start the program, sequentially perform a specified number of tasks and then stop.

Decision based control requires sensors in the system that provide feedback signals, which give information essential in controlling the system. When the system is in operation, the sensors provide continuous signals that are monitored by the NXT. When the level of a feedback signal deviates sufficiently from a specified level, the power to one or more actuators (fans) is adjusted to maintain the sensor signal within a control band centered about the control level. For decision-based control to be effective, the microprocessor must be provided with instructions that establish the upper and lower levels of the control band. Sensor signals that are outside the control band cause the controller to energize actuators (fans), which adjust the parameters affecting the system so as to maintain control. Programming for decision-based control is more challenging as the program entails a more structure involving forks, jumps and loops and sometimes loops within loops.

6.2 FLOWCHARTS

A flowchart is a programming tool that is often used to graphically illustrate each step in a program. Flowcharts are usually prepared prior to writing a structured program or even assembling a program with a graphical interface. Before writing a program, it is essential that the strategy for executing the steps that control the mechanisms involved in the steering the hovercraft be clearly established. The primary advantage of using a flowchart to plan the strategy for a program is that it provides a graphical representation of the program making its logic easier to visualize.

The American Standards Institute (ANSI) has developed several flowchart symbols that are presented in Fig. 6.1

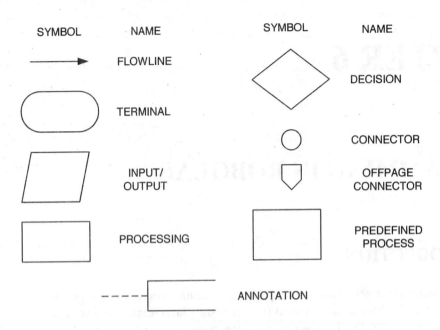

Fig. 6.1 Standard symbols used in preparing flowcharts.

The name and use of each ANSI symbol is listed in Table 6.1.

Table 6.1
Name and use of the flowchart symbols.

Symbol Name	Use
Flowline	For connecting symbols and showing the flow of the logic.
Terminal	For representing the beginning (start) and the end (stop) of the program.
Input/Output	For input and output operations such as reading sensor signals or displaying data.
Processing	For arithmetic and data manipulations operations.
Decision	For any comparison operation. The decision symbol has one input and two outputs.
Connector	For connecting two or more flowlines.
Off page Connector	For indicating that the flowchart continues on the next page.
Predefined Process	For representing a group of statements that perform a specific task or a subroutine.
Annotation	For providing additional information pertaining to one of the flowchart symbols.

Let's consider a few examples to demonstrate the use of flowcharts in programming.

EXAMPLE 6.1

When dinning in a nice restaurant, it is customary to tip the person serving your food by about 17%. Rather than counting pennies in determining amount of the final payment, it is common practice to add the tip to the initial bill and round the sum upwards to the nearest dollar. Let's prepare a flowchart for a program that will determine the amount of the final payment.

Solution: We begin with the terminal symbol with a command to start the program. Next a flowline connects the terminal symbol to an input/output symbol with a command to read the amount of the bill into the program. Flowlines connect to three processing symbols where sequential commands are given for the program to calculate the tip, add the amount of the tip to the initial bill and to round this amount upwards to the nearest dollar. A flowline connects the last processing symbol to an input/output symbol where the program is directed to display the amount of the payment. Finally, a terminal symbol is placed at the end of the top down listing of sequential steps to stop the program.

The flowchart in Fig. 6.2 is representative of a sequential program in which we move from one step to the next without skipping any blocks (or lines of code). Decisions were not a factor; consequently, it was not necessary to introduce one or more branch points (forks) in the structure of the program. However, in many applications decisions are essential, and a decision-structured program with one or more forks is mandatory. The flowchart for a decision structure is illustrated in Fig. 6.3.

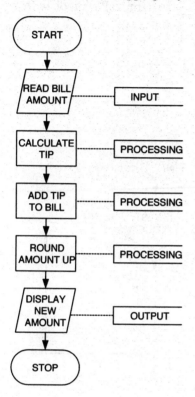

Fig. 6.2 Flowchart to determine the payment that includes an appropriate tip for an attentive waiter or waitress.

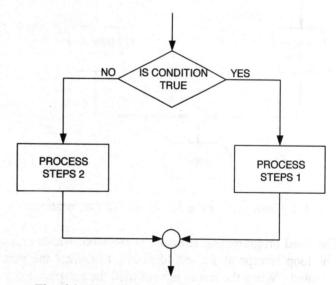

Fig. 6.3 Flowchart for a decision-structured program.

In a decision-structured program, we pose a question or a condition that is either true or false (yes or no). This gives rise to writing lines of code that state **IF** the specified condition is true **THEN** proceed to Process Steps 1 **ELSE** proceed to Process Steps 2.

EXAMPLE 6.2

Downtown San Jose, Costa Rico is laid out on a Cartesian grid. The Avenidas run east and west and the Calles run north and south. Avenida Central and Calle Central divide the city into quadrants. To the north of Avenida Central the avenidas have odd numbers and to the south they have even numbers. Similarly, Calles to the east of Calle Central have odd numbers and to the west they have even numbers. Prepare a flowchart for a program to determine if an avenida is on the north or south side of the city.

Solution: Again we begin the program with a terminal symbol with the command to start. A flowline connects this symbol to an input/output symbol with a command to read the number of the avenida as shown in Fig. 6.4. A flowline connects to a decision symbol, with a command to test if the number is even. **If** the number is even, **then** the flowline connects to an input/output symbol that displays the result "south". **Else** (if the number is odd, the decision condition is not satisfied), the flow line connects to the input/output symbol that displays "north". The flowlines from both input/output symbols go to a connector symbol. The flow line from the connector symbol goes to the terminal symbol with a command to stop the program.

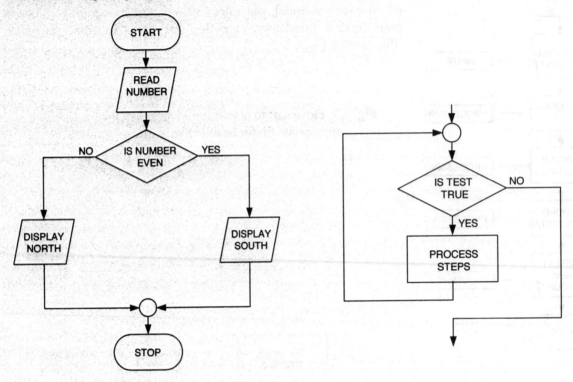

Fig. 6.4 Flowchart for the San Jose avenidas locations.   Fig. 6.5 Flowchart for a loop structure.

The third programming structure is the loop, which executes a set of instructions one or more times. The loop incorporates a test (decision) to control the number of times the instructions in the loop are repeated. When the test is not satisfied the program exits the loop. The flowchart symbols for a loop structure are illustrated in Fig. 6.5.

EXAMPLE 6.3

Suppose a class with 36 students takes an examination that is graded on a scale of 0 to 100. Prepare a flowchart for a program to determine the class average and to display this result.

Solution: Again we draw a terminal symbol with the start command to begin the program. Next, a process symbol is employed with instructions to set two variables defined as the count and sum both equal to zero as indicated in Fig. 6.6. Next in line is an input/output symbol with a command to read the first grade in a list of 36 grades. A connector symbol is inserted to indicate a junction point in the program. The flowline then proceeds to a decision symbol with logic to ascertain if there are more grades to be considered. If the answer is yes, the flowline connects to two sequential process symbols where the program adds one to the count and then adds the grade to the sum. An input/output symbol with the command to read the next grade follows. At this point, the program loops back to the connector symbol and the process is repeated until each grade is read into the program and processed. When all the grades have been accommodated, the test of the specified condition on the count in the decision symbol is negative. Then the flowline exits the loop and connects with another processing symbol. The average is then determined by taking the sum and dividing by the count. An input/output symbol is used with a command to display the result. Finally, the flowline connects with a terminal symbol to stop the program

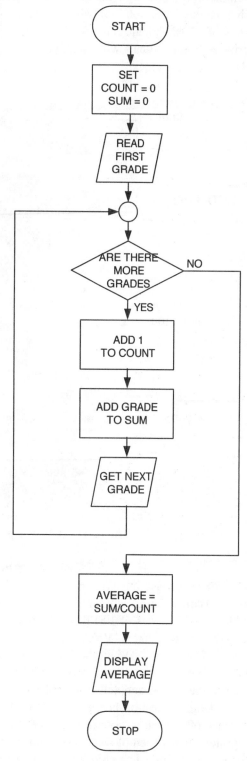

Fig. 6.6 Flowchart representing a decision and loop structure.

6.3   THE LEGO® NXT® CONTROLLER

The NXT, shown in Fig. 6.7, is a controller that contains an Atmel ARM processor (AT91SAM7S256), which incorporates a high performance 32 bit RISC architecture. This processor contains 256 Kbytes of flash memory (non-volatile) with single cycle access at 30MHz. It also contains an additional 64 Kbytes of SRAM (static random access memory). The ARM processor includes three different clocks to provide timing signals with frequencies that range from 22 to 42 kHz for the slow clock and 80 to 220 MHz for the PLL clock.

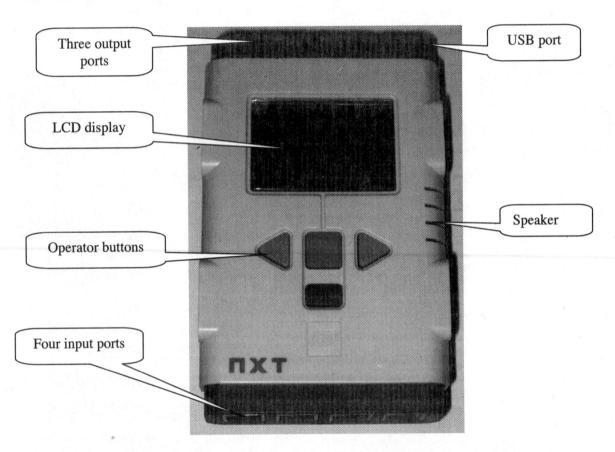

Fig 6.7 The NXT controls the actions of autonomous robots and vehicles.

The NXT also contains an 8-bit Atmel co-processor (ATmega48) that is used with the main processor for measuring voltage and controlling actuators. This processor contains 4 Kbytes of flash memory (non-volatile) with single cycle access at 30MHz. It also contains an additional 512 bytes of SRAM and 256 bytes of EEPROM memory. It has two 8-bit timer/counters, one 16-bit timer/counter and a real time counter. Voltage is measured with an eight channel 10 bit analog to digital converter (ADC). The 10-bit ADC in the NXT can acquire and store voltage measurements at frequencies up to 200 Hz. Six pulse width modulated (PWM) channels are available in the co-processor for supplying power to the motors and actuators, although the NXT only uses three of these channels.

Communication between the NXT and a computer is accomplished with either a USB cable or by Bluetooth wireless connection (radio). A CSR BlueCore™ 4 single chip radio and base band integrated circuit is employed to implement a Bluetooth 2.4GHz system, which provides enhanced data rates up to 3Mbps. The chip interfaces to 8Mbit of external flash memory and contains 47 Kbytes of RAM. When used with the CSR Bluetooth software stack, it provides a fully compliant Bluetooth system for data and voice communications. The BlueCore™ transmitter is powered by a 6-bit digital-to-analog-converter with a dynamic range greater than 30 dB. Its receiver incorporates channel filters and a digital demodulator for improved sensitivity and co-channel rejection. The USB 2.0 communication port is capable of 12 Mbits/s.

The number of programs you can store in the NXT is only limited by the memory available. The NXT microprocessor can perform both integer and floating point (decimal) mathematics. You can program the NXT to read and write files to its memory.

Input Ports and Sensors

There are eight ports on the NXT. The four input ports are located on the base of the unit below the display and control buttons, as shown in Fig. 6.7. These ports, numbered 1, 2, 3 and 4 are equipped with a six wire interface that can receive both analog and digital signals from sensors. Instructions in the LEGO User Guide for the NXT give the standard port locations for each of their sensors as:

Port 1: Touch sensor
Port 2: Sound sensor
Port 3: Light Sensor
Port 4: Ultrasound sensor

It is important to note that Port 4 is a high speed data connection that functions with an I^2C communication protocol, which is used with the LEGO ultrasonic sensor. The sensors are wired to the input ports using the 6-wire industry standard RJ121 connector with right side adjustment.

Sensors are classified as either active or passive. The passive sensors do not require power from the NXT to operate. The LEGO touch sensor, a simple momentary switch shown in Fig. 6.8, and the LEGO sound sensor, a microphone shown in Fig. 6.9, are examples of passive sensors. On the other hand, active sensors require power from the NXT to function. The LEGO light sensor, which contains a IR light source and a photodiode as shown in Fig. 6.10, is an example of an active sensor. The LEGO ultrasonic sensor, shown in Fig. 6.11, requires a communication channel to accommodate the timing signals required when bursts of sound are transmitted and received.

Fig. 6.8 The LEGO touch sensor (a contact switch) is a passive device that does not require power to function.

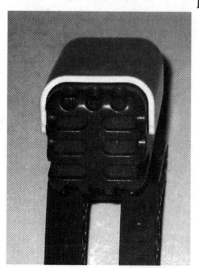

Fig. 6.9 The LEGO sound sensor (a microphone) is a passive device that does not require power to function. It measures sound in decibels (dB) or adjusted decibels (DbA).

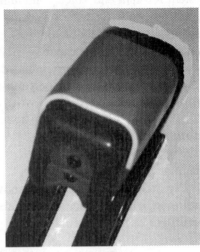

Fig. 6.10 The LEGO light sensor is an active device that requires power to light its IR source. Its readings range from 1 (dark) to 100 (bright).

Fig. 6.11 The LEGO ultrasonic (sonar) sensor requires a communication channel to send and receive signals in order to measure time for sound propagation to and from a target.

Output Ports

The three output ports, located on the top edge of the NXT above the LCD, are labeled as A, B, and C. These ports are also equipped with a 6-wire interface to provide power (voltage pulses) to motors, lights or sound making devices and to receive input from encoders that are installed in the LEGO servo motors. Instructions in the LEGO User Guide for the NXT give the standard port settings for the motors as:

Poet A: Motor for additional function
Port B: Motor for movement
Port C: Motor for movement

Each port may be operated in one of three modes—**on**, **off** and **floating**. When a port is programmed in the **on** mode, a motor or actuator connected to that port receives pulse width modulated power. When a port is programmed in the **off** mode, the motor actuator attached to that port is braked so it stops abruptly. When a port is programmed in the **floating** mode, a motor connected to that port does not receive power, but it is permitted to freewheel until it stops due to frictional forces.

The LEGO servomotor, shown in Fig. 6.12, contains a small motor, gearing and an encoder. The gearing increases the torque of the motor but decreases its speed. The motor is equipped with an encoder that provides a signal which enables measurement of the angle of rotation of the motor shaft to within ± 1°. Two of the six wires in the standard RJ121 connector are used to connect the encoder signal to the NXT.

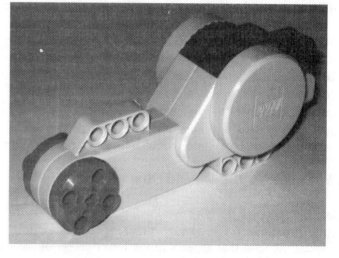

Fig. 6.12 The LEGO servomotor is connected to the output ports.

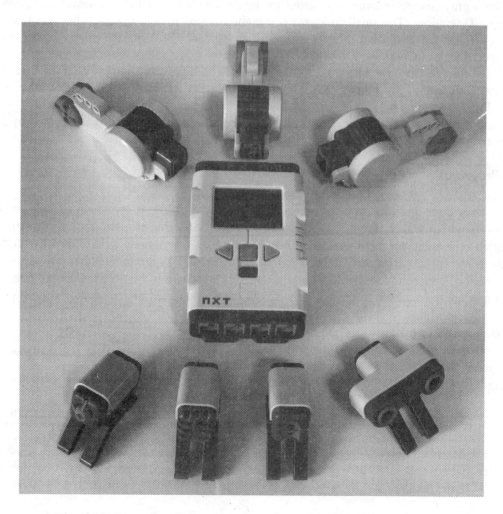

Fig. 6.13 Sensors and motors arranged around the NXT controller.

RCX motors and lights can be connected to the output ports of the NXT; however, an adapter cable is required to interface with its 6-wire RJ121 connector. In addition several companies[1] are producing special purpose sensors and output devices that are compatible with the NXT.

Communication with the NXT

Communication between the NXT and a computer is accomplished with either a USB cable or by Bluetooth wireless connection (radio). A CSR BlueCore™ 4 single chip radio and base band integrated circuit is employed to implement a Bluetooth 2.4GHz system. This chip provides a fully compliant Bluetooth system for data and voice communications at rates up to 3Mbps. The USB 2.0 communication port is faster with a transfer rate of 12 Mbps.

The Controller Buttons and Display

Near the center of the NXT, you will notice a liquid crystal display (LCD) and four buttons as shown in Fig. 6.7 and Fig. 6.13. The square orange button performs three functions—turns the NXT on, enters (selects) the icon that appears in the lower center of the LCD window and runs the program selected. The two light-gray triangular buttons are used for menu navigation as they to move the various icons across the LCD window. The small dark gray rectangular button is used to cancel the selection or to go back. It is pressed repeatedly until the "Turn-off" icon appears. Pressing the orange button when this icon is centered in the LCD shuts down the NXT.

When the NXT is turned on a short tune is heard and three icons appear near the bottom of the LCD, which include the "My Files", "NXT Program" and "Try Me" icons. These and other icons are defined in Fig. 6.14 to Fig. 6.17. The gray triangular button enables you move progressively among the various options available. The icon of interest is centered in the display and it is selected by pressing the square orange button.

Storing Programs

The NXT stores three different types of programs as indicated in Fig. 6.14, which include sound files, software files and NXT files. Programs written on a computer are downloaded into "software files" and songs written on a computer are downloaded into "sound files". Programs written directly on the NXT without using a computer are stored in the "NXT files". The NXT Program icon that provides entry into a large number of commands available within the NXT is presented in Fig. 6.15. The programs written on a computer are downloaded to the NXT by a USB cable connected to a computer's serial port. The NXT can also send and receive signals with its Blue tooth technology. When a program is successfully downloaded, the NXT responds with a burst of sound confirming its transmission.

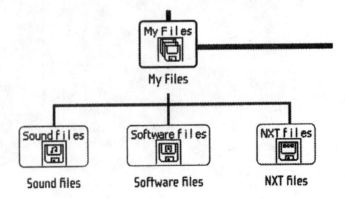

Fig. 6.14 The My Files icon and the three types of files that are options for storage in the NXT.

---

[1] These companies include: www.mindsensors.com; www.techno-stuff.com; www.hitechnic.com; and www.Imsensors.com.

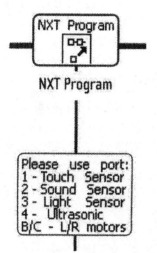

Fig. 6.15 The NXT program icon that enables programming to be performed directly on the controller.

The signals from the sensors are monitored by selecting the "View" icon that is accessed on the LCD by repeatedly pressing the triangular gray button. After the "View" icon is selected you have a choice of monitoring the signal from a large number of different LEGO sensors as indicated in Fig. 6.16. If a sensor is wired to a port, its output can be monitored on the LCD independent of a program running. In this sense, the NXT is functioning as a digital multimeter with its output converted to the units of measurement for the sensor involved.

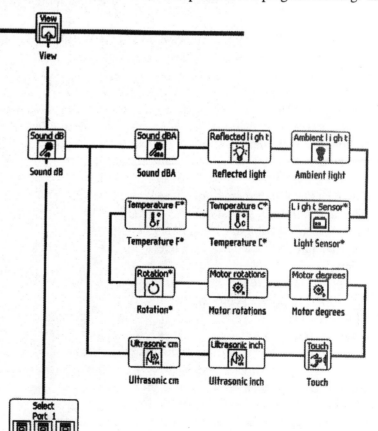

Fig. 6.16 The View icon enables measurements from the LEGO sensors to be monitored on the NXT display.

When a program is running, the output from the sensors connected to the four ports is displayed in a listing that is shown continuously on the LCD. The NXT display also shows the output of its internal clock in hour, minute, second and tenth of a second running time for the program. Turning the NXT does not reset its internal clock and the time displayed is cumulative for a given program.

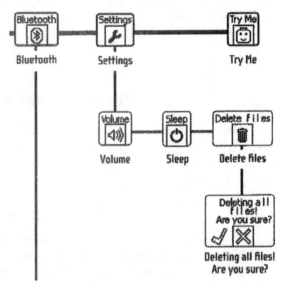

Fig. 6.17 The Bluetooth, Settings and Try Me icons that enable communication, NXT adjustments and a simple trial program.

The Bluetooth, Settings and Try Me icons are defined in Fig. 6.17. The Bluetooth icon provides access to several icons not presented in Fig. 6.17, which include Contacts, Connections, Search, On/Off and Visibility. The Settings icon enables access to Volume, Sleep and Deleting files. The Try Me icon issues a burst of sound when it is selected.

6.4   PROGRAMMING IN ROBOLAB™

ROBOLAB™ is an icon oriented programming language that is used to provide a list of instructions for the both the RCX and NXT microprocessors. It is based on LabVIEW™—a popular programming tool used by engineers for measurement and control in both academic and industrial settings. LabVIEW was developed by National Instruments in Austin, TX and is based on a graphical programming language known as G. As a programming language, G is similar to Basic, Fortran or the various versions of C. The major difference is that LabVIEW or G is based on graphics (icons) instead of many lines of written text. Hence, a LabVIEW program is comprised of a logical sequence of icons and is essentially independent of text except for comment statements made to explain various aspects of the program to a reviewer.

Dr. Chris Rogers and several others at Tufts University have developed several versions of ROBOLAB to permit people not yet trained in writing structured software to program the RCX or NXT controllers. With ROBOLAB, it is possible to create complex programs to control actuators, motors, relays or transistors on your team's hovercraft without spending a semester learning a structured programming language with several hundred commands.

The Opening Screen

This section describes the main screens in ROBOLAB, Version 2.9.3 that is available from the LEGO Educational Division, Pittsburg, KS. The opening screen for ROBOLAB, shown in Fig. 6.18, identifies the three professional groups that cooperated in the development of this icon based programming language.

In the center of the screen, observe three bars (buttons) labeled **Administrator**, **Programmer** and **Investigator**. The **Administrator** bar opens a program that enables you to adjust the settings on your NXT, define locations for your files, tests the communication from your computer to the NXT and hides the **Administrator** button. The **Programmer** bar activates the Main Menu screen that in turn enables you to select either the **Pilot** or **Inventor** levels of programming. The **Investigator** bar activates the Menu screen for the programming necessary to record measurements taken from LEGO sensors that include touch, light level, sound level, distance and angle of rotation of the servo motors.

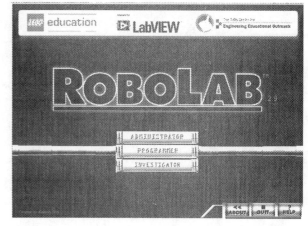

Fig. 6.18 Opening screen that appears as the files are loading ROBOLAB 2.9.3.

The Administrator Settings

The opening screen that appears when you click on the **Administrator** bar is presented in Fig. 6.19. This screen has three tabs positioned along its lower edge that switch the display as shown below.

1. **ADMINISTRATOR** screen is used to establish a method for communicating between your computer and the NXT.
2. **ROBOLAB Settings** screen is used to arrange file locations and to save or delete files.
3. **RCX/NXT Settings** screen is employed to monitor the battery level, firmware version, and to name the NXT and adjust the time before the power is automatically shut-off.

The screen displayed when the Administrator tab is activated is presented in Fig. 6.19. Four icons (buttons) are displayed. The **Select Com Port** button enables you to select the communication port where the cable from the NXT is connected to your computer. The **Download Firmware** button is to facilitate the downloading of firmware from your computer to the NXT. The downloading is accomplished with USB cable that transmits the firmware from the ROBOLAB program residing on your computer to the NXT. The firmware remains stored in memory in the NXT until the batteries are removed. If you are too slow when changing batteries, the NXT will prompt you to restore the firmware by repeating the downloading process.

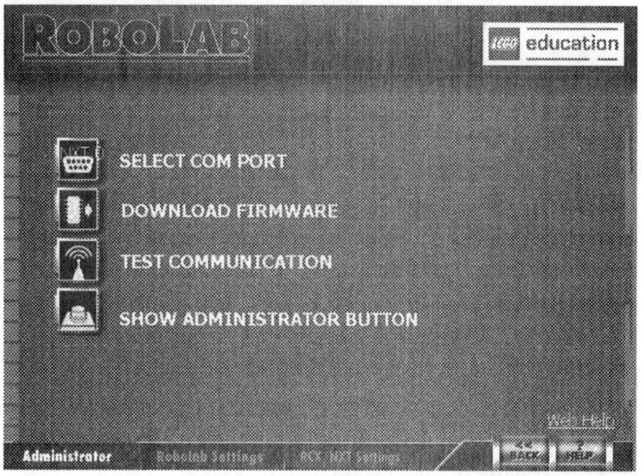

SELECT COM PORT

DOWNLOAD FIRMWARE

TEST COMMUNICATION

SHOW ADMINISTRATOR BUTTON

Fig. 6.19 Administrator screen enables selection of the controller, down loading firmware, testing communications and hiding or showing the administer button on the opening screen.

Click on the **SELECT COM PORT** button and if the NXT has already been selected as the controller of choice, the screen shown in Fig. 6.20 appears. Click on the check mark to initiate communication with the NXT.

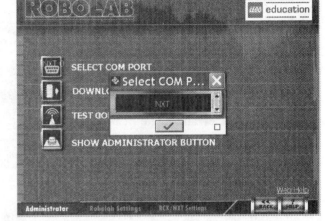

Fig. 6.20 Confirming the selection of the NXT as the controller connected to the communication port.

Next, click on the **DOWNLOAD FIRMWARE** button and the dialog box shown in Fig. 6.21 appears. Click on Ok and download the firmware from the ROBOLAB program on your computer to the NXT controller.

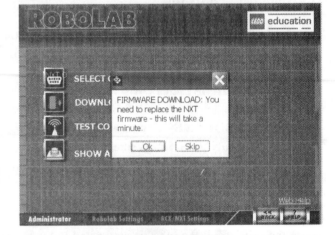

Fig. 6.21 Dialog box to aid in downloading firmware from your computer to the NXT.

After the download of the firmware is complete a dialog box appears, shown in Fig. 6.22, confirming successful communication between your computer and the NXT. The NXT also confirms a successful transmission with a little tune.

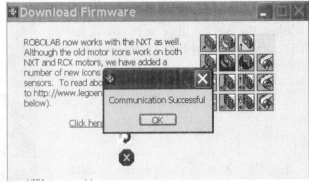

Fig. 6.22 Dialog box appears confirming the downloading of firmware from your computer to the NXT.

Click on the **TEST COMMUNICATION** button and a dialog box appears, as shown in Fig. 6.23, which confirms the communication link. The NXT also confirms the link with a short burst of sound.

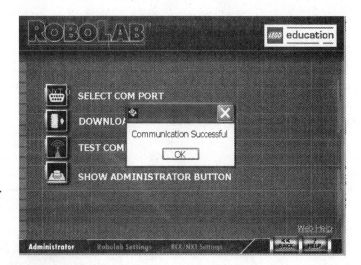

Fig. 6.23 Dialog box that confirms successful communication between your computer and the NXT controller.

Click on the **SHOW ADMINISTRATOR BUTTON** and a dialog box appears, as shown in Fig. 6.24, indication that the Administrator bar will be hidden on the opening screen. However, the administrator bar can be replaced on the opening screen by striking the F5 key.

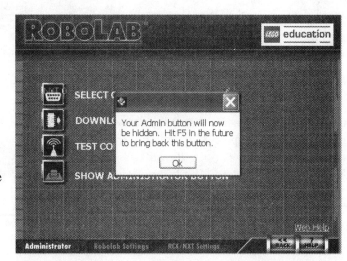

Fig. 6.24 Dialog box that confirms the removal of the Administrator bar from the opening screen.

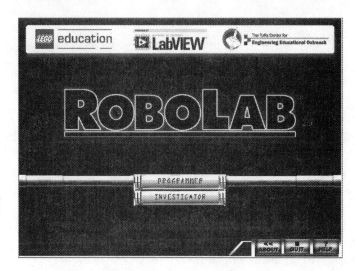

Fig. 6.25 The opening screen after the files are loaded and the administrator bar has been removed.

Robolab Settings

Clicking on the **RCX Settings** tab at the bottom of the **Administrator** screen activates the display presented in Fig. 6.26. To activate this screen, the NXT must be connected to your computer and be turned on. The settings stored in the NXT are interrogated and displayed on this screen. The path to the hard drive (usually C) for the programs stored in ROBOLAB is shown as well as the path for the user programs that are stored in the **Program Vault**.

Clicking on the icon with the + sign located on the left hand side below the **Theme Widow** creates a new folder that you can use to store any program that you write. The icon with the garbage can is used to delete programs selected from the theme window. Note that programs that are included with the ROBOLAB software are locked and cannot be deleted.

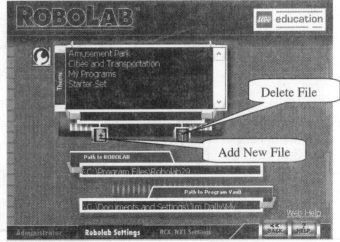

Fig. 6.26 Display for **ROBOLAB Settings** showing paths to the Robolab program and to Program files stored in the "vault".

RCX/NXT Settings

NXT Settings tab accesses the screen shown in Fig. 6.27, which enables you to name the NXT, name a program and select a power downtime. The changes are made by marking the entry and typing a substitute value. The battery level is indicated as well as the version of the NXT firmware that was downloaded previously.

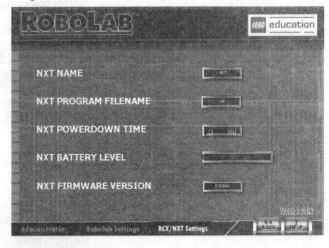

Fig. 6.27 The screen displayed for the RCX/NXT settings.

6.5 PROGRAMMING IN THE INVENTOR MODE

Programming in the **Pilot Mode** is simple because the programs are sequential—one command follows the next from start to finish. One pass through and the program is complete. However, programs containing only sequential commands are very limiting because tasks that require control structures such as looping, forks or branching cannot be written in Pilot. If control structures are required to execute a program, it is necessary to program in the **Inventor Mode**. For this reason, programming in the **Pilot Mode** will not be described in this book.

To begin programming in the **Inventor Mode**, click on the **Programmer** button on the opening screen (See Fig. 6.18). The display that appears, shown in Fig. 6.28, enables you to select from either **Pilot** or **Inventor** modes of programming. In this case, double click on **Inventor Level 4** in the **Level Window**, which is the most advanced level in the **Inventor Mode** of programming. **Note that Inventor Level 4 is the only level that can be used to program the NXT**.

When the Inventor program opens, two windows and the **Functions** palette appear as illustrated in Fig. 6.29. The upper window is the **Front Panel** and the lower window is the **Block Diagram**. Note that the stop and start signal lights are already positioned on the block diagram. To program in **Inventor**, you will drag icons from the **Functions** palette and its sub-palettes onto the **Block Diagram**.

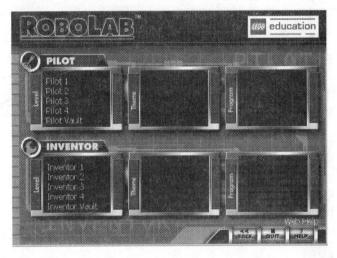

Fig. 6.28 Programming screen allows selection of four levels of programming in both Pilot and Inventor.

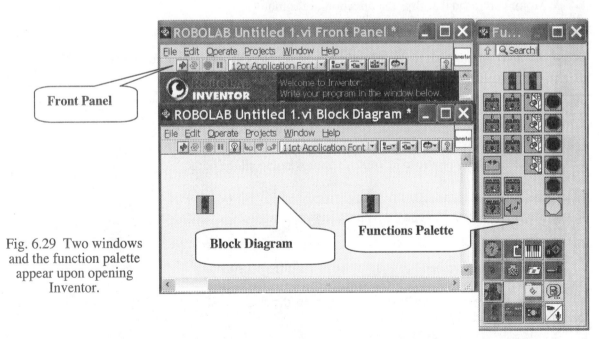

Fig. 6.29 Two windows and the function palette appear upon opening Inventor.

Inspection of the **Front Panel** and the **Block Diagram** shows that their menu bars are identical with each listing File, Edit, Operate, Project, Window and Help. Click on them and note the drop down menus that appear. If the **Functions** palette disappears, click on Window in the menu bar and then click **Show Functions Palette** that appears on the drop down menu.

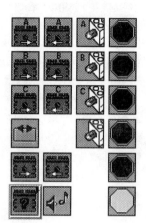

Fig. 6.30 Icons in the Functions palette used in programming the output ports.

6.5.1 The Functions Palette

The upper portion of the Functions palette, shown in Fig. 6.30 contains the icons for the commands used to control the output ports. The icons and their functions are listed below:

 Motor command for forward motion for motor wired to port A.

 Motor command that turns on all ports at power level 5.

 Motor command that flips the direction of the motors.

 Lamp command that turns on power to port A. Default setting is at full power.

 Stops power to motors or lamps wired to port A.

 Stops power to motors or lamps wired to all the ports.

 Stops power to motors so they float to a stop.

 Musical note command that plays one of six different sounds.

The yellow stop sign floats the output to the motors; whereas, the red stop sign produces a braking action suddenly stopping them.

The **Motor** icon ▯ permits access to a sub palette titled Advanced Output as shown in Fig. 6.31. To determine the function of each of these advanced output icons, point to the icon of interest and type Control + H. The **Context Help** window appears describing the function of this output command and shown in Fig. 6.32. The **Context Help** window, in this instance, describes the function of the

Random Direction icon, which sets the motor direction randomly. To use this command, the icon is dragged onto the Block Diagram and then the **Begin** and **End** terminals are wired into the chain of commands. Finally, the Output Ports which are affected by this command are identified.

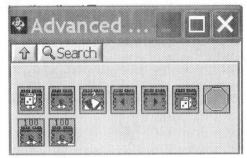

Fig. 6.31 The Advanced Output palette used to provide additional commands for the motors.

The **Context Help** window can be locked in the on position by single clicking on the small lock icon found in the lower left corner of the window. This lock is a toggle and the window can be unlocked by another click of the mouse.

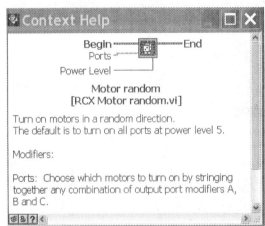

Fig. 6.32 The **Context Help** window is accessed by typing Control + H.

In the lower part of the Functions palette, you will find the group of 18 icons shown in Fig. 6.33. These icons provide an entry to several sub-palettes which contain additional commands essential to programming in Inventor Level 4. Several of these icons are described below.

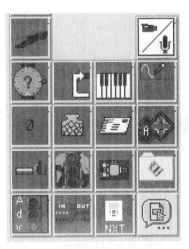

Fig. 6.33 Icons in the Functions palette that provide entry to other commands in Inventor Level 4.

 Wait For commands. Left clicking on this icon provides entry to the **Wait For** commands that are listed in a sub-palette presented in Fig. 6.34.

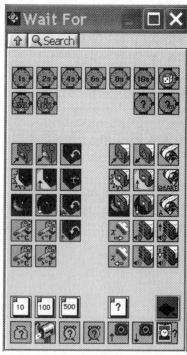

Fig. 6.34 The Wait For sub palette with its icons.

Structure commands. Clicking on this icon gives access to a sub-palette shown in Fig 6.35.

Clicking on one of these icons  gives access to additional sub-sub-palettes for **Forks**, **Jumps**, **Loops** and **Events**.

Fig. 6.35 Sub palette providing icons for structure commands.

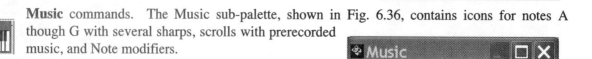

Music commands. The Music sub-palette, shown in Fig. 6.36, contains icons for notes A though G with several sharps, scrolls with prerecorded music, and Note modifiers.

Fig. 6.36 Sub palette providing icons for different notes and pre-recorded music.

Modifier commands. The Modifier sub-palette, presented in Fig. 6.37, enables specification of input and output Functions, with timers, containers and container values. It also permits Numeric constants to be defined.

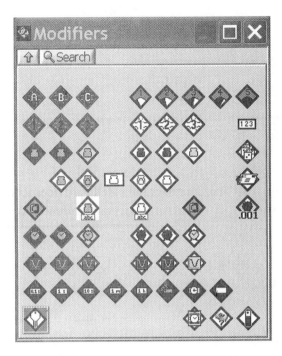

Fig. 6.37 The modifiers sub palette.

 Reset commands. This sub-palette, shown in Fig. 6.38, enables you to zero the light, sound and angle of rotation sensors, empty containers and mailboxes, zero timers and reset the clock on the RCX or NXT.

Fig. 6.38 The Reset sub palette used to zero readings from sensors, timers and containers.

Container commands. The Containers in the sub palette, shown in Fig. 6.39, provide storage locations in your program. Containers are available to collect data from all of the sensors and the timer, to perform simple arithmetic operations and to compute with simple formulas.

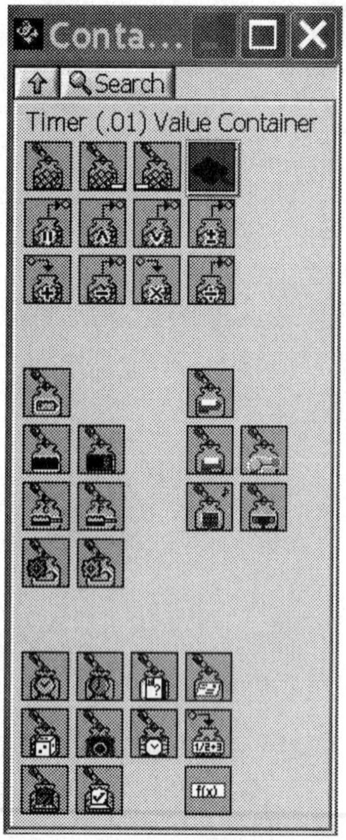

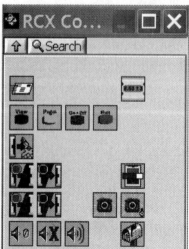

Fig. 6.39 The Container sub palette provides icons for in program storage and for elementary arithmetic operations.

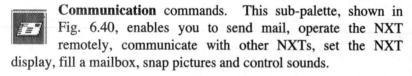

Communication commands. This sub-palette, shown in Fig. 6.40, enables you to send mail, operate the NXT remotely, communicate with other NXTs, set the NXT display, fill a mailbox, snap pictures and control sounds.

Fig. 6.40 The Communication sub palette provides icons for communicating with the RCX.

All of the icons presented on the sub-palettes and in some instances the sub-sub-palettes are too numerous to describe in this chapter. However, as an example, the sub-palette associated with the **Wait For** function is presented in Fig. 6.34. The cluster of clock faces located at the top of the sub-palette permits you to select the **Wait For** time in seconds. Some time delays are fixed, but others can be specified. The delay time can be specified to 1 millisecond. A random time generator is available that generates delay times, which vary randomly from 0 to 5 seconds. An icon is also available permitting the delay time to be set in minutes.

The icons near the center of the **Wait For** sub-palette are associated with the touch, light, temperature, distance, sound and angle sensors (servo motors). The delay in executing the program is in response to the output from the sensor to exceed or to drop below a certain prescribed value.

The icons, at the bottom of the sub-palette, control the delay time by waiting for a container, timer, clock or camera value to be achieved. Similarly, the delay time may be control by the numerical value of the mail received.

6.5.2 The Tools Palette

The **Tools Palette**, presented in Fig. 6.41, is accessed by clicking on Windows located on the menu bar and selecting **Show Tools Palette** from the drop down menu. Six of the ten icons on this palette are used extensively in preparing a program in the Inventor mode. Each of these six icons is described in the following paragraphs.

 The **Operator** tool is used to change numeric values.

 The **Select** tool is used to select icons from the various palettes and to drag them onto the Working Diagram. It can also be employed to resize text boxes.

Fig. 6.41 Icons representing the tools found in the **Tools** Palette.

 The **Text** tool is used to type text into text boxes and to change numeric values.

 The **Wiring** tool is used to connect wire from one icon Begin or End terminal to another and to connect modifier icons to their terminals.

The **Replace** tool is used to substitute one icon for another after a program is written. Icons can be deleted and then replaced with the Select tool; however, this process usually results in broken wires. The icons can be changed with the Replace tool without the need to correct wiring errors.

The Placement tool is used to move icons to different positions on the Working Diagram. It is also used to select icons from palettes and drag them onto the Working Diagram, a task that is also performed with the Select tool.

6.5.3 Wiring Icons and Modifiers

After determining the strategy for a specific program and preparing a flow chart, the next step in programming is to select the required icons representing the various commands and to drag them onto the working diagram. The order in which the commands are addressed is determined by the wiring which connects one icon to the next, etc. An example of the **Sound Type** icon is presented in Fig. 6.42, shows a heavy (pink) wire from the left that connects to the **Begin** terminal. A second heavy (pink) wire is attached to the **End** terminal. The chain of commands in a program moves from left to right. A third thin (blue) wire is attached to the **Modifier** terminal at the base of the icon. This wire is used to connect a modifier icon, which for this command controls the sound emitted when the speaker is activated.

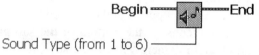

Fig. 6.42 Wiring an icon into a program.

The wires used in ROBOLAB are color coded. The beginning and the end terminals of the primary functional icons are connected with thick pink wires. Thin blue wires are used to connect integer Numeric modifiers. Thin green wires are used to connect the Port modifiers. Thick brown wires identify the sensor Port connected to Containers terminals.

Auto-wiring is a feature of Version 2.9.3 of ROBOLAB. When dragging an icon onto the Block Diagram, it attempts to automatically wire to adjacent icons. In some arrangements, the connections are made to the wrong terminals and the wiring must be removed by marking and deleting the misplaced wires. If you have placed your icons and are ready to begin wiring, hitting the space bar will automatically convert the **Select** tool into the **Wiring** tool.

EXAMPLE 6.4

Consider a simple program to demonstrate a few of the programming techniques utilized in Inventor Level 4. Suppose you are to write a program for your hovercraft to follow a straight line. The line is 50 mm wide and is formed using black tape applied over a relatively light surface. Your strategy is use a single light sensor to measure the reflected light from the surface. Based on the measurement, which enables feedback control, the power to one of the two propulsion fans used to drive the hovercraft will be adjusted. You start the hovercraft in the center of the black line and plan to follow its left edge.

Step 1: To begin let's prepare a flow diagram with a feedback loop that operates continuously. Continuous operation is required because it is expected that the hovercraft propulsion fans will have to be turned on and off many times to keep the vehicle close to the edge of the tracking line. The flow diagram will contain two conditional commands—the first to turn off the starboard fan (SF) when the intensity of the light sensor becomes greater than 55 indicating the hovercraft has moved off the black line and is over a reflecting light surface. The other conditional command is to turn off the port side fan if the light intensity measurement is less than 45 indicating that the hovercraft has moved over the black line. The program is sequential, although it does employ a loop that operates continuously because it is necessary to measure the reflected light intensity at all times. The flow diagram for the program required to control the course of the hovercraft is presented in Fig. 6.43.

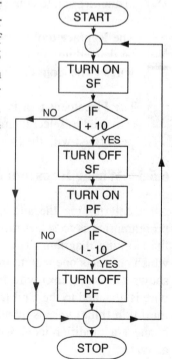

Fig. 6.43 Flow diagram for line following using feedback from a single light sensor.

Step 2: Let's prepare a program in Inventor Level 4 that will execute the program shown in Fig. 6.43.
Open Inventor 4 and from the **Functions** palette drag the **Motor A** icon 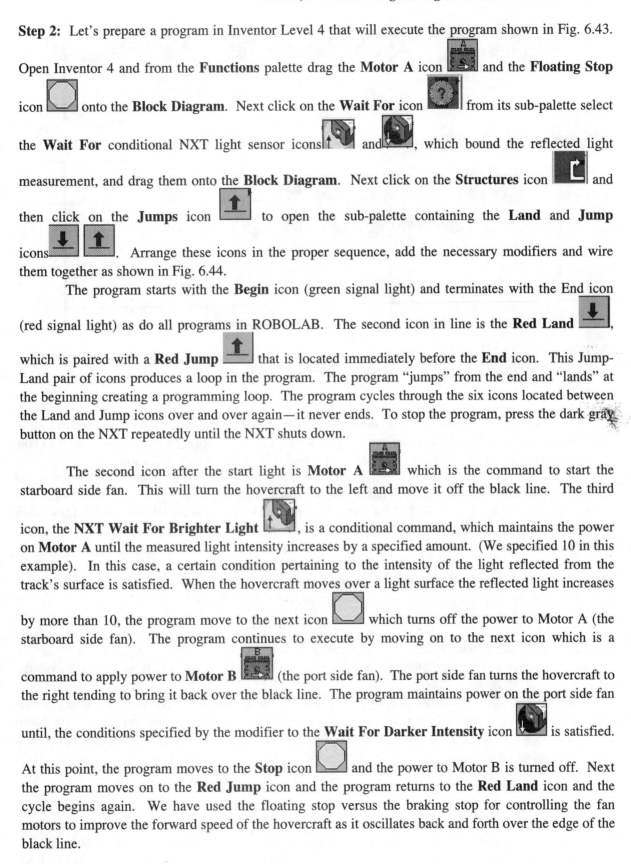 and the **Floating Stop**
icon onto the **Block Diagram**. Next click on the **Wait For** icon from its sub-palette select
the **Wait For** conditional NXT light sensor icons and , which bound the reflected light
measurement, and drag them onto the **Block Diagram**. Next click on the **Structures** icon and
then click on the **Jumps** icon to open the sub-palette containing the **Land** and **Jump**
icons . Arrange these icons in the proper sequence, add the necessary modifiers and wire
them together as shown in Fig. 6.44.

The program starts with the **Begin** icon (green signal light) and terminates with the End icon
(red signal light) as do all programs in ROBOLAB. The second icon in line is the **Red Land** ,
which is paired with a **Red Jump** that is located immediately before the **End** icon. This Jump-
Land pair of icons produces a loop in the program. The program "jumps" from the end and "lands" at
the beginning creating a programming loop. The program cycles through the six icons located between
the Land and Jump icons over and over again—it never ends. To stop the program, press the dark gray
button on the NXT repeatedly until the NXT shuts down.

The second icon after the start light is **Motor A** which is the command to start the
starboard side fan. This will turn the hovercraft to the left and move it off the black line. The third
icon, the **NXT Wait For Brighter Light** , is a conditional command, which maintains the power
on **Motor A** until the measured light intensity increases by a specified amount. (We specified 10 in this
example). In this case, a certain condition pertaining to the intensity of the light reflected from the
track's surface is satisfied. When the hovercraft moves over a light surface the reflected light increases
by more than 10, the program move to the next icon which turns off the power to Motor A (the
starboard side fan). The program continues to execute by moving on to the next icon which is a
command to apply power to **Motor B** (the port side fan). The port side fan turns the hovercraft to
the right tending to bring it back over the black line. The program maintains power on the port side fan
until, the conditions specified by the modifier to the **Wait For Darker Intensity** icon is satisfied.

At this point, the program moves to the **Stop** icon and the power to Motor B is turned off. Next
the program moves on to the **Red Jump** icon and the program returns to the **Red Land** icon and the
cycle begins again. We have used the floating stop versus the braking stop for controlling the fan
motors to improve the forward speed of the hovercraft as it oscillates back and forth over the edge of the
black line.

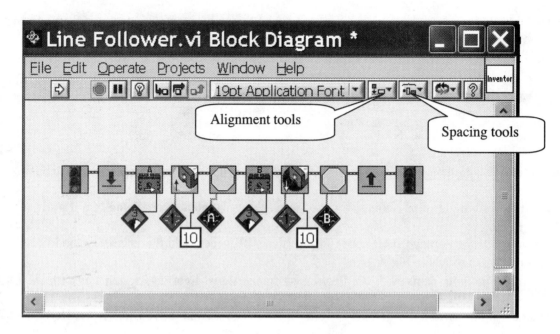

Fig. 6.44 Program employing feedback and a continuous loop to steer the hovercraft along the left edge of a black line on a light surface.

Programming in Inventor requires modifiers to identify Ports, conditional limits and power levels. For instance, the light sensor is connected to Port 1 as indicated by the modifier wired to both of the conditional icons. The light intensity change is also specified by wiring a Numeric indicator to both of these icons. Finally, the power level for Motors A and B is specified with still another modifier.

Another important step in the programming is its appearance. We have used the alignment and spacing features to align all of the icons and to space them uniformly. Finally the program is downloaded to the NXT through the USB cable. To download the program turn on the NXT and click on the small white **Run** button located just below File and Edit menu items. When the program is transmitted to the NXT successfully, it confirms its receipt with a beep and stores the program.

To save the program, click on File and from the drop down menu click on Save As. Name the program and save it under the My Projects folder in your computer.

6.5.4   Programming Structures

Working in **Inventor Level 4** enables a programmer to select from a number of different commands in developing a program to accomplish one or more tasks. These commands, which are accessed from the

Structures icon ![icon] on the **Functions** palette, **Forks**, **Loops**, **Jump** and **Land**, and **Events**, permit programs with multi-tasking, non-sequential commands, repeats and continuous cycling. The **Jump** and **Land** icons, which were previously used in Example 6.4, are identical to the **Go To** command in traditional programming.

Conditional Forks

Forks are employed when two or more tasks are involved. Forks split the command line at each fork icon that is placed in the program. The division of the command line into two paths is often dependent upon the output of some sensor. Sensor based icons that split the command line are shown in Fig. 6.45.

These include the touch, temperature, light and angle of rotation sensors. Forks can also be controlled based on a container count, the timer or clock count, the value of mail received, etc.

Fig. 6.45 The **Fork**s palette provides icons that control task splits. This sub-sub-palette is accessed from the **Structures** icon then the **Forks** icon.

Let's consider a few examples to demonstrate programming with some of the commands found in the Structures sub-palette.

EXAMPLE 6.5

Let's write a program that controls two tasks. The first task is to turn on a motor located at port A on the NXT when a switch is open. The second task is to turn on another motor located at port C when the same switch is closed. Motor A is to turn through and angle of 270°, stop and wait for 6 seconds. Motor C is to turn through an angle of 180°, stop and wait for 4 seconds. The program is to run continuously.

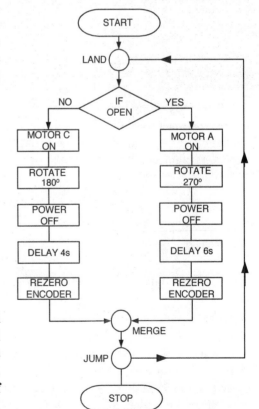

Fig. 6.46 Flow diagram with a conditional command.

Step 1: To begin, prepare a flow diagram with a command that causes it to operate continuously. The flow diagram will be centered about a conditional command that determines whether the switch is open or closed as shown in Fig. 6.46. A line of commands follows on either side of the **If** condition, which controls the behavior of the two motors. These two lines are merged before the program moves on to the **Go To** (Jump) command.

Step 2: The program written in Inventor Level 4 is presented in Fig. 6.47. The program is divided so that it can accomplish two tasks by using the **Touch Sensor Fork** icon. The choice of the task executed will depend on whether the **Touch Sensor** switch is open or closed. If the touch sensor switch is closed (on), the program takes the lower path and operates Motor C at a power level of 2 until the motor rotates through 180° , stops for 4 seconds , rezeros the encoder in the servomotor and then the process repeats until the touch sensor switch is opened. When the touch sensor switch is open (off), the program takes the upper path and operates motor A at a power level of 2 until the motor rotates through an angle of 270°, stops for 6 seconds, rezeros the encoder and then the process repeats. For both branches of the program, it was necessary to reset the encoder output from the servomotors to zero after each rotation. Note that a **Fork Merge** icon is used to merge the two program paths. **Red Jump** command is placed before the **End** icon and **Red Land** icon is placed after the **Begin** icon causing the program to loop continuously. The program will begin to execute as soon as the Orange button on the NXT is push when the stored program is positioned in the center of the LCD. To stop the program, press dark gray button until the NXT shuts down.

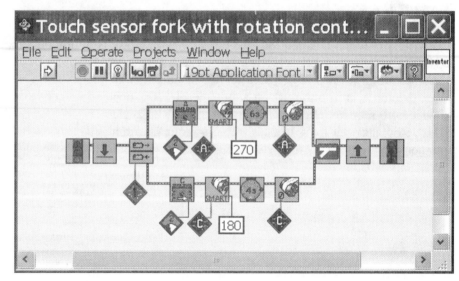

Fig. 6.47 A program that uses a **Touch Sensor Fork** to control motors connected to Ports A and C.

EXAMPLE 6.6

Modify the program shown in Fig. 6.47 so that it cycles 5 times and then ends.

Solution: The solution is relatively simple. Replace the **Jump** and **Land** icons, which were causing the program to cycle endlessly, with the **Start Loop** and **End Loop** icons as shown in Fig. 6.48. The number of cycles is specified by using a numeric modifier wired to the **Start Loop** icon. Note that the **Text** tool was used to identify the location of the **Loop** icons and to place the number 5 in the numeric modifier of the **Start Loop** command and 180 and 270 in the numeric modifiers for the

NXT Wait for Location icon .

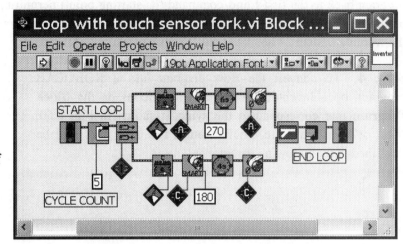

Fig. 6.48 Controlling the number of cycles using the **Loop** commands.

EXAMPLE 6.7

Prepare a program to operate a traffic safety system. The program is to perform two tasks, under certain environmental conditions, which are described below:

1. Operate a motor that actuates a valve that controls the flow of a deicing solution onto an ice covered bridge. The motor turns clockwise for two complete rotations to open the valve then stops with the valve open for 30 seconds. The motor then turns counterclockwise for two complete rotations to close the valve. The entire operation of the motor is to occur if and only if the temperature is less than 24 °F and if a control switch is turned on.

2. When the temperature is less that 24 °F, a flashing warning light is activated. The warning light flashes on for 3 seconds then off for 2 seconds. The warning light flashes on and off continuously when the temperature is less than 24 °F; otherwise, it ceases to operate at temperatures higher than this limit.

Step 1: The system described above is a warning system. It involves a flashing light and an active system that operates a valve that controls the flow of some substance, which is to alleviate the potential danger. From the system description, it is apparent that two conditional statements will be required. The flashing light, which is to operate whenever the temperature is less than 24 °F, is programmed with a sequence of four commands following the **Yes** branch of the **If/Then** statement shown in Fig. 6.49. A **Land** icon that receives a (**Go To**) command is placed prior to the flashing light commands to enable the warning signal to operate even when the switch is open.

Step 2: The operation of the motor actuated valve is dependent on two conditions—temperatures lower that 24 °F, and the closed position of a sensor switch. These two conditions are programmed by placing a second **If/Then** command on the right hand branch of the first **If/Then** statement, as shown in Fig. 6.49. A sequence of seven commands is programmed into the **Yes** branch of this conditional statement. These commands control the opening and closing of the valve. A **Jump** (**Go To**) is programmed into the **No** branch of the Switch Closed conditional command. Its purpose is to cycle that portion of the program which operates the flashing light when the temperature is less than 24 °F.

Step 3: **Merge** commands are placed to combine the two branches of the program from conditional statements. After the second **Merge** command another **Jump** command is inserted. This takes the program back to the first **Land** command (its starting point) permitting it to recycle continuously. The program will never reach the **Stop** or **End** command and will therefore operate continuously to maintain the warning system 24 hours per day 7 days a week.

Step: 4   To convert the flow diagram into a ROBOLAB program, open Inventor Level 4 and drag icons on the **Block Diagram** that correspond to the symbols shown in Fig. 6.50. Most of these icons are familiar, but a few new ones will be described in the following paragraphs.

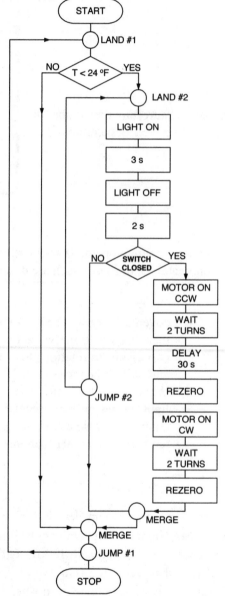

Fig. 6.49 Flow diagram for the traffic warning system.

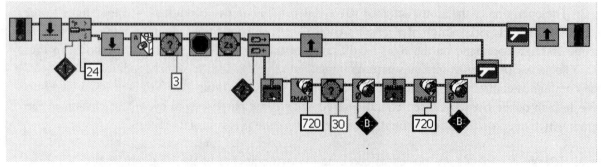

Fig. 6.50 A program in Inventor 4 to operate the traffic warning system.

The first conditional command the **Fahrenheit Fork** , found on the **Forks** sub-palette, depends on the temperature that is measured with a LEGO temperature sensor. Note the two **IF** options > and ≤ that are found at its two output terminals. If the temperature is greater than 24 °F, the program follows the **No** (upper) branch of the program. On the other hand, if the temperature is equal to or less than 24 °F the program follows the **Yes** (lower) path and executes the command causing the light to flash.

The second conditional command the **Touch Sensor Fork** depends on whether or not the switch is closed. The lower terminal corresponds to a closed switch and provides the entrance to the **Yes** (lower) branch of the program. The upper terminal corresponds to an open switch and provides the entrance to the **No** (upper) branch of the program.

The two branches from each of the conditional commands must be merged for continued progress of the program. The merge is achieved with the **Fork Merge** icon that is found on the **Forks** sub-palette. The two different **Go To** commands used in the program are inserted with the **Red Jump** and **Land** and the **Blue Jump** and **Land** icons. Proper execution of the program is dependent upon the correct placement of these Jump and Land icons[2].

Control of Motor B that opens and shuts the valve is achieved with the **NXT Wait for Location** icon . It is necessary to rezeros the encoders that provide the feedback signal for this command by using the **Rezero** icon .

6.5.5 Containers

There are two types of containers available for programming. The first type is for storing numbers and/or performing arithmetic operations on integer numbers. The second type combines the **Container** with **Structure** programming functions such as **Container Wait For**, **Container Loop** and **Container Fork**. The container icons are accessed on the by clicking on the **Container** icon found near the bottom of the **Functions** palette. The **Container** sub-palette, shown in Fig. 6.39, provides many useful commands to enhance programming capabilities.

The **Fill Container** icon is used to store a specified integer value. To store a numeric integer in a program, it is necessary to wire the **Fill Container** icon into the command sequence. Then the container is identified with a modifier such as Red, Blue or Yellow Container. Finally an integer number to be stored in the container is specified with a **Numeric Constant** modifier.

A group of eight container icons presented in Fig. 6.51 are used for arithmetic operations such as addition, subtraction, multiplication, etc. A second group of eight icons shown in Fig 6.52 are used in conjunction with sensors to record their values. For instance, the **Light Container** icon stores the value of the light intensity recorded by the light sensor. A final group of seven icons, shown in Fig. 6.53, are used for counting sensor response, elapsed time, event states, to enter formulas, etc.

---

[2] The Red Jump and Land icons cause the program to recycle continuously, which is a requirement. However, in some application the motor actuated valve may be required to operate only one time. If this is the case, it will be necessary to either design a switch for the motor so that it opens before the next cycle of the program or to develop a different program that limits the motor actuated valve to a single open-shut cycle.

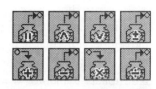

Fig. 6.51 Container icons used for arithmetic operations.

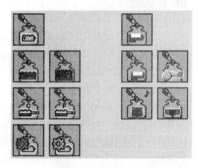

Fig. 6.52 Container icons used to record sensor output.

Fig. 6.53 Container icons used for counting sensor response, elapsed time, event states, to enter formulas, etc.

The container commands are frequently used because they provide rich programming options. We will demonstrate some of these options in the following two examples.

EXAMPLE 6.8

Your team is to measure the speed of a vehicle on a straight track as it travels between two fixed stations 500 meters apart. You have mounted a trip wire at each station that senses the passage of the vehicle by closing a switch as the vehicle moves over the trip wire. Prepare a program that performs the following tasks:

1. Compute the time required for the vehicle to travel the 500 meters between the two stations and display this result for 4 seconds on the NXT.
2. Compute the vehicle's velocity, and display this value for 6 seconds.
3. Comment on the accuracy of the measurement.

Step 1: Begin by preparing a flow diagram that outlines the commands necessary to calculate the velocity of the vehicle as it transverse the 500 meter course as shown in Fig. 6.54.

Step 2: We convert the flow diagram into a program in Inventor Level 4 by dragging icons on the **Block Diagram**. In this example, the program is relatively easy because it is sequential. Forks, Loops, Jumps and Lands are not required. However, a timer is necessary to measure the time for the vehicle to travel 500 meters. Also this example provides an opportunity to become familiar with the shortcomings of integer numbers when computing velocity. Begin the program shown in Fig. 6.55 by zeroing

(resetting) the timer. The Zero **Timer** icon ![icon] is found on a sub-palette accessed with the **Reset** icon

, which in turn is found on the **Functions** palette. The timer provides a count of one for every 0.1 second measured from the time the program begins to run. The timer count is measured at both stations

by the **Wait For Push** icon  , and stored by the **Touch Container** icon  . At this stage, we have measured the time (in counts) that the switches at stations #1 and #2 were closed by the passage of a vehicle.

Fig. 6.54 Flow diagram showing the commands needed to determine velocity from time measurements.

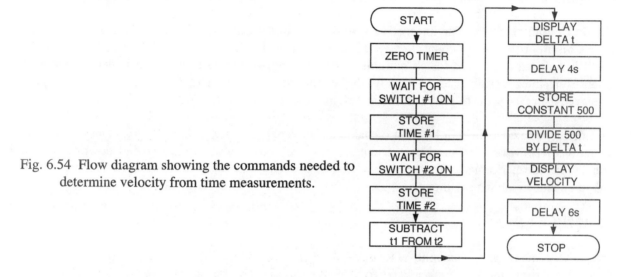

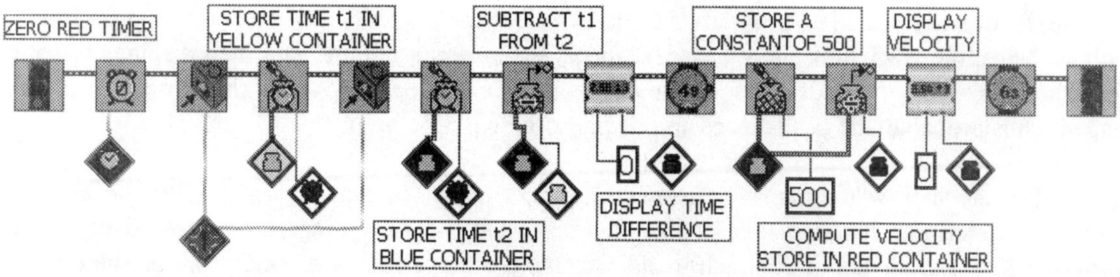

Fig. 6.55 The program to measure and display vehicle velocity in Inventor 4.

Step 3: The time for the vehicle to travel from station #1 to station #2 is determined by subtracting time t_1 from time t_2. The subtraction is accomplished with the **Remove From Container** icon  . This container is wired with the **Blue Container** modifier, which contains t_2 and the **Value of the Yellow Container**, which is t_1. The subtraction is performed and the count difference is stored in the **Blue Container**. The **Value of the Blue Container** is then displayed on the NXT for 4 seconds by using the

Set Display icon  with the appropriate multipliers. Note, that a zero is specified for the number of decimals used in the display. Because the containers use only integers, we are displaying the number of timer counts (one count for each 0.1 second). The Set Display icon is accessed on the sub-palette that appears when the **Communications** icon is activated. To maintain the reading on the display of the NXT for a sufficient time to read, the **Wait For 4s** icon is placed next in line.

Step 4: To determine the velocity over the distance of 500 meters between the two stations, the **Fill Container** icon is placed in the program sequence. A constant integer of 500 is input as a numeric modifier and this integer is stored in the **Red Container**. The velocity is computed using the

Divide Container . The integer 500 stored in the **Red Container** is divided by the **Value** of the **Blue Container** (the delta t count) to give the vehicle's velocity. In the division of the two integers, the result is rounded downward for all non-integer results. For example, 11/4 = 2.75 is determined as 2 when using integer mathematics. Finally, the number stored in the **Red Container** (the velocity) is displayed using the same procedure as indicated previously in Step 3.

Comment: The containers in Inventor 4 enable the storage of integers and some elementary arithmetic operations. However, the numbers are integers and they are truncated and rounded downward after calculations are performed. The result is a loss of accuracy that can be significant for experimental measurements involving small counts.

EXAMPLE 6.9

Program a game of chance using Inventor 4 and the RCX unit. In this game, the player places a bet, and then selects some integer number between 0 and 9. The house generates a second integer number between 0 and 9 using a random number generator. The player wins when the two numbers coincide; otherwise, he or she loses. The program is to determine if the player is a winner (one chance in ten). If the player wins, the NXT unit plays a tune, displays the winning number and then displays the amount of the winnings. Note the odds given by the house are one in nine. (How can they lose?) If the player loses, which is probable, the NXT beeps and displays the winning number.

Step 1: Prepare a flow diagram that shows the sequence of commands as well as the conditional statement to determine if the player or the house wins. The flow diagram, presented in Fig. 6.56, employs an initial sequence of commands that accepts the player bet, the number he or she chooses to play, and a number for the house obtained with a random number generator. The program also contains a conditional command that compares the player's number with the randomly generated house number. When the two numbers are equal, the player wins and a sequence of commands plays a tune, computes the winnings and displays the winning number as well as the amount of the winnings. If the player loses, the RCX beeps and the winning number is displayed.

Step 2: Prepare a program in Inventor 4 that will use the NXT to serve as a simple gambling machine. **Containers** are the key functions employed in assembling the icons necessary to program the NXT gambling house. Three different containers are used in the programming to store numbers as illustrated in Fig. 6.57.

The first icon in the sequence is the **Fill Container**. A **Numeric** modifier is added, which accepts the player's bet of 2 units. The amount of the bet is stored in the **Yellow Container**. The house

bet is established with the **Random Fill Container** , which is modified with a **Numeric** modifier to generate a number from 0 to 9. The winning number generated is stored in the **Red Container** modifier. Note the player has one chance in ten of winning. The **Container Fork** icon 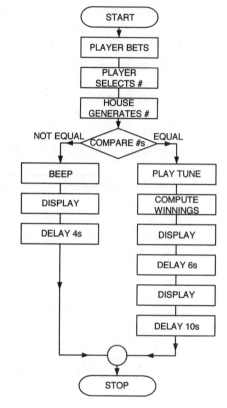 is employed to ascertain the winner—the house or the player. A **Numeric** modifier is wired to the **Container Fork** icon to enter the player's number. If the house number and the player's number are equal, the upper path of the program is executed; otherwise, the **Container Fork** command directs the program along the lower path.

Fig. 6.56 A flow diagram for the NXT gambling house.

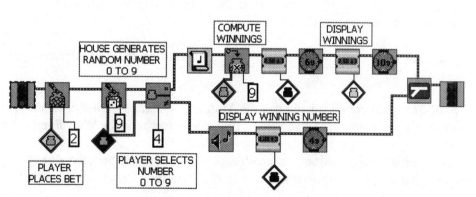

Figure 6.57 The program for gambling with numbers ranging from 0 to 9.

When the player loses, (the house number and the player's number are not equal), the **Play Sound** icon causes the NXT unit to beep. Next, the **Set Display** icon is used to display the winning number for 4 seconds. The **Value Of The Red Container** modifier is wired to the **Set Display** icon to provide the winning number.

When the player wins (equal numbers), A song is played by using the **Yellow Scroll** which plays "Twinkle Twinkle Little Stars" to announce the glad tidings. The **Multiply Container**

 is used to calculate the winnings. The house odds[3] (9 to 1) are provided with a **Numeric** modifier. The player's bet is entered into the calculation by wiring the **Yellow Container** to the **Multiply Container**. After performing the calculation, the product $9 \times 2 = 18$ replaces the number formerly stored in the **Yellow Container**. Next, the winning number (4) is displayed on the NXT by wiring the **Value Of The Red Container** to the **Set Display** icon. The **Wait For 6** seconds command enables the operator to read the winning number. The amount of the winning is displayed with the **Set Display** icon, which is wired with the **Value of the Yellow Container** as a modifier.

Finally the **Fork Merge** icon is employed to bring the two programming paths together and take the program to the **End** command. The program runs once for each time the Orange button on the NXT unit is activated. The time to execute the program depends on the path taken and the delay times used to monitor the display on the NXT unit.

6.5.6 Communication

The NXT is able to communicate in two different ways—by sounds and by Bluetooth (or USB) transmissions. We have already demonstrated both methods. The NXT beeped and played a tune in the execution of the program written for Example 6.9. We have used the USB cable repeatedly when downloading programs from our computer. However, the NXT is capable of sending or receiving

simple messages to or from other NXT units by using the **Communication** function ![icon]. Click on this icon and the **Communication** sub-palette appears. There are several **Communication** icons found on this and other sub-palettes that are useful in writing programs involving two or more NXT units performing tasks that are related and often dependent upon one another. Several of these icons are described below:

 The **Send Mail** command found on the **RCX Communication** sub-palette is used to send a message from one RCX to another.

 The **Fill Mailbox** command is used to reset the **Mailbox** to another value specified with a **Numeric** modifier.

 The **Wait For Mail** command empties the **Mailbox** and waits until the number received from another RCX is equal to a specified value. This icon is found on the **Wait For** sub-palette.

 The **Mailbox Fork** command selects a path depending upon whether the value of the mail is > or ≤ to a specified number. This icon is found on the **Structure/Forks** sub-sub-palette.

 The **Loop While The Value Of Mail Is Less Than** command starts a loop that repeats while the value of mail in the **Mailbox** is less than a specified value. A similar command **Loop While The Value Of Mail Is Greater Than** is also available. These icons are located on the **Structure/Loops** sub-sub-palette.

---

[3] The player's odds are 1 in 10 while the house odds are 1 in 9. Who will win in the long run?

The **Mail Container** command sets the **Container** to the value of the mail in the mailbox. This icon is found on the **Container** sub-palette.

 The **Empty Mailbox** command resets the value of the mail in the **Mailbox** to zero. This icon is found on the **Reset** sub-palette.

 The **Value Of Mail** modifier provides the numeric value of the mail in the **Mailbox**. This icon is found on the **Modifiers** sub-palette.

6.5.7 The NXT Detective

It is possible to interrogate the data being received and stored in the NXT by using a feature of ROBOLAB called **Detective**. To open **Detective**, select it from the **Projects** menu on the Block Diagram as shown in Fig. 6.58. In addition to being enabling you to select the communication port, test communication and download firmware, **Detective** allows you to view and change hardware settings, view sensor readings, view container values and logged data.

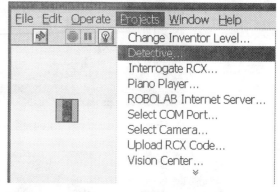

Fig. 6.58 Open NXT Detective feature from the Projects menu on the Block diagram.

Clicking on Detective opens the dialog box shown in Fig. 6.59, which has buttons for Hardware Settings, Sensor Scope, Container Scope and Data Logged.

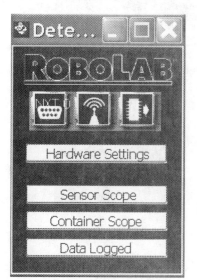

Fig. 6.59 Options available for interrogating the NXT using Detective.

Click on Hardware Settings and a listing of the current settings for the NXT appears as illustrated in Fig. 6.60. After selecting an item from the list, a short description of it appears in the window to the right. It is possible to change the setting of many of the items on this list. .

Fig. 6.60 Hardware settings accessed through the Detective feature in ROBOLAB 2.9.3.

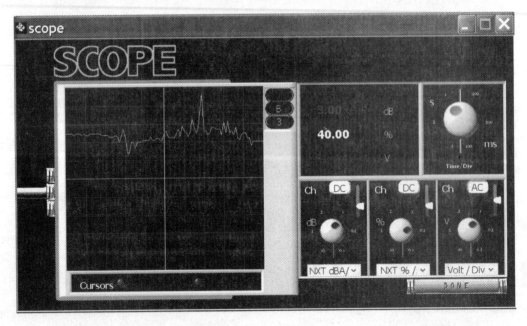

Fig. 6.61 The virtual oscilloscope available with the NXT and ROBOLAB.

Click on the Sensor Scope button in the Detective dialog box and a virtual oscilloscope appears, as shown in Fig. 6.61. The color coded traces on the sensor scope display input sensor information from ports 1, 2, or 3 or rotation sensor data from ports A, B and C. To switch between the rotation sensors and input sensors, left click on the port number. The default setting for the readout gives the raw sensor voltage. To readout the calibrated values such as centimeters for the distance measured by the ultrasonic sensor or degrees for the rotation sensor, select the appropriate sensor type from the drop-down menu for the channel associated with a specific sensor. The DC/AC button toggles the trace between DC and AC mode. In the AC mode the average will automatically be subtracted from the signal. The slider allows you to shift the entire trace up or down on the face of the virtual oscilloscope.

Click on Containers Scope button on the Detective dialog box and a listing of the 24 containers appears as shown in Fig. 6.62. The number stored in the container is listed in the table to the left. Click on a specific container that is being filled by a running program and its value is presented as a function of time in the chart to the right.

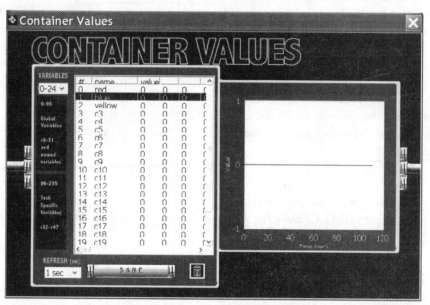

Fig. 6.62 Detective Container Scope enables you to check a program as it runs.

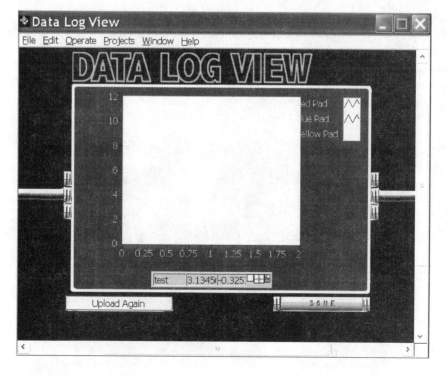

Fig. 6.63 Detective Data Logging Scope enables you to graph data uploaded from a running program.

The Data Logging Scope, accessed through the Detective dialog box, is presented in Fig. 6.63. This feature of ROBOLAB and the NXT enables you to view the data collected and stored in the NXT. With the USB connection between the NXT and your computer the uploading of the data is nearly instantaneous.

6.5.8   Inventor Summary

Section 6.5 provides a brief introduction to programming using **Inventor Level 4** in ROBOLAB. The **Function** and **Tool** palettes were described and most of the icons found on these palettes were identified. The procedure for accessing the very useful **Context Help Window** was discussed.

The **Inventor** mode of programming differs in many ways from the **Pilot** mode. The requirement to wire the command icons and their modifiers in the Inventor mode adds complexity to the programming task. To assist in the wiring task, a description of wire sizes and colors was provided. Several examples were also introduced to demonstrate programming with **Structures** and **Containers**. These examples provide you with opportunities to study control loops, conditional commands and data storage and manipulation.

The features of NXT Detective were described by introducing the Hardware settings, Sensor Scope, Container Scope and the Data Logging View. It is recognized that this Section provides only an introduction to programming in the Inventor mode; however, a much more detail treatment may be found in Eric Wang's book titled **Engineering with LEGO Bricks and ROBOLAB**, 3rd Edition.

6.6   PROGRAMMING IN INVESTIGATOR

6.6.1 Introduction

Investigator follows **Pilot** and **Inventor** as the third mode of programming within ROBOLAB. It is used to enable the NXT to acquire data taken with sensors, store this data and then to transmit it to your PC. Let's explore the screens encountered when opening ROBOLAB in the **Investigator** mode. The first screen, shown in Fig. 6.18, provides three buttons, which enable you to open ROBOLAB in the **Pilot**, **Inventor** or **Investigator** modes. Select the lower button, open **Investigator** and the screen presented in Fig. 6.64 appears.

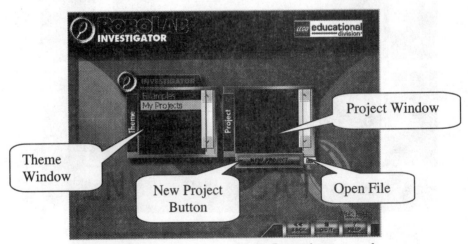

Fig. 6.64 The opening screen in the **Investigator** mode.

Let's click on **My Projects** in the **Theme Window** and then double click on the **Project** titled **Investigator Example 1** and note the **Navigator** and **Programming Windows** that appear as indicated in Fig. 6.65.

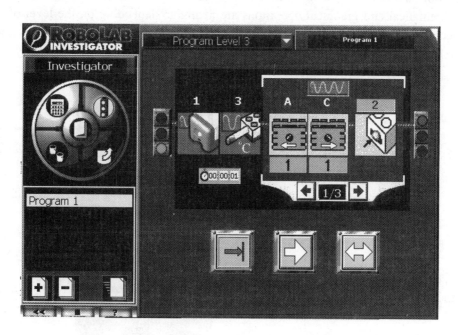

Fig. 6.65 The **Navigator Window** (left) and **Programming Window** (right) in Investigator Level 3.

The Navigator Window

For the moment, let's focus on the **Navigator Window** that controls the programming tasks. Investigator is more complex than either Pilot or Inventor because Investigator divides the programming task into five different areas each of which is accessed by buttons in the **Circle** of the **Navigator Window**. These buttons, as well as others, are defined in Fig. 6.66. Because of the need to switch from one task to another while programming, comparing data, analyzing results or documenting a project, the Navigator Window remains open as you switch from task to another.

Each of the five areas that you will use as you program the NXT to acquire, store and manipulate data is defined in the paragraphs below:

 The **Program Area** is similar to that found with the Inventor mode. You write your programs by dragging icons into this area. These icons (commands) permit the acquisition of data for a predefined period. You are basically providing instructions to the NXT to acquire measurements using its data acquisition system. You may chose from five different levels of complexity for programming. This area also contains a large white arrow for downloading the program from your PC to a RCX unit.

 The **Upload Area** enables you to transfer data stored on a page (file) in the NXT to your PC or (Mac). Remember that each **Data Set** is stored on an individual page and is transferred one page at a time from the NXT to a new page on the Upload Area on your PC.

The **Compute Area** permits you to perform **mathematical operations** on your data after it has been transferred to your PC. The operations can be simple such as addition, subtraction, multiplication and division, or more involved operations such as differentiation or integration. Many mathematical and graphical operations are available depending on the level of computing tools selected. Level 5 provides the complete range of mathematical operations available in ROBOLAB. You may also write ROBOLAB code to manipulate data in almost any manner you wish.

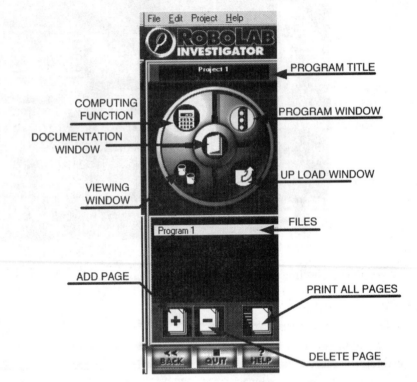
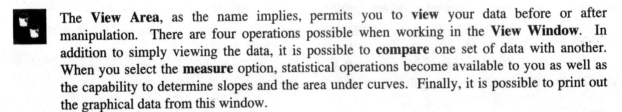
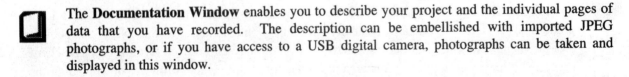

Fig. 6.66 The **Navigator Window** in **Investigator**.

The **View Area**, as the name implies, permits you to **view** your data before or after manipulation. There are four operations possible when working in the **View Window**. In addition to simply viewing the data, it is possible to **compare** one set of data with another. When you select the **measure** option, statistical operations become available to you as well as the capability to determine slopes and the area under curves. Finally, it is possible to print out the graphical data from this window.

The **Documentation Window** enables you to describe your project and the individual pages of data that you have recorded. The description can be embellished with imported JPEG photographs, or if you have access to a USB digital camera, photographs can be taken and displayed in this window.

6.6.2 The Programming Area

Click on the **Begin** ion on the **Navigator Wheel** and the screen associated with Project 1 appears, as shown in Fig. 6.67. The **Program Title** given here as "Light and Angle Sensor" can be edited to provide a description of your project. Also **Program Level 3** is shown in the upper left window. Five different programming levels are embedded in the ROBOLAB code when working in the Investigator mode. The programming window shown in the center of Fig. 6.67 contains the icons that form your program. Many of these icons have already been introduced, but several of the new ones will be described in the next subsection. Finally, the download button used to transfer the program from your PC to the RCX by the IR transmission tower is identified.

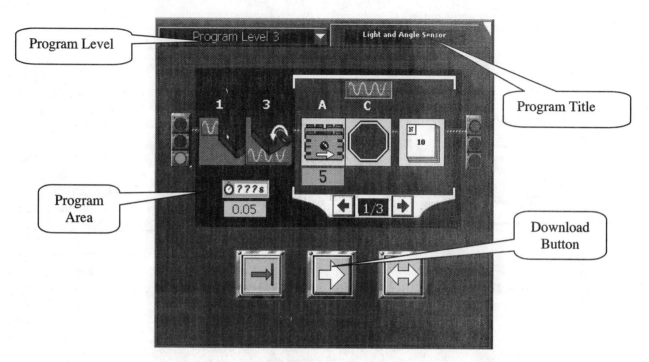

Fig. 6.67 The Programming Window associated with Level 3 programming.

6.6.3 Programming in Investigator Level 3

Programming in **Investigator Level 3** is at an intermediate level. It affords the ability to record data from two different sensors at different sampling rates. You can also power devices such as motors that are attached to the output Ports of the NXT. Programming in **Investigator Level 3** is like programming in the Pilot Mode because wiring is not necessary. In Level 3, the icons are arranged in a pre-wired, linear sequence. Loops, forks and jumps in the program are not viable options in Level 3.

Let's open the **Programming Window** associated with **Program Level 3** by clicking on the ∇ symbol to the right side of the Program Level display to obtain the screen shown in Fig. 6.68.

Specifying the Sensors

First consider sensor selection. It is clear from Fig. 6.68 that the sensors are connected to input Ports 1 and 3. The choice of sensors for the two ports differ slightly as indicated by the icons shown on the pull down menus presented in Fig. 6.69. The sensors in the top row are identical for both Ports 1 and 3. The icons identify (in order) the following sensors: RCX touch or contact, RCX light intensity, temperature °C, and temperature °F. The first three sensors in the second row, which are identical for Ports 1 and 3 include the RCX angle, RCX and rotation and the generic sensor. The last icon in the second row for Port 1 is a clock and for Port 3 it is the NXT touch sensor. The third row for Port 1 includes a digital (USB) camera and the NXT touch, light and distance sensors. The third row for Port 3 identifies NXT light, distance, sound and rotation sensors. If you select the NXT rotation sensor, it will collect data for the encoder in either Motor A or Motor C. The two choices in the bottom row for Port 1 include the

NXT sound and rotation sensors. **The motors should not be connected to either Port 1 or Port 3.** An icon in the bottom row for Port 3 indicates no sensor is attached to it.

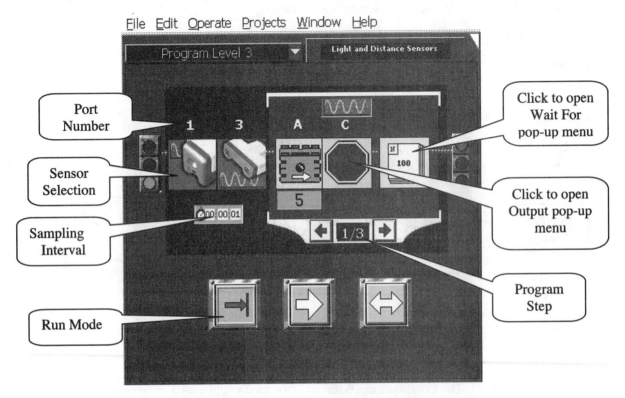

Fig. 6.68 The **Programming Window** associated with **Investigator Level 3** programming.

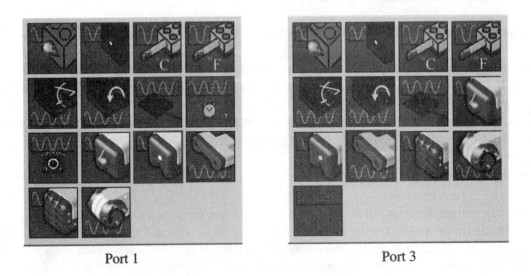

Fig. 6.69 Admissible sensor selections at Ports 1 and 3.

Select the **Generic Sensor** icon for Port 1. A button labeled "Gene ∇" appears under this icon. Click on this button and the pull down menu shown in Fig. 6.70 appears. This listing enables you to adapt Port 1 to acquire data from a wide variety of sensors. We have selected the voltage option from this listing and the **Generic Sensor** icon is labeled as shown in Fig. 6.71. With this setting, the output voltage from the sensor will be measured and stored in the memory (RAM) of the NXT.

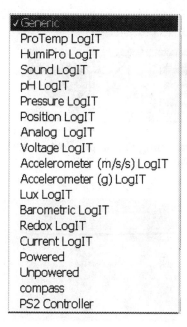

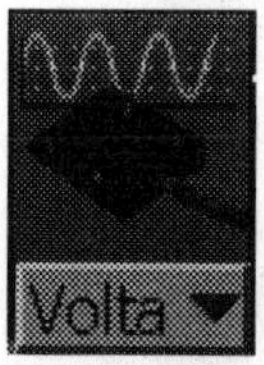

Fig. 6.71 Specifying the signal input
for Port 3 as a voltage.

Fig. 6.70 Pull down menu listing types of sensors
that can be accommodated by the RCX.

Specifying the Data Logging Interval

The sampling rate is specified with the Data Logging Interval icon ![icon] located below the sensor icons (See Fig. 6.67). Click on it and a row of five icons that provide a wide choice of data logging rates appears as shown in Fig. 6.72. The first three of these icons (left to right) permit you to sample the output from the sensor every second, minute or hour. The fourth icon enables you to specify the time between readings in seconds. The last icon in this row represents the touch sensor. When this selection is made, the output from the sensor will be recorded each time the contact on the touch sensor is depressed. Note that the touch sensor is connected to Port 2 when operating in this mode for data sampling.

Fig. 6.72 Icons employed to specify data logging rate.

Programming the Output Ports

A three step program is mandatory when working in **Investigator Level 3** to control the output Ports A and C and to specify a **Wait For** command. The output Ports and the **Wait For** command are programmed in the Window shown in Fig. 6.73. Click on the icon under either the designations for Port A or C and a pop up menu showing motors operating in the forward and reverse directions, a lamp and a stop sign.

　　　The number below the **Motor** icon shown in Fig. 6.73 adjusts the power level. Click on this number and a popup menu appears as shown in Fig. 6.74. You have a choice of five power levels—1 to

5. The voltage on the motors is always constant. The power level is adjusted by using pulse width modulation where the voltage to the motors is turned on and off at a high frequency. At a power setting of 5, the voltage is applied continuously. At a power setting of 1 the voltage is applied for only 1/5 th of a cycle.

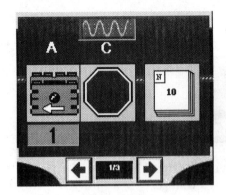

Fig. 6.73 Icons available for programming the output Ports A and C and the **Wait For** command.

Fig. 6.74 Icons used to control the power level applied to the motors.

In the lower right hand corner of Fig. 6.74, there is an icon that identifies Port 1. When selecting this icon, you are specifying that the motor operate at a power level controlled by the output of the sensor connected to Port 1. The Lamp, shown in Fig. 6.73, can be switched on and off if it is connected to either Port A or C. On the other hand if the stop sign is selected, the output voltage at that Port will be turned off.

Programming the Wait For Command

The third and last icon in Fig. 6.73 is the **Wait For** command. Click on this icon and a pop up box appears with 16 different icons as illustrated in Fig. 6.75. None of the NXT sensors are included in this box, although the NXT sensors may be added in future releases of ROBOLAB. Thus, we are limited to either using the RCX sensors with an adapter cable (Table 1.1) or restricting the programming to the first two rows of **Wait For** commands. There is one exception: you can use the NXT touch and light sensors on Port 2 with the RCX **Wait For** command. The only difficulty is that the red LED on the light sensor will not be turned on to illuminate the target.

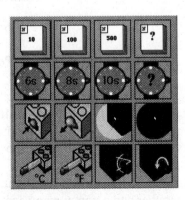

Fig. 6.75 Icons used to specify the **Wait For** command.

These icons permit you to **Wait For** the following events:

- The number of measurements recorded—10, 100, 500 or a specified number.
- The number of seconds between readings—6, 8, 10 or a specified number of seconds.
- Until either the NXT or RCX touch sensor is activated—contact released or depressed.
- Until the NXT or RCX light sensor detects a higher or lower light intensity.
- Until the RCX temperature sensor measures a temperature > or < than a specified value.
- Until the RCX rotation sensors measures an angular displacement > or < than a specified value.

Data Logging

In some situations, logging data may not be required. In these situations, it is possible to avoid recording the sensor output by clicking on the icon 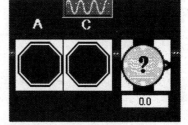 located above Port C in Fig. 6.73. Clicking on this icon provides you with the two options shown in Fig. 6.76. The display with a sine wave indicates that you will record data during this step in the program. The display without the sine wave is selected if data logging is not required.

Fig. 6.76 Data logging icons for turning the logging process on or off.

Three Step Programming

The **Level 3** programming in **Investigator** requires three steps—whether you need them or not. The program step is identified below the program icons for the output Ports as indicated in Fig. 6.77. The 1/3 shown in this illustration indicates that the program in this Window represents the first step in a sequence of three. We advance from one step to another by clicking the right arrow and decrease the program step by clicking the left arrow. You can effectively eliminate a step by programming the output ports and the time delay by using the sequence of icons shown on the right side of Fig. 6.77.

Fig. 6.77 (Left) Program step and control buttons. (Right) A program to eliminate a step.

Downloading and Running the Program

Buttons that control downloading and running the program, located below the Programming Window, are shown in Fig. 6.78. The white single ended **Run** arrow is used to download the program from a PC to the NXT with the USB cable. To run the program in the NXT, it is necessary to depress its orange Run button. The white double-ended **Run** arrow downloads your program in the direct mode. When operating in the **Direct mode**, the program is transferred from the PC to the NXT and the program begins to run immediately. It is not necessary to press the orange button on the NXT to initiate the program. The NXT will immediately begin to send the sensor measurements to the computer via the USB cable where it is displayed as it is received. Of course for the data to be transmitted while the program is running, the NXT must remain connected through the USB cable at all times. The **Direct**

mode is useful in debugging a program because you can observe the data as it is collected and determine if it is correct.

Fig. 6.78 Control buttons for downloading and running your program.

When you click on the **Red** arrow button, a pop up menu appears, which is shown on the right side of Fig. 6.78. These icons provide you with two possibilities for running your program. The **Red** single ended arrow button on the left side is for running the program once and stopping. The **Red** double looping arrows on the right side enable you to run the program continuously.

Handling Several Programs

In some applications, it may be necessary to write several programs in Investigator. It is possible to create a new program by clicking on the page icon with the **+** sign that is located at the lower left hand corner of **Navigator Window**. When you create a new page for a program it is numbered and displayed in the **File Listing Area** as indicated in Fig. 6.79. More descriptive titles can be added by typing new program names in the title block of the **Programming Window**. It is possible to move from one program to another simply by clicking on the program title in the **File Listing Area** shown in Fig. 6.79. Similarly, you can delete a page (or program) by marking it and clicking on the page icon with the **–** sign.

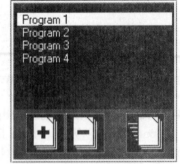

Fig. 6.79 A listing of program files on the area below the Navigator wheel.

6.6.4 Uploading Area

Suppose that you have programmed your NXT to acquire data from one or more NXT sensors. To transfer this data set to a PC, click on the **Upload** icon on the **Navigator Circle** and the Window shown in Fig. 6.80 appears. Six features associated with this window are identified in Fig. 6.80 including: (1) The title bar to identify the data set, (2) Upload tools, (3) Upload button, (4) Graphing tools, (5) Graph area and (6) Page (data set) listing area.

Click on the **Upload** button (the white arrow in Fig. 6.80) and the data set is uploaded from the NXT to your PC via the UBS cable. The uploading process is almost instantaneous and the NXT chirps when the process is complete. If you are transferring one set of data, it will occupy one page and be stored in one of the colored buckets used to designate memory locations. If you have data from several sensors, the data from each sensor is uploaded onto a separate page and stored in a different colored bucket.

To generate the illustrations in Fig. 6.81, we uploaded two sets of data—one from a NXT light sensor and another from a NXT sound sensor. In the uploading process, ROBOLAB stored the data in two different files and labeled them as Data Set 1 (light intensity) and NXT Sound Sensor. Data Set 1 was identified with the **Red Bucket** and the sound sensor data was placed in a **Blue Bucket** as shown in Fig. 6.81.

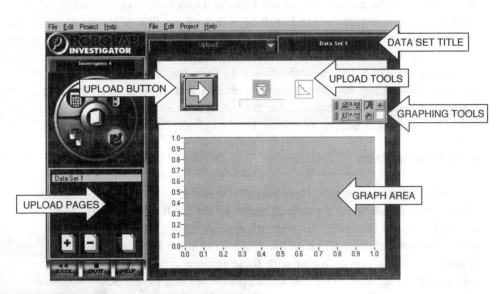

Fig. 6.80 The
Upload Window.

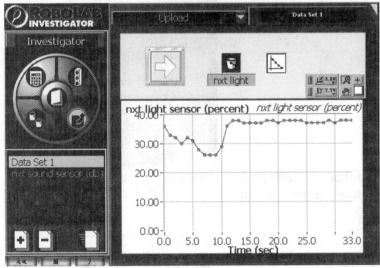

Fig. 6.81 Data presentation from
the NXT light sensor after
uploading.

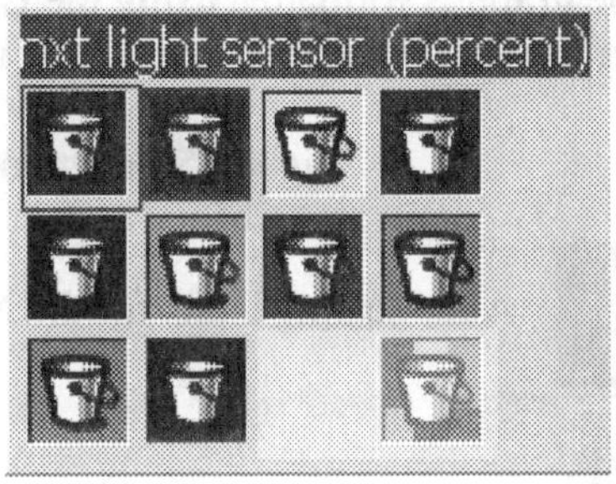

Fig. 6.82 Pop-up menu identifies the
buckets for storing data.

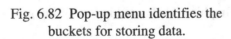

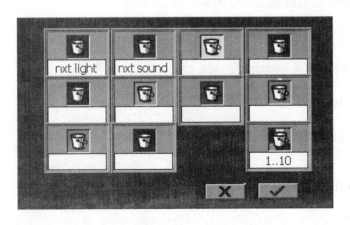

Fig. 6.83 Pop-up box showing the colored
buckets with titles.

By clicking on the **Bucket** icon in the **Upload Tools** area (See Fig. 6.81), a pop up menu with all of the buckets appears as indicated in Fig. 6.82 Click on the title of the bucket and the entire bucket set with titles appears, as shown in Fig 6.83. You can edit the title by marking it and typing a new title.

If you plan to collect several sets of data, it is necessary to add a page (open a new file for each uploading event). If you are uploading data from two different sensors, ROBOLAB will automatically open a second page for the second set of data and store it in a different colored bucket. After the uploading process is complete, the bucket color can be changed clicking on the bucket icon and selecting a different bucket color from the ten colors presented in Fig. 6.82.

Graphing the Data

A graph of the data set (e.g. light intensity versus time) is automatically displayed in the graph area of the **Upload Window**. In the default mode, ROBOLAB automatically plots the points and draws a curve through them. However, several options are available for graphing each data set. Click on the graph icon in the **Upload Tools** area and a pop up menu with these options appears as illustrated in Fig. 6.84. The first three options are self evident from the icons. The fourth option produces a bar chart instead of a curve and the final option is for a tabular display of the data that is also presented in Fig. 6.84.

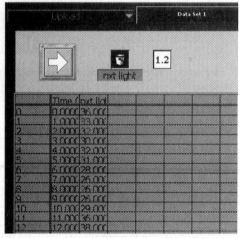

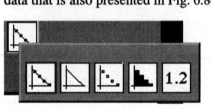

Fig. 6.84 Icons enabling different methods for representing data sets and a tabular display of the measurements recorded.

Formatting the Graph

The graph is formatted using the **Graphing Tools** located on the right side of the **Upload Window** (See Fig. 6.80). These tools, presented in Fig. 6.85, are described below:

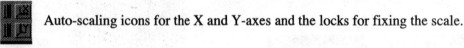

Fig. 6.85 Palette of Graphing Tools used in formatting graphs.

Auto-scaling icons for the X and Y-axes and the locks for fixing the scale.

Icons for formatting axes by adjusting precision, mapping mode, grid color, etc.

Icon for zoom control enables the adjustment of the size of the graph and to crop (trim).

Icon for pan control permits movement of the graph in any direction.

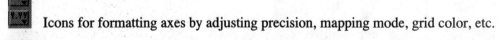

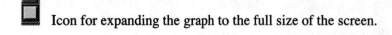

Icon for expanding the graph to the full size of the screen.

An example of the full screen display of a set of data together with two cursors is shown in Fig 6.86. Cursor readout boxes are located below the full screen graph.

Fig. 6.86 Full screen display of the graph with two cursors for determining intermediate measurement values.

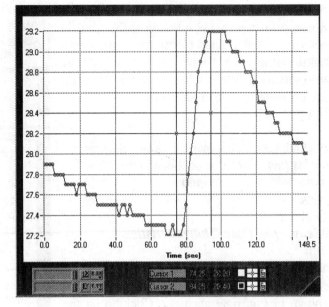

6.6.5 The View Area

When operating in the **View Area** it is possible to:

1. View the graph representing each data set.
2. Compare one data set against another.
3. Measure the geometric and statistical characteristics of a data set.
4. Print out the results.

Of these four functions, **Compare** and **Measure** are the most useful. **View** and **Print** may also be performed in the **Upload Window**. The **View Window** is shown in Fig. 6.87.

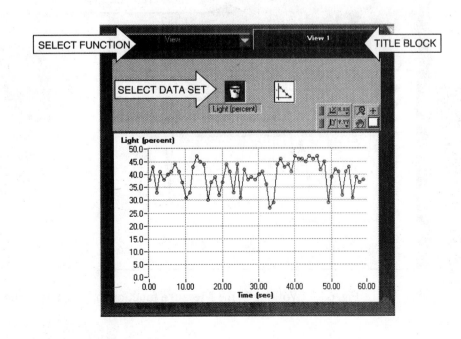

Fig. 6.87 The **View Window**.

Compare Data Sets

The **View Window** is used in the compare mode when the results of two experiments are presented on the same graph. A comparison is made by clicking on the ∇ symbol on the View bar. A menu drops down from which the function to be used is selected. We selected **Compare** to provide a comparison of light intensity versus temperature. The results of this comparison are presented in Fig. 6.88

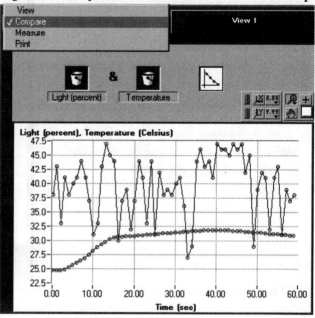

Fig. 6.88 A comparison of one data set against another is obtained in the **View Window**.

Measuring Data Characteristics

When selecting the **Measure** function from the menu on the **View Window**, the screen shown in Fig. 6.89 appears. Clicking on the icon with the oscillating curve produces an array of icons that enable you to select from six different measurement options, which include: (1) Minima, (2) Maximum, (3) Mean, (4) Standard Deviation, (5) Slopes and (6) Area under a curve. In Fig. 6.89, which shows random changes in light intensity, the mean (average) of the light intensities are represented. The mean value is printed in the box to the right of the **Measure** icon.

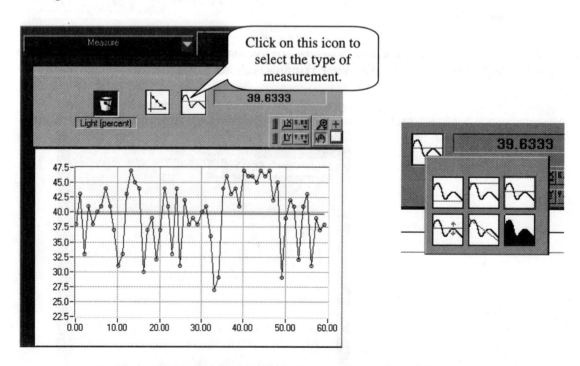

Fig. 6.89 Working in the **Measure** function in the **View Window**.

Printing Output from the View Area

When selecting the **Printing** function from the menu on the **View Window**, the screen shown in Fig. 6.90 appears. Click on the print icon and the file you have created in the **View Area** is printed.

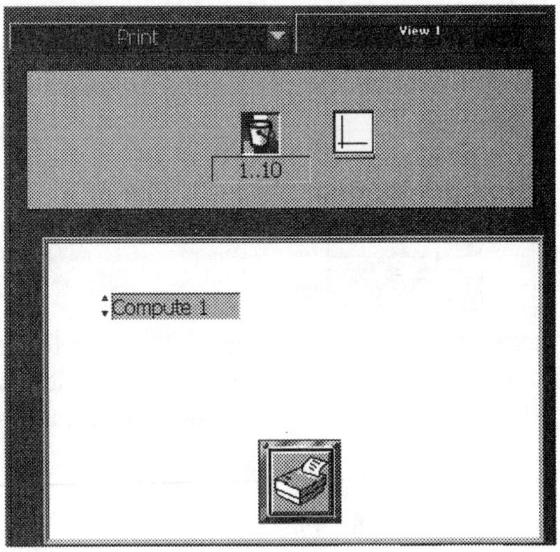

Fig. 6.90 The **Print Window** used to print pages from both the View and Compute Areas.

6.6.6 The Compute Area

The **Compute Area** is used to manipulate or process data that has been uploaded into a colored bucket. There are five different tools available for this purpose; however, this description is limited to tools 1, 2 and 3, which is consistent with the programming in **Investigator Level 3**. Tools 1, 2 and 3 are pre-wired in a simple series sequence with a limited number of pre-selected operations (functions). Tools 4 and 5 are more complex, and like the **Inventor** mode, these tools require the user to select functional icons from a palette and to wire them together in the correct sequence to create a program.

Computing with Tool 1

Begin with Tool 1 by clicking on the ∇ symbol on the tool bar, and the screen shown in Fig. 6.91 appears. There are six sequential icons associated with Tool 1 that are located at the bottom of the screen. The first icon, the **Red Bucket**, is used to select the data set that is to be manipulated. Click on the second or the fourth icon (from the left side) and the pop up menu, shown in Fig. 6.92, appears providing a choice of nine mathematical operations. These operations include adding, subtracting, multiplying and dividing as well as determining the sine, cosine and tangent when using each data point as an argument of these trigonometric functions. Also, you can determine the natural log and raise the exponential e by the value of each data point.

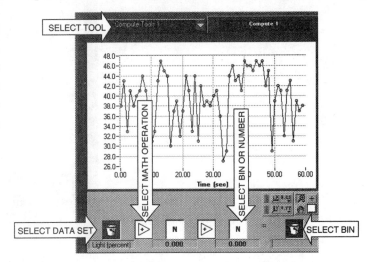

Fig. 6.91 Mathematical operations associated with Tool 1.

Click on the third or fifth icon from the left and the pop up menu that is shown on the right hand side of Fig 6.92 appears. This menu provides icons for 10 buckets or bins and one floating-point number N. The buckets or bins are file locations that may contain data or alternatively be ready to receive a new set of measurements or a file that has been generated by using one or more mathematical operations. The last icon in this row (a bucket) represents the file location in which the manipulated data set resides. After the mathematical operation has been performed, the transformed data is displayed graphically.

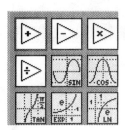

Fig. 6.92 Icons for selecting mathematical operations to the left, and bucket colors and a floating point number to the right.

New pages can be added so that several different computations can be performed on one or more data sets. In each case, the transformed data is displayed in a graphical format.

Computing with Tool 2

Load **Tool 2** by clicking on the ∇ symbol on the tool bar and select the second of the five tool options. The screen shown in Fig. 6.93 appears. The icons for manipulating the data are shown at the bottom of this screen. The two **Bucket** (bin) icons provide us with the opportunity to load one data set in the Y-axis and another in the X-axis. The data in the Buckets are plotted against another. Click on the ∇ symbol to the right of the bucket associated with the X-axis and the drop down menu appears that provides a choice of three quantities—X, Y and N. These three quantities are measured when data is acquired by the NXT. X represents the time in the event when the data point was recorded, Y is the quantity measured in appropriate units or percentages and N is the number of the data point. We have shown Y versus N in Fig. 6.93. Note that 60 data points were recorded (0 to 59).

Fig. 6.93 The window associated with **Tool 2**.

Computing with Tool 3

The window associated with **Tool 3** is presented in Fig. 6.94. This tool enables the user to perform several different mathematical operations on a specified data set. The data set is selected by chosing the Bucket (bin) containing the results of an experiment. Next two math icons are available that permit sequential mathematical operations on this data set. The icons, also shown in Fig. 6.94, enable the following operations:

First row:
- No change (multiplying the value of each data point by 1.000).
- Minima (determine the data point with the minimium value).
- Maxima (determine the data point with the maximum value).
- Mean (determine the average value of all the data points).

Second row:
- Standard deviation (determine the statistical quantity known as the standard deviation).
- Slopes (determine the average slope of the curve generated by the data set).
- Areas (determine the area under the curve).
- Differentiate (differentiate the data set at each point).

Third row:
- Integrate (integrate the data set point by point).
- Average lines (provide an average curve for two or more curves).
- Fit line (fit a line through the data points and provide the equation for this line).

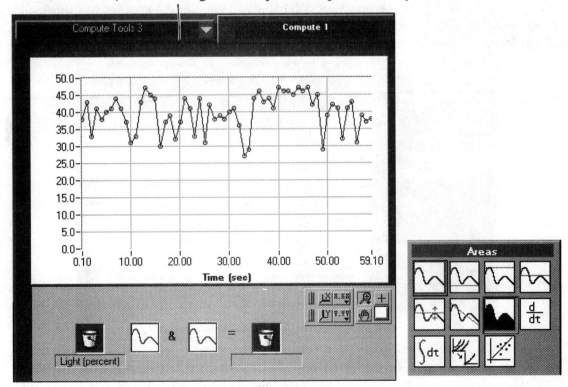

Fig 6.94 The window for **Tool 3** and the icons describing the available mathematical operations.

Exporting Data

When you have completed the processing of the data, it can be exported as a text file by clicking on File on the menu bar. A drop down menu appears as illustrated in Fig. 6.95.

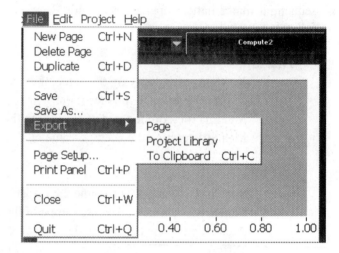

Fig 6.95 Procedure to export data from the
Compute Area.

6.6.7 The Journal Area

The **Journal Window**, presented in Fig. 6.96, is employed to prepare a report for your project. The report usually includes a statement of the problem or alternatively the objective of the experiment conducted. Next, you describe the expected outcome of the experiment, and then discuss the sensors used to collect the data and the sampling rate. Digital photographs of the experimental apparatus may be inserted into one of the journal pages. Results drawn from one or more data sets are described and conclusions drawn. Several pages may be used in documenting the project.

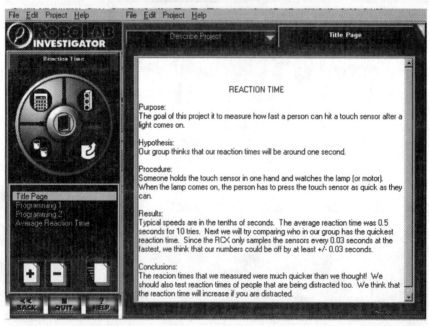

Fig. 6.96 The **Journal Window** that is used to describe and document the results of your experiments.

6.7   SUMMARY

The NXT incorporates a microprocessor that can be programmed to control the operation of your hovercraft. The degree of control possible is dependent upon the number and type of sensors that you connect to the NXT. Programming consists of downloading a series of commands in a compatible language that direct the microprocessor to execute a number of specific actions. This chapter deals in detail with programming in a graphical language known as ROBOLAB. However, before going into the details of ROBOLAB, the use of flowcharts was described. Flowchart symbols were defined and flowcharts for several different programming examples were presented.

The software used with the NXT for controlling our sensors and actuators is an icon based language known as ROBOLAB. Tufts University, National Instruments and the LEGO Group worked together to developed it. It is based on the popular program LABVIEW™—software used by many engineers for laboratory measurements and process control. ROBOLAB is partitioned into three main divisions—**Administrator**, **Programmer** and **Investigator**. The **Administrator** division enables you to download firmware, establish file locations, specify a communication port, etc. The **Programmer** division allows you to program the NXT in either the **Pilot** or **Inventor** modes. Programming in the **Pilot** mode is the easiest to learn and use; however, its capabilities are limited to sequential tasks. Programming in the **Inventor** mode is more challenging; however, this approach enables you to structure more complex programs with forks, jumps, loops, events, tasks, etc.

An extended description of programming in the **Inventor Level 4** has been included. The important palettes and sub-palettes such as the **Functions**, **Tools**, **Structures**, **Communications** and **Containers** were introduced. Techniques for wiring functional icons and their modifiers were described. Several examples were employed to demonstrate programming methods. Flow charts were used in several cases to better illustrate the strategy upon which the program was based.

Significant emphasis on programming involved the **Investigator** mode because of its inherent capability for data logging. Initially programming in **Investigator Level 3** was described. This level involves relatively simple programming with linear sequential structures similar to that found in the **Pilot** mode.

REFERENCES

1. Wang, E. L., <u>Engineering with LEGO Bricks and ROBOLAB</u>, 3rd Edition, College House Enterprises, Knoxville, TN, 2007.
2. Baum, D., <u>Definitive Guide to LEGO Mindstorms</u>, Apress, San Francisco, CA, 2000.
3. Ferrari, M., Ferrari, G. and R. Hempel, <u>Building Robots with LEGO Mindstorms</u>, Syngress Publishing, Rockland, MA, 2002.
4. Erwin, B., <u>Creative Projects with LEGO Mindstorms</u>, Addison-Wesley, Boston, 2001.
5. Dally, J. W., Riley, W. F. and K. G. McConnell, <u>Instrumentation for Engineering Measurements</u>, John Wiley & Sons, New York, NY, 1993.
6. Baum, D., Gasperi, M., Hemple, R. and L. Villa, <u>Extreme Mindstorms: An Advanced Guide to LEGO® Mindstorms</u>, Apress, San Francisco, CA, 2000.
7. Hardy, Q., "Son of Lego," Forbes, September, 4, 2006, p. 108.
8. Manes, S., "BYOB Build Your Own Bot," Forbes, September, 4, 2006, p. 110.

LABORATORY EXERCISES

6.1   Write a program to measure the voltage on Ports A and C with all five power level settings. Download the program to the NXT and using a digital voltmeter measure the voltage at each Port for each power level. Did you note a significant difference in voltage from one power level to another? If not, how is the power level adjusted?

6.2   With your instructor's assistance, measure the voltage output from Port A as a function of time for each of the five power levels. A storage oscilloscope should be used in making these measurements to display the time voltage trace associated with the pulse modulation. Measure the width of the pulses and establish the duty cycle for each power level.

6.3   Write a program that will enable you to determine the forward speed of a wheeled vehicle as a function of the power level applied to the LEGO servomotors. Download the program to the NXT and then conduct a test to measure the velocity of your vehicle at each of the power settings. Prepare a graph of speed versus power level. Is this a linear function? Record the results in your laboratory journal.

6.4   Write a program that will enable turning a wheeled vehicle as quickly as possible. Download the program to the NXT and then conduct a test to measure the time required for your vehicle to rotate 360° in either direction when it is operating at full power. Repeat the exercise to determine the time required to turn through 45°, 90°, 180° and 270° angles.

6.5   Using the meter in the LCD display of the NXT and a LEGO NXT light sensor measure the light intensity reflecting from a white surface. Also measure the light intensity reflecting from black tape. Is this difference between the two measurements significant? Why?

EXERCISES

6.1   State, county and city governments collect taxes from travelers visiting their area when they rent hotel rooms. Prepare a flowchart for a program to compute a hotel bill that accommodates the taxes collected by these three different governments.

6.2   Prepare a flowchart for a program to determine if a calle is east or west of Calle Central in San Jose, Costa Rico.

6.3   Suppose a rich uncle deposits $10,000.00 in a Roth Account (tax free) for you to use after its value reaches $1,000,000. The account earns 6% interest compounded annually. Prepare a flowchart for a program that will determine when you will become a millionaire and be able to withdraw funds from this account.

6.4   Write a program for measuring the intensity of light reflected from the surface of four different materials being considered for the hovercraft's track. Down load the program, conduct the experiment and upload the data. Completely document the experiment and its results.

6.5   Devise an experiment to measure the velocity of the hovercraft as a function of the power applied to its propulsion fans. Record your measurements of velocity with respect to time over a one - minute period at each power level. Prepare a graph of the velocity-time relation for each power level.

6.6   Describe the experiment and the results achieved in Exercise 6.5 using the Journal Area in ROBOLAB.

6.7   Design another experiment to measure the speed of the fans used in Exercise 6.5. We suggest employing a light sensor together with the NXT for this measurement. Based on the signals from this sensor, derive an equation to convert the pulses recorded to the speed of the fan.

6.8 Conduct the experiment described in Exercise 6.7 measuring the angular velocity of the fan in terms of RPM. Did you check to make sure that the data represents the true waveform and is free of aliasing effects?

6.9 Design an experiment to measure the variation in light intensity as a function of position in a classroom. Based on the data obtained during the scan of the classroom, prepare a graph that shows the results over the footprint of the room.

6.10 Write a program to determine the mean light intensity its standard deviation for the data collected in Exercise 6.9.

6.11 Prepare a Journal window that documents the experiment and its results that were presented in Example 6.7.

6.12 Prepare a Journal window that documents the experiment and its results that were presented in Example 6.8.

6.13 Prepare a Journal window that documents the experiment and its results that were presented in Example 6.9.

6.14 Determine the angular velocity in revolutions per second (RPS) of the new LEGO servomotor with its gear reduction when operating at power levels of 2 and 5. Prepare a Journal window that documents this experiment and its results.

Notes

PART II

ENGINEERING GRAPHICS

CHAPTER 7

THREE-VIEW DRAWINGS

7.1 INTRODUCTION

Engineering graphics is a broad term used to describe a visual means of communication. Most people normally think of communication in terms of writing and speaking because they are more commonly used in the normal course of life. However, when trying to communicate design ideas, writing and speaking are not always the best way to express complex concepts. A more visual way to communicate is needed. It is more effective to present design concepts and details by means of drawings, sketches, pictures and many different types of graphs. Visuals aids, such as drawings, convey ideas quickly and with remarkable precision. Also, it is easier to transmit and to receive design information if it is conveyed in the form of drawings, sketches and graphs.

The objective of this part of the book is to introduce you to visual methods of communication. You should understand the advantages of presenting information in drawings, sketches and graphs because they are very common techniques used in communicating engineering concepts. In this part of the textbook, methods for preparing orthographic projections, three-dimensional drawings, sketches and several different types of graphs are presented.

There are two general approaches used in preparing drawings and graphs. The first is a manual approach where the visuals are drawn by hand using a few simple drawing instruments. The second approach makes use of a computer and suitable software programs that greatly facilitate the preparation of visual representations. The manual methods for preparing two-dimensional drawings will be described in this chapter. Methods for constructing pictorial drawing will be described in Chapter 11. In a different textbook, a computer-aided design (CAD) software program used in preparing engineering drawings is introduced.

7.2 VIEW DRAWINGS

Let's begin by considering the block-like object shown in Fig. 7.1. You could describe this object as a rectangular block with a slot cut from its top. However, this written description is vague because you have not conveyed the relative proportions of the block, the precise location of the slot or the size of its features. In preparing an engineering drawing, it is essential to communicate proportion, exact location and size of every feature of the object that you are describing. The objective of the drawing is to convey sufficient information for the object to be produced by anyone capable of reading a drawing. You may use pictorial drawings and/or multi-view drawings to quickly and accurately convey this information to the reader. The drawing shown in Fig. 7.1 is a pictorial that is used to improve visualization of features specified in multi-view drawings.

The pictorial (isometric) drawing in Fig. 7.1 is shown with three arrows, which represent the directions of observation of the object showing the front, top and side views. When considered individually, each view is a two-dimensional rendering of the three-dimensional object. Of course, two-dimensional drawings are incomplete, because they show only the information observed in one of many possible views. Nevertheless, the two-dimensional views are important because it is possible to show objects to a true scale on a sheet of ordinary paper or a computer monitor. The problem of incomplete information inherent in two-dimensional drawings is avoided by presenting a sufficient number of views to completely locate and size each feature of the object.

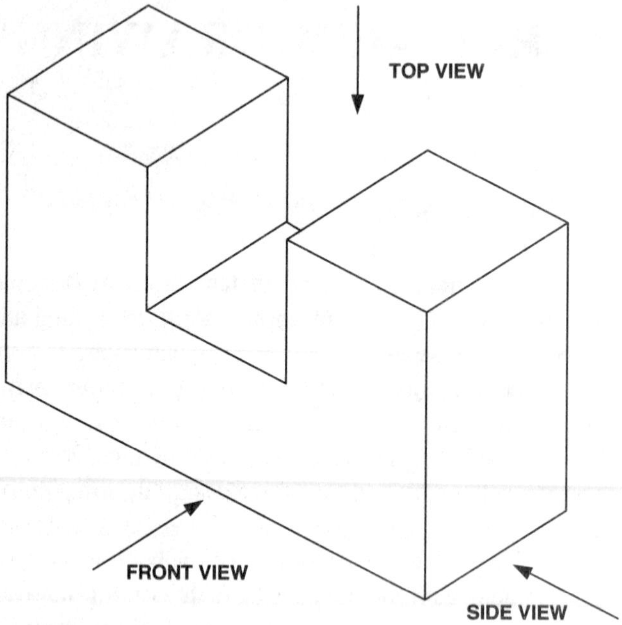

Fig. 7.1 Pictorial drawing showing a rectangular block with a slot. The three viewing directions define the front, top, and side views.

A three-view drawing of the rectangular block presented in Fig. 7.1 is illustrated in Fig. 7.2. The single pictorial drawing of the block is now represented with three different two-dimensional drawings representing the front, top and side views. These three drawings completely define the proportions of the rectangular block, its size and the location and size of the slot.

The arrangement of the views is important. The front view is placed in the lower left corner of the paper, the top view is directly above it and the right side view is to the right of the front view. This arrangement is important because it permits one to prepare the three-view drawing using orthographic projection, where dimensions are projected from one view to another. For example, you measure width and height of the block, draw the front view and project construction lines upward to the top view. The width of the object is identical in both the top and front views. You also project construction lines to the right from the front view to define feature heights in the side view. The front and side views both show the height dimensions of the block. You may also project from the top view to the side view by drawing a 45° construction line from the right hand corner of the front view as shown in Fig. 7.3. Then you project construction lines to the right from the top view until they intersect the 45°-construction line. Next, turn the lines downward to show the depth of the block on the side view. The construction lines used to prepare Fig. 7.2 are illustrated in Fig. 7.3. Orthographic projection is very important because it saves you time in preparing your drawings. It reduces the number of time-consuming measurements that must be made in preparing the different views necessary to completely describe any given part. The construction lines also help you in forming visual images of the locations of each feature on the adjacent views.

One more point to consider before leaving Fig. 7.2. Did you notice the dashed line on the side view? Why were all of the lines solid except that one? A convention is followed for the use of line styles and weights in preparing engineering drawing. Solid lines represent edges of features that are observed in a particular view. As you observe the top of the block shown in Fig. 7.1, you can see six lines—the four sides of the rectangle and the two edges of the slot. Because all of these edges are visible, solid lines are used to represent all six edges in the top view. However, when the object is viewed from the side, only the four edges outlining the rectangle are visible. When viewing the block from the right side, the slot cannot be observed. However, when the pictorial or the top and front views are examined, it is evident that a slot exists. To show the slot on the side view, a dashed line representing a hidden line is employed. Two-dimensional drawings differ from photographs because hidden features can be shown on the appropriate view with dashed lines.

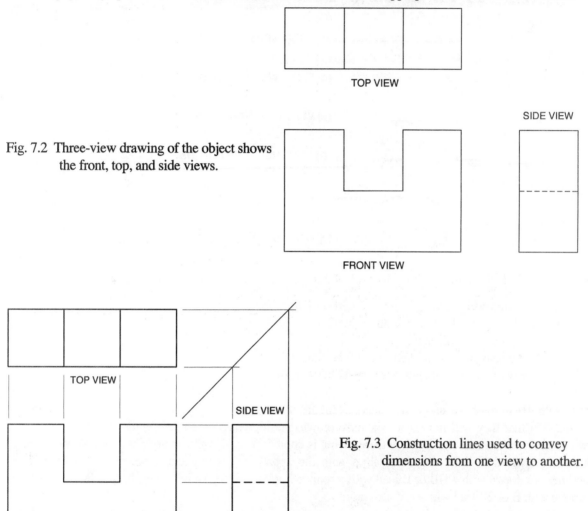

Fig. 7.2 Three-view drawing of the object shows the front, top, and side views.

Fig. 7.3 Construction lines used to convey dimensions from one view to another.

7.3 LINE STYLES

Different line styles and different line weights (thickness and darkness of the lines) are used in preparing engineering drawings as illustrated in Fig. 7.4. The weight and the thickness of the lines are controlled by the selection of the pencil lead and the sharpness of the point. The degree of hardness is governed by variable amounts of clay that is mixed with graphite to form pencil lead. Hard leads have a small amount of clay while soft leads have higher clay content. The scale from hard and light to soft and dark is given below:

Hard and Light ⇐ 6H–5H–4H–3H–2H–H–F–HB–B–2B–3B–4B–5B–6B ⇒ Soft and Dark

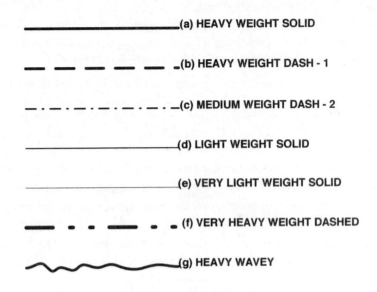

Fig. 7.4 Different line styles used in view drawing.

a. Heavyweight, solid-lines representing edges
b. Heavyweight, dashed-lines representing hidden lines.
c. Medium-weight, long-dash then short-dash lines for centerlines.
d. Lightweight solid-lines for dimensioning.
e. Very thin, light-lines for construction.
f. Very heavyweight, dashed-lines for section cuts.
g. Heavy wavy-lines to indicate a break in the view.

Pencils with 4H to 6H leads and a very sharp point are used to draw construction lines. These lines are so light and thin that they will not show when the drawing is reproduced. Light-weight dimension lines are drawn so they will remain visible when the drawing is copied. F lead with a sharp point and a very slightly rounded tip is used to draw dimension lines with the appropriate width and darkness. Medium-weight centerlines are drawn with a HB or B lead with a rounded point. Heavy-weight lines, both solid and dashed, are drawn with B or 2B lead with a rounded point.

The very soft leads—3B to 6B are usually employed by artists to prepare sketchings, but not by engineers. Engineers handle a completed drawing much more than an artist. In handling, the soft lead smears degrading the appearance of the drawing and detracting from the quality of the copies made from the originals.

It is recognized that many of you will be short on pencils, but perhaps team members can share resources. Do not use ballpoint pens for drawing. There are two problems with ball-point pens. First, you have no control over the width of the line because the line width is determined by the diameter of the ball in the pen. Second, you may make errors in preparing the drawing. The lines drawn with a ballpoint pen cannot be erased; your drawings are often sloppy and difficult to read. Work in pencil and use a clean, white, vinyl eraser to eliminate all traces of your errors. Strive to produce an accurate, clean and error-free drawing.

While on the topic of pencils and erasers, below is a list some other simple tools that you should consider acquiring.

1. Transparent ruler with scales in both inches and millimeters.

2. Two triangles—30/60 and 45/45 degrees.
3. Protractor for measuring angles.
4. Compass for drawing arcs and circles.
5. Templates for drawing circles, ellipses, etc.

This is a short list. A complete set of drawing instruments or a drafting board with a T-square have not been included. The purpose of this course is not to teach drafting, but to introduce you to the essentials of engineering graphics. A complete set of drafting tools is not necessary to learn the early lessons in graphics. You are encouraged to invest the small sum needed to acquire the items listed above. You will find them useful in many other courses during your program in engineering.

Finally, a suggestion is made for the paper to use in preparing your drawings. National Brand engineering paper 8-1/2 by 11 in. in size is recommended. It is light green in color to relieve the strain on your eyes. It also has a grid (five squares to the inch) on the reverse side of each sheet. The grid lines show through to the front side of the sheet providing guidelines helpful in preparing many different drawings, sketches and charts. The paper erases well, produces good copies and is pre-punched for a three-ring binder. Equipped with the proper pencils, eraser, drafting tools and paper, you are ready to draw several different features normally encountered in preparing multi-view drawings of engineering components.

7.4 REPRESENTING FEATURES

The techniques for drawing various features in three-view drawings will be demonstrated in this section. The examples include a block with a slot and a step, a block with a tapered slot and a block with a step and a hole.

Block with Slot and Step

The rectangular block in Fig. 7.1 has one feature—a slot. Many different geometric features are required in depicting parts that incorporate holes, curved boundaries, steps and tapers. A pictorial drawing of an object with both a slot and a step, presented in Fig. 7.5, serves as the next example. Let's prepare a three-view drawing of this object using the method of orthographic projection. Examine the pictorial starting with the front view.

As you examine the front view in Fig. 7.5, visualize the outline of the block. Even with the step and the slot, the outline of the block (the outside edges) is a rectangle. Draw the rectangle, representing the outline of the front view, in the lower left hand corner of your quadrille paper. Do not worry about exact dimensions of the rectangle at this stage; however, try to maintain the proportions shown in Fig. 7.5. When drawing the rectangle in the front view, extend the vertical lines upward into the region of the top view and the horizontal lines to the right into the region of the side view. These are construction lines; you should keep them thin and light. Now examine the step in the upper right of the front view of the block. Note the intersection of the planes defining the step with the front plane of the block. These intersections produce two new edges that are visible in the front view. Draw the location of these edges with the horizontal and vertical lines shown in the upper right hand corner of the rectangle (see Fig. 7.6). Extend the vertical line used to locate the step in the front view upward and the horizontal line to the right. Finally, locate the slot on the left side of the pictorial drawing in Fig. 7.5. Again note the planes defining the slot and observe that they intersect the front plane to form two vertical edges. Draw two vertical lines at the correct location to represent these edges. Project these lines upward into the region of the top view with construction lines. You have now completed the front view and are ready to draw the top view.

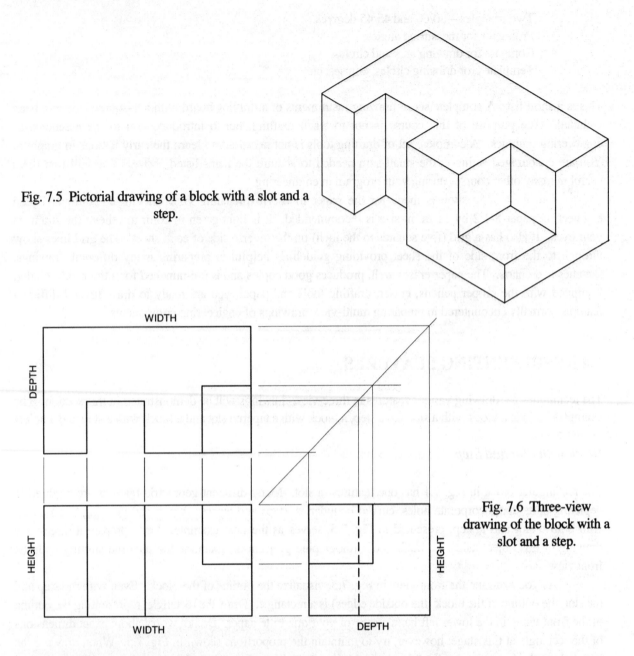

Fig. 7.5 Pictorial drawing of a block with a slot and a step.

Fig. 7.6 Three-view drawing of the block with a slot and a step.

WIDTH

DEPTH

HEIGHT

WIDTH

HEIGHT

DEPTH

Look downward at the top of the pictorial drawing. Again, the outline observed is rectangular in shape; however, the rectangle is not perfect because the slot interrupts it. Looking downward, you will observe that the edge defining the outline of the rectangle across the width of the slot is missing. It is possible to observe the bottom plane at the location of the slot. Draw the outline of the rectangle in the location of the top view of Fig. 7.6. Use the construction lines that exist at the location of the top view to establish the width dimensions. An open section in the rectangle is evident at the location of the slot. Next, show the depth of the slot in the top view by drawing the three lines that represent the edges formed by the three vertical planes of the slot with the top plane of the block.

Next, examine the step. The step has two vertical planes that intersect the top plane to form its defining edges. Draw two lines representing those edges in the drawing to complete the top view. If you made full use of the construction lines projected from the front view into the region of the top view, it was easy to draw the top view. It was only necessary to locate the position of three horizontal lines that defined the depth of the slot, step and rectangular block. These depth dimensions are carried to the right with construction lines.

The final view is of the right side. It is also possible to draw a left side view, but the accepted convention is to draw the right view. After having completed the front and top views, with the orthographic projection of construction lines, the side view is easy to prepare. Draw a construction line at 45° so that it intersects the horizontal construction lines from the top view. Then draw a second set of construction lines downward from these intersection points into the region of the side view. The side view now contains seven construction lines. The three horizontal lines projected from the front view give the height of the step and the rectangular block. The four vertical lines projected from the top view give the depth of the slot, step and block. The location and dimensions of all of the lines in the side view have been established from your orthographic projections. No measurements are necessary. To complete the side view, darken select portions of the construction lines that represent edges in the side view. Again, examine the side of the pictorial drawing in Fig. 7.5 and observe that the rectangular outline of the block is completed. Also, note the edges produced by the intersections of the planes forming the step with the plane of the right side of the block. Draw these two edges in the upper left corner of the side view (see Fig. 7.6). From a visual perspective, your drawing of the right side is complete; however, a multi-view drawing often carries more information than you can directly observe. You know from the front and top views that a slot exists. Your projection lines from the top view to the side view show the depth of the slot. Even though you cannot "see" the slot in the side view, its presence and its location is represented with a dashed (hidden) line. The three-view drawing of a rectangular block containing a slot and a step is complete as illustrated in Fig. 7.6.

Block with Tapered Slot

Let's consider drawing a slightly different feature—a block incorporating a tapered slot as illustrated pictorially in Fig. 7.7. Begin a three-view drawing with the front view in the lower left corner of your drawing paper as shown in Fig. 7.8. Examine the front view in Fig. 7.7 and observe the edges produced by all the planes intersecting the front plane. Drawing these edges gives four vertical lines and three horizontal lines. The tapered surface intersects the front plane giving an edge shown as a horizontal line in the front view. The top surface is represented by still another edge visible in the front view. All of these lines are projected into the remaining two views with construction lines.

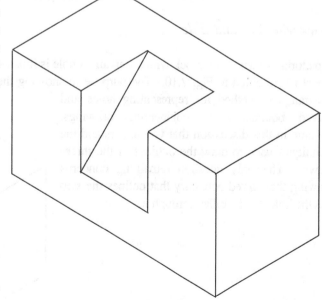

Fig. 7.7 Pictorial drawing of a block with a tapered slot.

Examine the top view and observe that the rectangular outline of the block remains intact. Observe three edges inside the rectangular outline. The edge due to the intersection of the taper plane with the top surface produces a horizontal line, and the two vertical planes, defining the width of the slot, intersect the top surface to produce two vertical lines in the top view. Project the horizontal lines in the top view to the right so that they intersect the 45° construction line shown in Fig. 7.8. Then project these three lines downward from the intersection points on the 45° line into the side view.

The six construction lines projected into the side view provide the dimensions needed to complete this view. Clearly, it is evident from Fig. 7.7 that the rectangle outlining the side view is intact. In fact when you view the right side, only a simple rectangle formed by the four edges is visible. The information regarding the tapered surface, evident in the pictorial drawing and the front view, is not visible in the side view. Information, showing the location of the tapered surface on the side view, is added to the right-side view by using the dashed (hidden) lines.

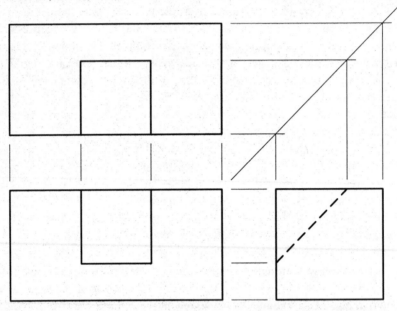

Fig. 7.8 Three-view drawing of the block with a tapered slot.

Block with Step and Hole

A pictorial drawing of a block with a step and a hole is illustrated in Fig. 7.9, and a three-view drawing of this object is presented in Fig. 7.10. The purpose in showing the fourth example of a three-view drawing is to illustrate the method for representing holes and curved boundaries in engineering drawings. Assume in this discussion that you understand the techniques used to draw the outlines of the three-views. The only question remaining concerns drawing the curved boundary that defines the step and the hole that is drilled through it.

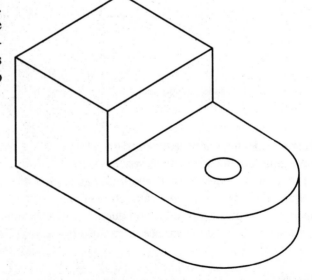

Fig. 7.9 Pictorial drawing of a block with a step and a hole.

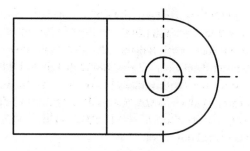

Fig. 7.10 Three-view drawing of the
block with a step and a hole.

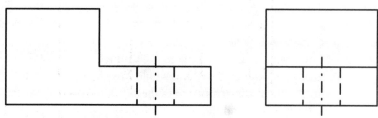

The circular boundary outlining the
step is evident only in the top view
(see Fig. 7.10). The front and side
views give no evidence of the presence of a curved boundary because the curved surface does not produce
intersections (edges) on the front or side planes. The curved surface in the top view is produced by using a
compass with its point located in the center of the hole. The radius is set on the compass so that the circular
arc is tangent to the horizontal lines forming the outline of the block in the top view. It is advisable to draw
the arc prior to drawing the horizontal lines that are tangent to the ends of the arc.

A compass is also employed to draw the hole in the top view. Note the two orthogonal lines defining
the center of the hole. These are centerlines that locate the position of the hole relative to other features on the
block. The presence of the hole is also depicted in the front and side views even though it is not visible in
either of these views. Use dashed (hidden) lines to locate the edges of the holes. The centerline of the hole in
the front and side views is also shown. Note the long-dash, short-dash line used to denote the centerlines in all
three-views. The centerlines are very important when you dimension the drawings, because holes are located
relative to their centerlines, not their edges.

7.5 DIMENSIONING

In the previous section, three-view drawings were prepared maintaining proportionality, but without concern
for the dimensions. This approach was a simplification taken to clarify the discussion of three-view drawings.
In actual practice, the dimensions are critical because they control the size of the component and the location
of all of its features. Let's examine dimensions from two different points of view. First, as the individual
preparing the drawing, you must know (or decide) upon the dimensions required to completely define the
component. If you are the designer of a new component, it is necessary to start with a blank sheet of paper
and assign the dimensions that you believe will optimize the design of the component. If you are redesigning
an existing component, it is often possible to start with the original drawing of that component and modify
existing dimensions to refine the design. The person preparing the drawing must know all of its dimensions.
It is your responsibility to include on your drawing all dimensions necessary for some stranger, perhaps living
in another country, to fabricate the component.

Next consider the second point of view—as the person using the drawing. This person may
manufacture the part, may assemble the product that incorporates the part or may repair the product.
Engineering drawings serve many purposes and are used by many people not involved in the design. The
three-view drawing defines this component without ambiguity. The dimensions give its size precisely. The
component, as defined by the drawing and its dimension, can be made by anyone in the world. The

component must be interchangeable with another one manufactured previously by a different plant in another country.

Let's begin to learn about dimensioning by adding dimensions to the drawing of the block with a slot, as shown in Fig. 7.2. This drawing has been copied and dimensions added as indicated in Fig. 7.11. Examine the front view of this drawing note the width of the block has been defined as 3. This dimension is inserted in a break in the dimension line. No units are given except for a notation in the drawing block that all dimensions are in inches, mm, ft, etc. Arrowheads that point to the two extension lines terminate the dimension line. The arrowhead is long (about 1/8 in.) and thin. The extension lines are separated from the view drawings by a small gap (about 1/16 in.).

In the example presented in Fig. 7.11, the block was dimensioned using inches as the unit of measure. The fact that the width was defined as 3 indicates to the person manufacturing the block that the width may be $3 \pm 1/64$ in. The tolerance on the width is implied or is given in the drawing block. If the designer finds it necessary to specify a block with a higher dimensional accuracy, the dimension would be specified using decimals. For example, if you provided the dimension as 3.00, the person manufacturing the part would understand that the width of the block would have to be between 3.00 ± 0.01 in. If you require still more accuracy, specify the dimension as 3.000; the block must be produced with a width between 3.000 ± 0.005 in.

Remember—there is a great difference between specifying the dimension as 3, 3.00 and 3.000 in the accuracy of the block and tooling required to manufacture it. There is also a significant difference in the cost. As you decrease (tighten) the tolerances, the cost of manufacturing a component increases markedly.

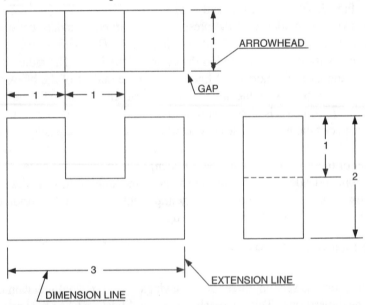

Fig. 7.11 Example showing the dimension line, extension line, arrowhead and gap.

Let's return to Fig. 7.11 and examine the dimensioning of the top and side views. The width on the front view has already been shown, but the dimensions for the depth and height of the block and the location and size of the slot must be provided. Note, the top view is used to give dimensions for the width and depth, and the side view is used to give dimensions for the height and depth. In the top view of Fig. 7.11, the depth of the rectangular block has been specified as 1. The dimensions locating the slot and defining its width on this view have also been given. The dimensions of the height of the block and the height of the slot are provided on the side view.

The dimensioning of the block and its slot has been completed. Every dimension necessary to manufacture the component has been specified in the drawing. Also the drawing has not been over-dimensioned. (The same dimension has not been specified more than one time). The choice of where you place the dimensions is somewhat arbitrary. You may dimension the width on either the front or top views, the depth on the top or side views and the height on the front or side views. Two choices for the location of each dimension are possible. The selection between the two views in defining a particular dimension is usually made to reduce drawing clutter and maintain clarity.

Dimensioning Holes and Cylinders

To demonstrate techniques for dimensioning either holes or cylindrical boundaries, let's add dimensions to the drawing shown in Fig. 7.10. Again start with the front view and locate the centerline of the hole as 3.20 from the left edge of the block. A hole is always located by dimensioning between suitable edges to its centerline. Use centerlines for dimensioning because the drill employed to produce the hole is inserted into the component at the location of its centerline.

In this case the dimension for the width is not shown in the front view because it is preferable to specify the radius of the curved (cylindrical) boundary in the top view. Examine the top view and note the radius of the curved boundary is specified as 1.20 R. The R is added to the dimension to indicate that it is given in terms of the radius and not the diameter. In specifying the radius R, you have in effect defined the width of the block as the sum of 3.20 + 1.20 = 4.40. If you had specified the total width of the block in the front view, the drawing would have been over-dimensioned.

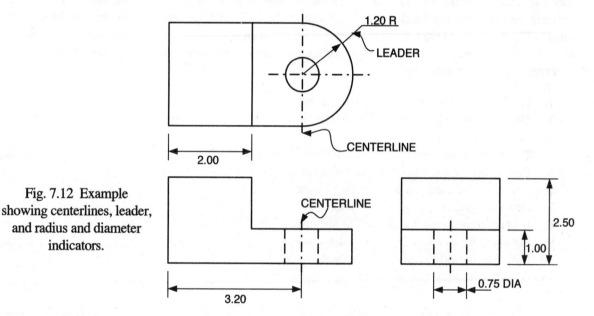

Fig. 7.12 Example showing centerlines, leader, and radius and diameter indicators.

The location of the step from the left edge of the block has been specified as 2.00 in the top view. The height dimensions of the features are given in the side view. The total height of the block is given as 2.50 and the height of the step as 1.00. Both heights are measured from the bottom edge of the block. The diameter of the hole is dimensioned on the side view as 0.75 DIA. Holes are dimensioned in terms of their diameter because the drills used in drilling the holes are specified by their diameter, not radius.

Are the dimensions as shown in Fig. 7.12 complete? You might question if the depth of the block has been described because it has not been shown in the side view. However, the depth was specified with the radius of the curved boundary in the top view. The depth is equal to two times the radius R or 2.40.

Again, units for the dimensions have not been shown directly. The units are defined in the drawing block along with other information pertaining to manufacturing the component. The use of two standards for units, the U. S. Customary and the SI systems may cause some confusion in preparing and reading drawings. Some drawings are prepared using U. S. Customary units of in. or ft, and others are prepared using the SI system with units given in mm, cm or m. Sometimes, when the drawing is to be used by many people from different countries, both systems are employed in the dimensioning. An example of dual dimensioning is given in Fig. 7.13. In this case, a break the dimension line is not made. U. S. Customary units are placed above the line and SI units are placed below the line and enclosed with parentheses.

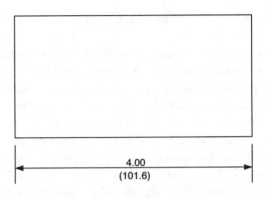

Fig. 7.13 Illustration of a convention used for dual dimensioning.

7.6 DRAWING BLOCKS

The drawing block serves a very important function on an engineering drawing. In the previous section, reference was made to the fact that the units for the dimensions are specified in the drawing block. The unit of measurement used in preparing the drawing is only one of the many facts presented in the drawing block. As illustrated in Fig. 7.14, the drawing block is located in the lower right hand corner adjacent to the border of the drawing. A typical drawing block conveys important information pertaining to all drawings produced by a certain company and to the individual drawing. Information commonly shown in the drawing block includes:

1. The name of the company issuing the drawing.
2. The name of the part that the drawing defines.
3. The scale used in preparing the drawing.
4. Tolerances to be employed in manufacturing the part.
5. Date of the completion or the release of the drawing.
6. Material to be used in manufacturing the part.
7. Heat treatment of the part after manufacturing.
8. Units of measurement to be used in manufacturing the part.
9. Initials of the individual preparing the drawing.
10. Initials of the individual checking the drawing.
11. A unique drawing number to identify the drawing.

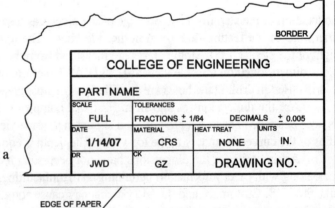

Fig. 7.14 Example of information presented in a drawing block.

Let's examine each of these items. The name of the company is self-evident; it is illustrated in Fig. 7.14 as the College of Engineering. In designing a part, identify it with some name such as bracket, support, shaft, etc. and include this name on your drawing. The part name should be brief (no more than two or three words).

Next you will indicate the scale used in preparing the drawing. The example shown in Fig. 7.14 indicates that full scale was used in drawing the three-views of the part. The scale that you select depends on the size of the part and the size of the paper that is used for the drawing. In a typical college class, you are usually constrained to standard size copy paper (8-1/2 by 11 inches) because of size limitations imposed by low-cost laser printers. Suppose the part that you are designing is 12 inches wide, 6 inches high and 1 inch deep. Will the three-view drawing of this part fit on this standard size paper? The answer to this question is a clear NO. You need to scale down the part dimensions on the drawing so that it will fit on a single sheet of paper. In this case, you would probably use ¼ scale. With ¼ scale, the front view of the part would be 3 inches wide by 1-½ inches high. You would dimension the drawing using the actual size of the part (i.e. you would show the width as 12 inches although it is only 3 inches on the scaled drawing). You indicate to those individuals reading the drawing that the size of each view has been reduced by a scaling factor of ¼ with a notation in the drawing block.

Tolerances that are to be employed in manufacturing the part are specified in the drawing block. Many companies have standardized their tolerances and these limits are preprinted in drawing blocks that are incorporated on the company's drawing paper. The date of completion of the drawing is also shown. In some instances, the drawing release date is used instead of the completion date. The release date identifies when the design engineers turn over the ownership of the design of a product to the operations (production) function, which is responsible for producing the product.

The material from which the part is to be fabricated is often identified in the drawing block. In the example shown, the abbreviation CRS indicates that cold rolled steel will be used in manufacturing the component. Sometimes the heat treatment of the component, if required, is identified in the drawing block. Heat treatment is a multi-step process employed to enhance the strength and/or hardness of a component fabricated from certain metals. Because heat treatment is usually performed after the part is machined; it is standard practice to indicate whether or not heat treatment is required. This information assists those in charge of operations in controlling the flow of parts in the production process.

The units used in the drawing are given in the drawing block. You may use U. S. Customary units, SI units, or dual units. You cannot use mixed units. (Note the difference between mixed and dual units). U. S. Customary units are in terms of inch (in.) or foot (ft). SI units are in terms of millimeter (mm) or meter (m).

Those responsible for the drawing are identified. The individual who has prepared the drawing signs the block with his or her initials. The drawing is checked for completeness and accuracy wit the individual checking the details initialing the drawing block. Finally, the drawing is given a number. In a large company, which may release thousands of drawings each year, the engineering records department controls the drawing numbers. This department maintains a numbering system that insures that each part or component has a unique drawing number. The engineering records department also organizes the numbering system to group together all of the drawings needed to build a specific product.

7.7 ADDITIONAL VIEWS

Some of the parts used in products are complex and additional views may be needed to ensure that the drawings are properly interpreted. You are not restricted to using only three-views in preparing detail drawings. For very simple parts, you can adequately describe the object with one or two views. There is no need to waste time drawing additional views. For complex parts that are more difficult to visualize, you can provide additional views to clarify the drawing. These extra views may show either external surfaces or internal sections. If it helps to visualize the object, you may add back, bottom and left side views to the more commonly employed front, top and right side views. To draw these additional external views, simply rearrange the layout of the views on the drawing as shown in Fig. 7.15. This drawing, which shows five views of a multiply tapered block, illustrates the proper arrangement to show additional external views. It also maintains the advantages of orthographic projection. Clearly, the extra views help in visualizing the complex shape of this block.

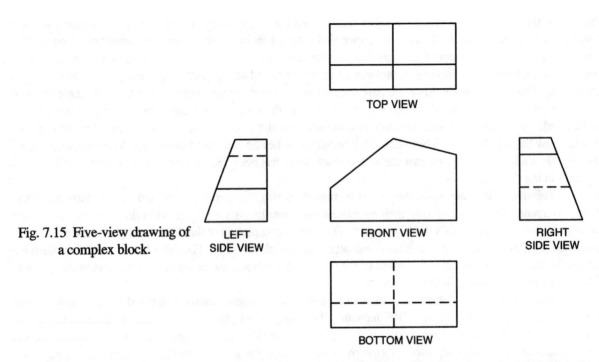

Fig. 7.15 Five-view drawing of
a complex block.

An addition internal (sectional) view is often more useful in clarifying a drawing than an additional external view. The concept of a sectional view requires you to mentally slice the object into two pieces with a cutting plane. You then open the part and view the internal section revealed by the cut. An example of the sectioning technique is illustrated in Fig. 7.16 where adding a sectional view has modified Fig. 7.8. On the front view

drawing in Fig. 7.16, a very heavy dashed line is shown to indicate the location of the cut. The cut line is turned through 90° at both ends and arrowheads are drawn to indicate the direction of the view. The arrowheads are each labeled with the letter A. In this case, you are viewing the section cut from the right side. Accordingly, you should expect the section view to closely resemble the right side view. You then draw the section that is revealed by the cut.

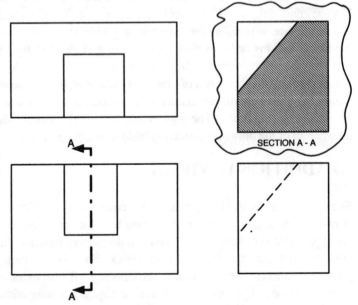

Fig. 7.16 Section cut and the
corresponding section view.

The section view may be placed at any convenient location on the drawing. In Fig. 7.16, it has been placed it in the open space in the upper right hand corner of the drawing. The section view is usually encircled with a wavy line to distinguish it from the usual three-views. The section view is identified with the section cut by using the letters A-A in a notation. In some drawings, you might make several cuts each revealing a different section with each sectional view identified by A–A, B–B, C–C, etc. In this view, the section revealed by the cut is shown with cross-hatching. The remaining part of the view (outside the cut section) is drawn without cross-hatching. Prepare a section view of the component shown in Fig. 7.12. Make the section cut along the centerline through the hole in the top view, and then examine the section from the right side.

7.8 SUMMARY

Engineering graphics have been introduced by describing techniques for preparing multi-view drawings. Multi-view (usually three-view) drawings are used to communicate the relative proportions of a component and the exact size of all of its features. Drawing techniques are covered in detail with a few simple examples. It is important that you invest the time necessary to learn these techniques because acquiring skills to draw and read three-view drawings is an essential means of communicating in engineering.

   Dimensioning is an important task in preparing an error free drawing. All of the dimensions must be included so that the part can be manufactured from the drawing. Often the part is manufactured in another city, or state or even another country. The person manufacturing the part does not know you and cannot ask questions to clarify the ambiguities in the drawing. The drawing and all of its dimensions must completely define the component. An important implication of the dimensioning is the tolerances required in fabricating the part. As you write the numbers for the dimensions, you implicitly assign the tolerances. There is a significant difference between 3, and 3.000 in the precision required in manufacturing and in the cost of the component.

   Drawing blocks were described to illustrate the type of information that they normally convey. Remember drawing blocks are not unique and you will observe differences in the information presented by different companies. The concept of scale was also introduced in this section. The scale used in preparing a drawing is your decision. Select the scale factor so that the three-views fit the paper without crowding. Sometimes you will scale down the view drawings so they will fit on the sheet. Other times you will scale up the views of a very small part so that very small features may be clearly represented.

   The coverage of three-view drawings presented in this chapter is very brief. If you need additional information, more complete textbooks are listed in the references.

REFERENCES

1.   Earle, J. H., Engineering Design Graphics, 4ᵗʰ edition, Addison Wesley, Reading, MA 1983.
2.   Bertoline, G. R., and E. N. Wiebe, Fundamentals of Graphics Communication, 3ʳᵈ Ed., McGraw Hill New York, NY, 2002
3.   Lockhart, S. D. and C. M. Johnson, Engineering Design Communication, Prentice Hall, New York, NY, 1999.
4.   Jensen, C. H. and J. D. Helsel, Engineering Drawing and Design, Glencoe McGraw Hill, New York, NY, 1997.
5.   Sorby, S. A., Manner, K. J., Baartmans, B. J. and S. S. Sorby, 3-D Visualization for Engineering Graphics, Prentice Hall, New York, NY, 1998
6.   Giesecke, F. E. et al, Engineering Graphics, 7ᵗʰ Ed., Prentice Hall, New York, NY, 2001.
7.   Krulikowski, A. Geometric Dimensioning and Tolerancing, Delmar Publishers, Delmar Publishers, Clifton Park, NY, 1998.

EXERCISES

7.1   Prepare a three-view drawing similar to the one shown in Fig. 7.2 except change the width of the notch from 1 inch to 1–1/2 inch.

7.2   Prepare a three-view drawing similar to the one shown in Fig. 7.2 except change the width of the notch from 1 inch to 1-3/4 inch.

7.3   Prepare a three-view drawing similar to the one shown in Fig. 7.2 except change the depth of the object from 1 inch to 1-3/4 inch.

7.4   Take a piece of clay or foam plastic and use a razor knife to manufacture a block with the shape shown in Fig. 7.6.

7.5   Prepare a three-view drawing of a block with a taper notch like that shown in Fig. 7.8. Select the dimensions yourself, but be consistent from one view to another with these dimensions.

7.6   Prepare a three-view drawing similar to that shown in Fig. 7.10 except increase the diameter of the hole to 1.0 inch.

7.7   Prepare a three-view drawing similar to that shown in Fig. 7.10 except decrease the diameter of the hole to ½ inch.

7.8   Prepare a three-view drawing similar to that shown in Fig. 7.10 except increase the depth of the object from 2.4 to 3.0 inch.

7.9   Dimension the drawing that you prepared for Exercise 7.1.

7.10 Dimension the drawing that you prepared for Exercise 7.2.

7.11 Dimension the drawing that you prepared for Exercise 7.3.

7.12 Dimension the drawing that you prepared for Exercise 7.5.

7.13   Dimension the drawing that you prepared for Exercise 7.6.

7.14 Dimension the drawing that you prepared for Exercise 7.7.

7.15 Dimension the drawing that you prepared for Exercise 7.8.

7.16 Design a drawing block that your team will employ to identify their drawings.

7.17 Prepare a five view drawing of the object shown in Fig. 7.5.

7.18 Prepare a five view drawing of the object shown in Fig. 7.7.

7.19 Prepare a five view drawing of the object shown in Fig. 7.9.

7.20 Prepare a drawing with a section view for the block defined in Fig. 7.1.

7.21 Prepare a drawing with a section view for the block defined in Fig. 7.5.

CHAPTER 8

PICTORIAL DRAWINGS

8.1 INTRODUCTION

Pictorial drawings are three-dimensional illustrations of a machine component, structure or any object. For a person trying to visualize an object with some complexity, the pictorial drawing is the most effective means to convey its size and shape. Some people have difficulty placing the three standard (front, top and side) views together to "see" the object. Pictorials drawings assemble the three-views in a single image providing a three-dimensional rendering facilitating visualization. Because pictorials are so easy to visualize, they are often used to illustrate catalogs, maintenance manuals and assembly instructions.

Three different types of pictorials are in common usage:

- Isometric
- Oblique
- Perspective

A simple cube with these three different types of pictorials is illustrated in Fig. 8.1. The isometric pictorial is drawn with its three axes spaced 120° apart. The term isometric means "equal measurement" indicating that the three sides are all scaled by the same factor relative to their true length. Parallel lines defining the edges of the cube are also parallel on the isometric drawing. Drawing paper with isometric axes is available in well-stocked office and/or drafting supply stores. You are encouraged to use this special paper because it greatly facilitates the preparation of an isometric pictorial.

Oblique pictorials are drawn with the front view in the x–y plane. The oblique lines, which represent the z-axis (depth), are projected at some angle—often 45°; however, the angle used for the oblique lines may vary from 0 to 90°. Parallel lines defining two or more edges on the cube are also parallel on the oblique drawing. If the true length of the lines is employed in scaling all three sides, it is known as a cavalier oblique pictorial. The cavalier oblique style is frequently used, but the resulting pictorial is distorted. The distortion is due to the depth dimension, which appears to one's eye to be too long.

The perspective is a pictorial drawing that represents what you appear to "see". Artists usually draw or paint using the perspective style. Engineers sometimes represent their designs in this style. Unfortunately, perspective is the most difficult of the three types of pictorials to master. In perspective drawing, you do not have a well-defined coordinate system. Parallel lines tend to converge to a vanishing point as they recede from the observer. Scales used on the different axes are different in order to foreshorten the lines located some distance from the picture plane. The use of converging lines instead of parallel lines and the foreshortening of select dimensions give the drawing perspective. A prospective drawing is similar to a photograph of an object.

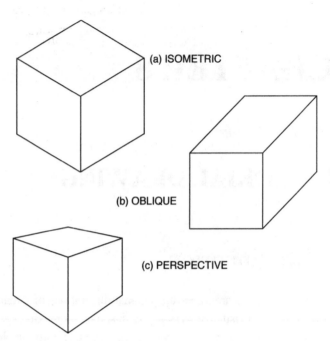

(a) ISOMETRIC

(b) OBLIQUE

Fig. 8.1 A cube represented with isometric,
cavalier oblique and perspective pictorials.

(c) PERSPECTIVE

8.2 ISOMETRIC DRAWINGS

In this discussion of the three forms of pictorial drawings, let's consider the following three characteristics of three-dimensional, isometric drawings:

1. The axes used to frame a drawing.
2. The direction of viewing the object.
3. The dimensions used to draw width, height and depth.

Let's begin with the isometric axes shown in Fig. 8.2. The isometric pictorial employs axes that make 120° angles with each other. The axes divide the paper into the three zones utilized to present three-views. If the axes form the letter Y with the vertical line oriented downward from the two branches, you are looking downward toward the object. From this perspective, you visualize the top view in the region between the branches of the Y as shown in Fig. 8.2. The front view is displayed in the region to the right of the vertical axis, and the left-side view is located in the region to the left of this axis.

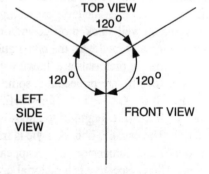

Fig. 8.2 Isometric axes at 120° angles give three regions for the
front, top and left-side views.

The axes and the direction of viewing the object have been defined for isometric coordinates in Fig. 8.2. Next consider the dimensions used on an isometric drawing for the width, height and depth. The word isometric means equal measurement; hence, it is clear that the same scale is used along all three axes. To illustrate the techniques used to prepare isometric drawings, consider a simple rectangular block with width W, height H, and depth D. An isometric pictorial of this rectangular block is presented in Fig. 8.3. To draw an isometric pictorial of a rectangular block, follow the procedure outlined below:

1. Use a 30°– 60°– 90° triangle to draw the isometric axes identified with the numbers 1, 2, and 3 as shown in Fig. 8.3.
2. Measure the length of H from the origin down the vertical axis to establish point A.
3. From point A, draw two additional lines (numbers 4 and 5) parallel to lines number 1 and 2.
4. Along line 5, measure the width W locating point B. Similarly measure the depth D along line 4 to locate point C.
5. From points B and C, draw the vertical lines 6 and 7 that intersect lines 1 and 2 and locate points E and F.
6. From point F, draw line 8 parallel to line 2, and from point E draw line 9 parallel to line 1.
7. Lines 8 and 9 intersect at point G to complete the isometric drawing.

The isometric pictorial that is illustrated in Fig. 8.3 shows the left-side view, the top view and the front view because the block is being viewed from above with the vertical edge bisecting the left side and the front side views. The origin of the isometric coordinates is positioned at the upper left hand corner of the rectangular block.

The procedure for preparing isometric drawings is easy to implement; the lines are all parallel to the isometric axes and the measurements are all made to the same scale.

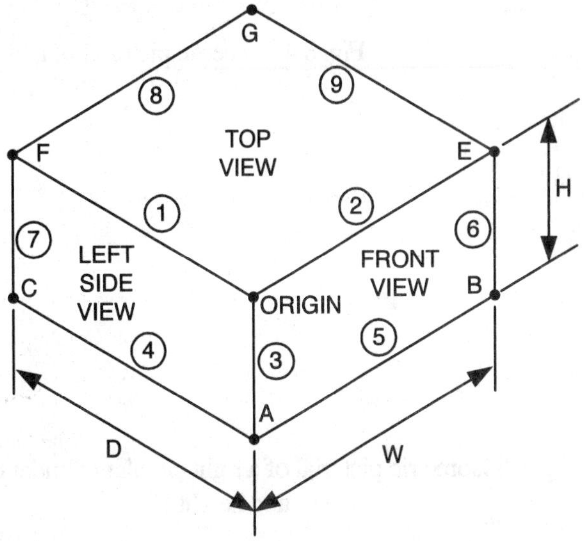

Fig. 8.3 Isometric pictorial of a rectangular block.

The example of a rectangular block was too easy. Let's try a more complex geometry as shown in Fig. 8.4. Examine the three-view drawing of an odd shaped block illustrated in Fig. 8.4. It shows an unusual three-view drawing because it presents the left-side view instead of the more conventional right-side view. The three-view drawing is presented in this manner because the isometric drawing shows the same three-views. For the simple rectangular block, presented in Fig. 8.3, the origin of the isometric axes was placed at the upper left hand corner of the front view. For the more complex geometry shown in Fig. 8.4, the origin is again placed at the same location.

The isometric pictorial, presented in Fig. 8.4, is drawn on isometric paper with only a straight edge to guide the lines. With a steady hand, you may sketch the lines without the straight edge. Isometric paper gives many evenly spaced lines parallel to the isometric axes eliminating the need for the 30° triangle. Begin at point B and draw plane 1 in the region for the left view. Do not draw the entire left-side view, because there is a taper to the block that complicates this view in its isometric representation. Defer completing the left-side view, and move to the top view and draw plane 2. Both planes 1 and 2 are clearly defined in the three-view drawing and are easy to construct on the isometric pictorial. Again the top view is not completed because the step and the taper complicate the geometry. Move next to the front view and add plane 3, which defines the depth of the step. Now return to the top view and draw plane 4, as shown in Fig. 8.4. Because the top view is now complete in the isometric drawing and the height of the step is established, it is easy to draw plane 5. Note plane 5 is on the surface formed by the taper; it does not lie in a plane formed by the isometric axes. This non-isometric plane is located by first drawing the four well defined isometric planes on the pictorial. The location of plane 5 is then clearly established by the position of planes 1 and 4.

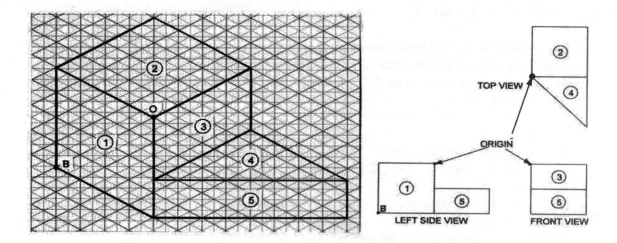

Fig. 8.4 Isometric pictorial of a block with a step and a tapered side.

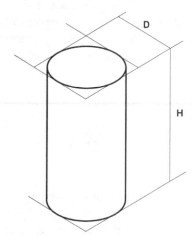

Fig. 8.5 Isometric pictorial of a right circular cylinder of diameter D
and height H.

You now understand the procedure for drawing isometric pictorials of simple rectangular blocks and more complex blocks with both a step and a taper. Let's next consider a cylinder of diameter D and height H and illustrate it in an isometric pictorial. To draw the cylinder, lay out two isometric axes located a distance H apart using light construction lines as indicated in Fig. 8.5. On the upper set of axes, draw a square in the top plane with the length of the sides of the square equal to the diameter of the cylinder. Then select an isometric ellipse (35° - 16') from an ellipse template, and draw an isometric ellipse so that it is tangent to each side of the square drawn in the top plane. (If you do not have one of these handy templates, sketch the ellipse in by hand). Note that the ellipse is tangent to the isometric axes at four points as indicated in Fig. 8.5. Move down to the second set of isometric axes, which represent the bottom plane. Take the template and draw an isometric ellipse again. This time draw only the front half of the ellipse, because only that portion is visible when viewing the cylinder from above. The two ellipses are joined with vertical lines at their outer most points to complete the cylinder. The construction lines in Fig. 8.5 have been retained to aid you in understanding the procedure used in preparing this drawing. To finish the isometric representation, erase these construction lines and shade the cylinder to enhance the visual effect of a three-dimensional object. Shading and shadows will be discussed later in this chapter when this isometric drawing of a cylinder will be referenced again.

8.3 OBLIQUE DRAWINGS

Oblique pictorials and isometric pictorials are similar, because they both utilize parallel lines in constructing the three views. The difference between isometric and oblique pictorials is in the definition of the axes. In oblique drawings, an x, y, and z coordinate system is employed as illustrated in Fig. 8.6. The three coordinate axes divide the sheet into three regions for drawing the front, top and right-side views. With the axes defined as shown in Fig. 8.6, the object is viewed from above and to the right. The z-axis, which is the receding axis in Fig. 8.6, is drawn with a 45° angle relative to the x-axis; however, other angles such as 30° or 60° are often employed to construct the receding (z) axis.

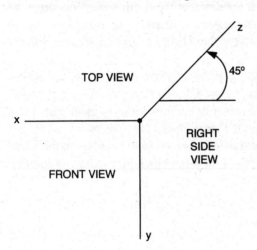

Fig. 8.6 Oblique axes use the x, y, and z coordinate system.

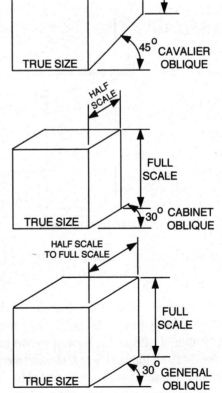

Fig. 8.7 A pictorial of a cube drawn with cavalier, cabinet and general oblique techniques.

Three different types of oblique drawings are frequently used for pictorials, as illustrated in Fig. 8.7:

1. Cavalier oblique is drawn with the receding axis at any angle from about 30° to 60°, but the measurements along all three axes are made with the same scale.
2. Cabinet oblique is drawn with the receding axis at any angle from about 30° to 60°, but the measurements along the z axis are half-scale.
3. General oblique is drawn with the receding axis at any angle from about 30° to 60°, but the measurements along the z axis vary from half to full-scale.

If you examine the cube represented by the three types of oblique drawings in Fig. 8.7, it is evident that full-scale (true length) measurements are used in the front view in all three types of oblique pictorials. The difference among them is the scale employed along the receding (z) axis. In cavalier oblique, full-scale measurements are made along the receding axis; however, this scale produces a drawing that is out of proportion. The cube does not look like a cube.

In cabinet oblique, half-scale measurements are made along the receding axis. The resulting drawing is in better proportion than the cavalier oblique, but sometimes it appears that the depth dimension along the receding axis is too short. The general oblique where the measurement on the receding axis can be varied from half-scale to full-scale is preferred. The scale is adjusted between these limits to give what appears to the eye to be the correct proportions.

To illustrate the procedure followed in drawing an oblique pictorial view, examine the three-view drawing of a pair of the connected rectangular blocks as shown in Fig. 8.8. To begin an oblique pictorial, draw the x, y, z-axes using point O, located at the lower left-hand corner in the front view, as the origin. First, draw part of the front view (plane 1) corresponding to the front of the large block. Next, draw part of the top view (plane 2) of the large block. In drawing the front view, use true lengths (on plane 1) to layout the width and height of the large block. On the top view, establish the depth of the large block maintaining proportion by using a scaling factor of ¾.

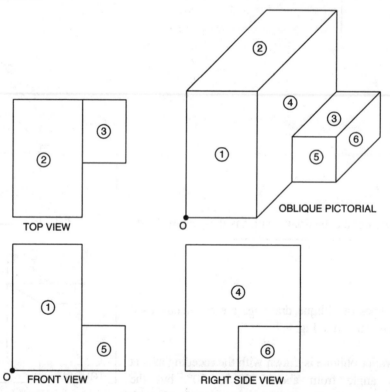

Fig. 8.8 Three-view drawing of a connected pair of rectangular blocks together with an oblique pictorial of the same object.

Note the smaller block obstructs the right-side view of the large block. This obstruction is handled by drawing the small block. Using the information in the three-view drawing, you can place plane 3 on the oblique pictorial. Plane 3 is located by working from the back edge in the top view and the top edge in the front view. The width and height measurements required to locate and size plane 3 in the oblique pictorial are true lengths, but the depth is again scaled by ¾. After you have drawn plane 3 locating the small block in the pictorial, it is easy to draw plane 4 by referring to the right-side view in the three-view drawing. Complete the drawing of the small block by dropping vertical lines down from the three corners in plane 3 and by closing the sides that form planes 5 and 6.

The resulting oblique pictorial clearly captures the relative proportions and positioning of the two blocks. A comparison of the pictorial with the three-view drawing demonstrates the advantages of the pictorial in visualizing the object—the pictorial is much more effective. The three-view drawing is employed for the precise definition of size and location of all of the features of a component or structure. As such it is dimensioned so that it can be used in the shop for manufacturing. The pictorial is not usually dimensioned because it is not a substitute for a detailed three-view drawing.

For two final examples of preparing oblique pictorials, consider the drawing of a cylinder and a block with a circular hole as shown in Fig. 8.9. It is easy to draw a cylinder or a circle on an oblique pictorial providing the required circles are placed on either the front plane or any plane parallel to the front plane, because both the width and height dimensions are true length. In addition the x-y axes are orthogonal on the front view; hence, the circle is not distorted into an ellipse. You can draw the circle with a compass or a circle template that is a significant advantage.

To represent the cylinder shown in Fig. 8.9a, draw two circles centered at points A and B. Both points A and B lie on the z (receding) axis, which has been drawn at a 45° angle to the x-axis. The circles are drawn with construction lines using full-scale diameters. The spacing of the two circles along the z-axis is ¾ scale relative to the length of the cylinder. Draw two lines parallel to the z-axis tangent to the circles for the sides of the cylinder. Draw the front circle with a heavy line, and darken the visible portion of the back circle. The resulting pictorial shows a somewhat distorted cylinder. The distortion is due to the fact that the back circle has been drawn with the same diameter as the front circle. When observing an actual right circular cylinder, the back circle would "appear" to be smaller than the front circle. The apparent difference in the size of features located in different planes will be addressed in the next section on perspective drawing.

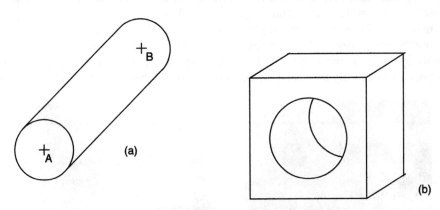

Fig. 8.9 Examples of oblique pictorials.

The final example of oblique pictorial, presented in Fig. 8.9b, shows a simple rectangular block with a central hole. The block has been drawn so that the circle of the hole is observed in its entirety in the front view. The question is how to draw the circle in the back plane? First draw the square locating the back plane in the pictorial with light construction lines. Then find the center of this square and draw the second circle with the same diameter as the front circle. Note, the two circles intersect and only a portion of the circle on the back plane is visible. Darken the construction lines on the back circle over that part of the arc that is visible through the hole. The resulting pictorial is an effective three-dimensional drawing showing the appearance of a hole through a rectangular block. Artists would critique the drawing, because it too distorts the visual image of the object. The scale used for both the back and front planes are the same producing distortion of the image. This distortion is usually acceptable in engineering drawing, but better pictorials are possible with perspective drawings.

8.4 PERSPECTIVE DRAWINGS

Perspective drawings are pictorials that represent observations similar to those made with our eyes or a camera. This method of illustration is critical to the success of an artist making either sketches or paintings. Engineers also use perspective drawings particularly when preparing visuals for those who are not trained to read more conventional three-view drawings. The drawing instruments used are the same as those described previously except for adding a thumbtack and a piece of string to the list.

There is one very significant difference between perspective drawings and isometric or oblique drawings. In both isometric and oblique drawings, the lines defining the edges of an object are parallel to the axes; however, in perspective drawing, the lines defining the receding edges are not parallel. Drawing parallel lines distorts the drawing, because parallel lines appear to converge as they recede in space. The best illustration of this fact is a pair of railroad tracks, shown in Fig. 8.10. Looking down the tracks, you see that the tracks converge to a point. Also the railroad ties and the trees and poles lining the track appear to become shorter in the distant view. Everyone knows that the tracks are parallel. What is going on? To understand what is happening, it is necessary to define four terms used in describing prospective drawing.

1. A picture plane is the surface (i. e. the sheet of paper) of the pictorial. The edges of the paper represent the window through which you "see" a three-dimensional object that is the subject of the perspective drawing.
2. A horizon line divides the sky and the land (or the sea) if you are outdoors. The horizon line is at the level of your eyes and will change with your elevation. In a room the true horizon cannot be located because the walls of the room block the view. In this case, it is assumed that the horizon line is at eye level.
3. A viewing point and direction of view depends on the location of eyes relative to the object. You can look directly at the object, from left to right, right to left, downward, upward, etc. What you see changes markedly depending on these parameters. Look at an object from a window, and change where you stand and the direction of your view. Does the view of the object change?
4. A vanishing point is where parallel lines converge to a point as they recede into the distance. You can clearly identify the vanishing point in Fig. 8.10 where the tracks appear to meet.

Fig. 8.10 Railroad tracks illustrate the visual effect of converging parallel lines
and foreshortening of distant lines.

One Point Perspective Drawings

Depending on the view, an object can be represented with a one, two or three-point perspective. Let's start with the simple one-point perspective and illustrate the approach by drawing a simple rectangular block. In one-point perspective, you place the front view in the picture plane and show its true width and height as illustrated in Fig. 8.11. Then a construction line is drawn to represent the horizon. The location of this line depends on the viewing point and the viewing direction. In Fig. 8.11, you are viewing the block straight-on (not from the right or the left); however, you are above the block. Your eyes look downward and observe the top surface of the block. The elevation of your eyes relative to the top of the block is taken into account by the location of the horizon line. Next, locate the vanishing point on the horizon line at the center point behind the front view, because you are looking straight-on at the block. Draw construction lines from the vanishing point to the top corners of the block in the front view as shown in Fig. 8.11. The back edge on the top view is drawn parallel to the front top edge to establish the depth of the block. Note that the back edge is much shorter in length than the front edge. The shortening of the lines on the recessed planes gives an illusion of the third dimension. The edges of the sides are darkened to complete the pictorial. Again these edges are converging giving an illusion of depth on the two-dimensional sheet of drawing paper.

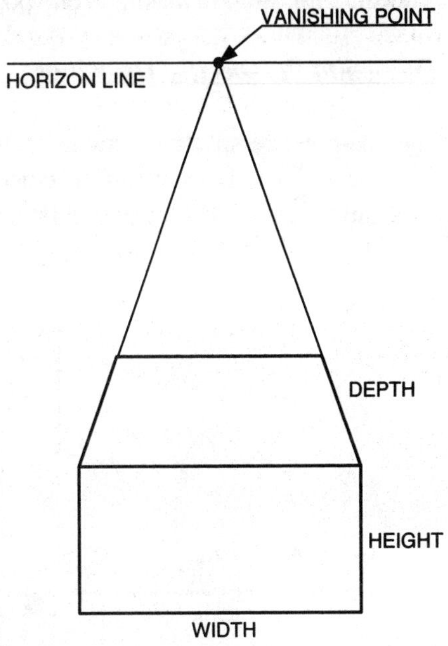

Fig. 8.11 An example of a pictorial drawing of a rectangular block with one-point perspective.

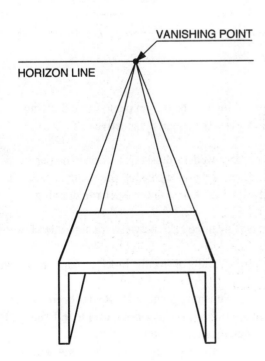

Fig. 8.12 A one-point perspective drawing of a coffee table.

Another example of one-point perspective is the drawing of a coffee table presented in Fig. 8.12. Again the front of the table is drawn to scale in the picture plane. The width and height dimensions are true lengths. Coffee tables are low, so when standing it is necessary to look downward to view the table. In this example, you are centered relative to the table and looking downward at its top surface. To develop a perspective view with this direction of viewing, draw a horizon line at eye level aligning the vanishing point with the center of the table. Next, draw light construction lines from the top outside corners of the table to the vanishing point. Also draw construction lines from the lower inside corners of the table legs as shown in Fig. 8.12. These construction lines form two triangles. The larger of these two triangles is used to provide the length of the back top edge of the table. Next, draw the edges of the tabletop to complete the perspective rendering of the top view. The smaller of the two triangles is used as a guide in drawing the bottom edges of the legs that are visible under the table.

In a normal perspective drawing, the construction lines, the vanishing point and the horizon line would be erased. These drawing aids have been retained in Fig. 8.12, because they were used to illustrate the procedure followed in making a one-point perspective drawing.

Two-point Perspective Drawings

The one-point perspective drawing is useful when viewing an object straight-on with its front view shown in the picture plane. However, if the object is rotated so that neither the front or side view is in the picture plane, as illustrated in Fig. 8.13, a two-point perspective is required.

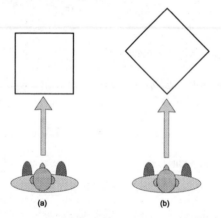

Fig. 8.13 Directions of viewing control the number of vanishing points used in perspective drawing.
a. Straight-on—one-point perspective.   b. Angle view—two-point perspective.

Again consider a rectangular block to illustrate the approach followed in drawing a two-point perspective pictorial. From Fig. 8.13, it is evident that only one vertical edge of the block lies in the picture plane. You start with this fact and proceed step by step to illustrate the block with a two-point perspective drawing.

- Draw a vertical line, as indicated in Fig. 8.14, to establish the edge between the left-side view and the front view.
- Draw the horizon line to reflect the fact that you are standing above the block looking downward onto the top surface.
- Place two vanishing points on the horizon line. The vanishing point (VP - R) has been located to the right of the vertical line farther from this line than the vanishing point on the left-side (VP - L), because the block is being viewed at a slight angle from the right toward the left.
- Measure the true length of the vertical line (line 1) and label its ends with the letters A and B.

- Draw four construction lines connecting points A and B with VP - R and VP - L.
- Draw vertical lines (2 and 3) to locate the back edges of the left side and the front of the block. The ends of these lines, located by the construction lines, are labeled (C, D) and (E, F).
- Draw top edge lines (4, 5) by connecting points F, A and D.
- Draw bottom edge lines (6, 7) by connecting points E, B and C.
- Draw construction lines from point F to VP - R and from point D to VP – L. Then locate point G at the intersection of these two lines.
- Draw edge lines (8, 9) by connecting points F, G, and D.

The two-point perspective drawing of the rectangular block is complete as shown in Fig. 8.14. The procedure may seem long and tedious, when it is described in a sequence of steps, but it is not difficult. When you understand the basic concepts of establishing the true length of the vertical line in the picture plane and the vanishing points on the horizon line, the remaining steps are routine.

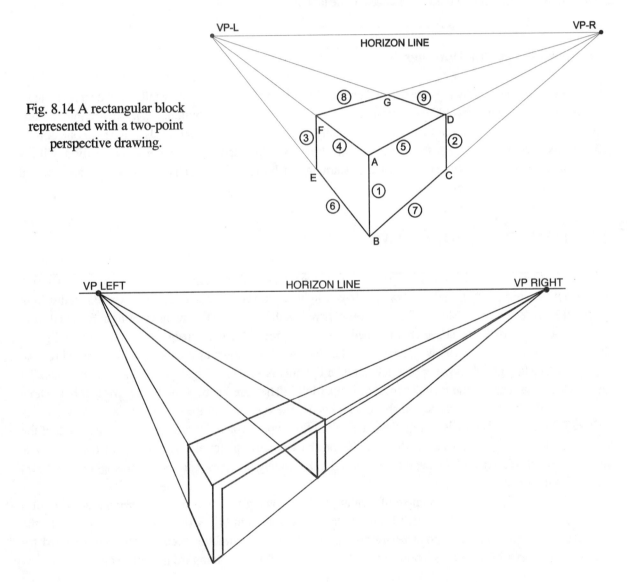

Fig. 8.14 A rectangular block represented with a two-point perspective drawing.

Fig. 8.15 A two-point perspective drawing of a coffee table.

To complete the discussion of the two-point perspective, let's again consider the coffee table. This time, a two-point perspective drawing will be used to illustrate this table. Begin by placing the vertical edge of a table leg in the picture plane as indicated in Fig. 8.15. The edge of the leg is shown in true length on the picture plane. Establish the horizon line and the vanishing points to provide the direction of the view that you are showing. Draw construction lines from the ends of the vertical line to the vanishing points. Vertical lines are then drawn to show the width and the depth of the table. The location of these vertical lines is not established by measurement, because both the width and depth dimensions are not true length in this two-point perspective. Place the vertical lines in a location that maintains—**to the eye**—the correct proportions of the table.

When the vertical lines that bound the left-side view and the front view are drawn, the remainder of the drawing is easy to complete. Construction lines are drawn from the ends of these vertical lines to the vanishing points. The construction lines outline the table. It is only necessary to darken those segments of the construction lines that define the visible edges of the table.

Three-point Perspective Drawings

Three-point perspective drawings are used when the object being illustrated is very tall. Architects drawing a city view with skyscrapers would use three-point perspectives and taper the building as it extends into the sky. Engineers usually deal with smaller objects that usually can be represented in pictorials with either one or two-point perspective drawings. For this reason, the methods used in preparing three-point perspectives will not be described. However, if you are interested in learning much more about perspective drawing, an excellent book by Powell [2] is recommended.

8.5 SHADING AND SHADOWS

Adding shading and shadows to the pictorial drawings makes them appear more realistic. Let's first distinguish between shading and shadowing. If you light an object, some surfaces are exposed to this light and other surfaces are in the shade. Consequently, there is a difference in the intensity of light reflected from these surfaces. Those in the shade are darkened to some degree. When an opaque object blocks the light, a shadow is produced. The shadow occurring on the floor plane is shown as a dark area in the drawing. An example of shading and shadowing of a right circular cylinder is shown in Fig. 8.16. In this example, parallel light rays are illuminating the cylinder from the upper left. (They are included on the drawing only to show the logic for determining the shade and shadow regions). The right half of the cylinder is in the shade and is darkened slightly. A shadow is cast on the plane of the floor upon which the cylinder rests. The depth of the shadow is the same as the depth of the cylinder on the isometric pictorial. The length of the shadow is dependent on the direction of the parallel rays of light. Extending the lines representing the light rays defines the edge of the shadow.

Let's consider another example of shading and shadowing in a two-point perspective drawing of a cube. The drawing, presented in Fig. 8.17, is similar to that shown in Fig. 8.14, so you may assume that the cube can be drawn with a two-point perspective pictorial. The construction technique is demonstrated for determining the exact size of the shadow in Fig. 8.17, when the light is coming from a point source.

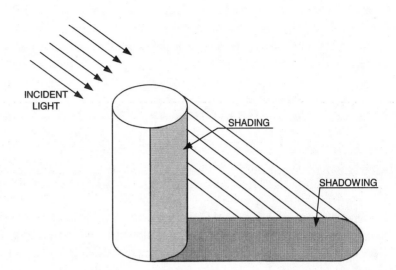

Fig. 8.16 Shading and shadowing of
an isometric pictorial of a right
circular cylinder.

Suppose you have completed the two-point perspective drawing of the cube and are ready to shade and shadow the drawing. To begin select the location of the light source. It is your choice, but you should place the light source well above the horizon either to the left or the right of the cube. Second, you must select the vanishing point for the shadow. Again it is your choice as long as it is directly below the light source and in the ground (floor) plane. The reasons for these two constraints are evident. The shadow must vanish when the light source is directly overhead, and the shadow must always lie on the ground (floor) plane. In Fig. 8.17, the vanishing point of the shadow was placed on the horizon line; however, it could have been located anywhere on the ground plane under the light source.

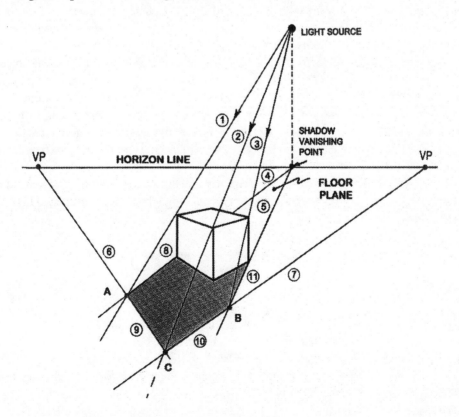

Fig. 8.17 Technique for determining shadow outline on a two-point perspective drawing.

Consider a step-by-step procedure for locating the shadow outline shown in Fig. 8.17:

1. Draw the light rays (1, 2, and 3) from the source through the three forward corners on the top of the cube.
2. Draw two construction lines (4 and 5) from the vanishing point of the shadow through the two bottom-side corners.
3. Locate points A and B at the intersections of the lines (1 and 4) and (3 and 5).
4. Draw lines 6 and 7 from the left and right vanishing points through points A and B.
5. The intersection of lines (6 and 7) locates point C.
6. Darken segments of the construction lines (8, 9, 10 and 11) to give the shadow outline.
7. Erase all of the construction lines and darken the shadow region within the outline.

The shadow has been added to the drawing, but the front and left-side views of the cube are in the shade. To complete the pictorial, these two sides should be darkened slightly indicating they are in a shadow.

8.6 SUMMARY

Three different pictorial drawings—isometric, oblique, and perspective—have been introduced. Both isometric and oblique pictorials are widely used in engineering to present three-dimension illustrations on a two-dimensional format (a sheet of drawing paper or a computer monitor). The isometric and oblique pictorials are easy to draw because lines are parallel, and most, if not all, of the measurements are to scale. However, exact scaling of dimensions and parallel lines tend to distort these three-dimensional illustrations. The eye or camera "sees" dimensions with different scales, and parallel lines as parallel only if the view is in the picture plane. If the view is on any other plane behind the picture plane, the dimensions are foreshortened. If the parallel lines are on views with receding axes, they tend to converge. These distortions are ignored in engineering drawings when preparing isometric or oblique pictorials.

Perspective drawings provide a more realistic representation of three-dimensional objects. Artists almost always use perspective concepts in their sketches or paintings. In engineering, one-point or two-point perspectives are usually employed to produce realistic three-dimensional drawings that are used in catalogs and maintenance manuals. The techniques of perspective drawing require understanding of four quantities: the picture plane, the horizon line, vanishing point or points, and viewing direction. The techniques for drawing perspectives with converging lines and foreshortened dimensions follow directly from the definitions of these four quantities.

Shading and shadowing is an added technique for making pictorial drawings even more realistic. Methods for determining the outline of shadows either from point light sources or collimated light sources that produce parallel light rays have been described.

REFERENCES

1. Franks, G. Pencil Drawing, Walter Foster Publishing, Laguna Hills, CA 1988.
2. Powell, W. F. Perspective, Walter Foster Publishing, Laguna Hills, CA 1989.
3. Earle, J. H., Engineering Design Graphics, Addison Wesley, Reading, MA 1983.
4. Bertoline, G. R., and E. N. Wiebe, Fundamentals of Graphics Communication, 3rd Ed., McGraw Hill New York, NY, 2002
5. Lockhart, S. D. and C. M. Johnson, Engineering Design Communication, Prentice Hall, New York, NY, 1999.

6. Jensen, C. H. and J. D. Helsel, <u>Engineering Drawing and Design</u>, Glencoe McGraw Hill, New York, NY, 1997.

7. Sorby, S. A., Manner, K. J., Baartmans, B. J. and S. S. Sorby, <u>3-D Visualization for Engineering Graphics,</u> Prentice Hall, New York, NY, 1998

8. Giesecke, F. E. et al, <u>Engineering Graphics</u>, 8ᵗʰ Ed., Prentice Hall, New York, NY, 2003.

EXERCISES

8.1   What are the three types of pictorial drawings? Why are pictorial drawings used in engineering applications?

8.2   Prepare an isometric pictorial drawing of a rectangular block 50 mm wide by 35 mm deep by 30 mm high. What scale should you use in preparing this drawing? Why?

8.3   Prepare an isometric pictorial drawing of a right circular cylinder 30 mm in diameter by 60 mm long. What scale should you use in preparing this drawing? Why?

8.4   Prepare an oblique pictorial drawing of a rectangular block 25 mm wide by 50 mm deep by 75 mm high. What scale should you use in preparing this drawing? Why?

8.5   Prepare an oblique pictorial drawing of a right circular cylinder 25 mm in diameter by 60 mm long. What scale should you use in preparing this drawing? Why?

8.6   Prepare an isometric pictorial drawing of a rectangular block with a centrally located circular through hole. State the dimensions of the block and the diameter and orientation of the hole.

8.7   Prepare an oblique pictorial drawing of a rectangular block with a centrally located circular through hole. State the dimensions of the block and the diameter and orientation of the hole.

8.8   Prepare an isometric pictorial drawing of a component for a component that your team is developing.

8.9   Prepare an oblique pictorial drawing of a component for a component that your team is developing.

8.10   Prepare a prospective pictorial drawing of the building that houses the Dean's offices. View the building from one of its front corners.

8.11   Draw a pictorial illustration of a sphere with shading and shadowing.

Notes

CHAPTER 9

COMPUTER AIDED DESIGN (CAD)

9.1 INTRODUCTION TO CAD

A product consists of components. In the design process, individual components are constructed first. Afterwards, the designed components are assembled. In this chapter, we introduce a CAD software system to create the individual components and assemblies. The selected CAD software system is Pro/ENGINEER Wildfire, Version 3.0.

There is a variety of hovercraft models, which have been designed and constructed. The following picture shows a hovercraft model. The 3 maximum dimensions, or the dimensions of the total height, the total width and the total length, are 206, 280 and 540 mm, respectively.

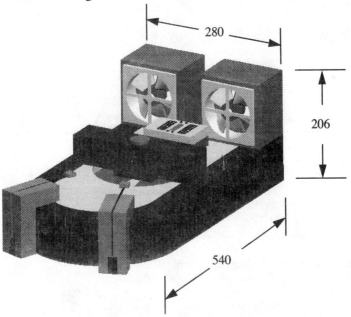

As illustrated in the above picture, the hovercraft model has a lift fan, which is located in the central part. There are 2 propulsion fans located in the rear part. The lift fan and propulsion fans are all attached to the base structure. To autonomously control the motion of the hovercraft, 2 sensors and a micro-computer or a RCX box are used. The sensors and the RCX box are also attached to the base structure.

In the following, we will present the steps and procedures for constructing this hovercraft model using Pro/ENGINEER. To facilitate the presentation, we decompose this hovercraft model into 4 sub-systems. These 4 sub-systems are:

1. The base structural system

2. The lift fan system where an axial fan or a centrifugal fan may be used.

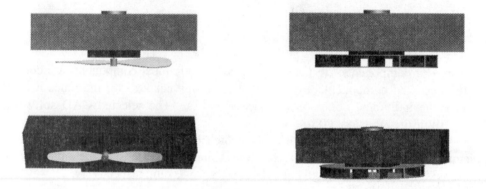

3. The propulsion fan system.

4. The sensor-based control system.

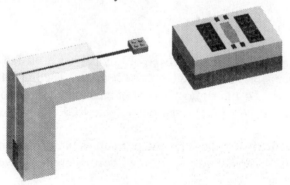

In the following sections, we present the procedure to create each of the 4 subsystems, starting with the base structural system. For each sub-system, we present an exploded view first, illustrating all components used in the sub-system. Afterwards, we present a step-by-step procedure to create a 3D solid model for each component. Finally, we assemble the created 3D solid models to construct a 3D solid model of the subsystem. When the 3D solid models of the 4 subsystems are available, we assemble the four (4) 3D solid models of subsystems to construct a 3D solid model of the hovercraft model.

9.2 CREATION OF THE BASE STRUCTURAL SYSTEM

Figure 9-1 is an exploded view of the base structural system. As illustrated, this subsystem has a top_plate component, a bottom_plate component, 3 support-bar components and a skirt component.

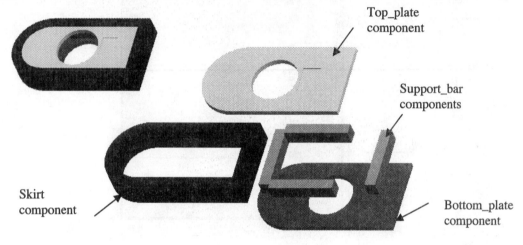

Fig. 9.1 An Exploded View of the Base Structural System Showing the 4 Components.

We begin with creating the support_bar component to introduce the Pro/ENGINEER design system. Afterwards, we construct the top_plate component, the bottom_plate component and the skirt component. Finally, we assemble them together.

Creation of the Support_Bar Component

To start the design process using Pro/ENGINEER, you need to launch the CAD system. Click the **Start** menu > **Program > PTC > Pro ENGINEER**, or just click the displayed icon of **ProE**.

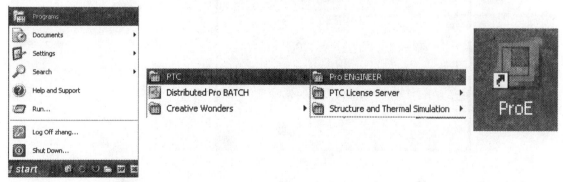

Step 1: select the icon, called "**Create a new file**", which is displayed on the menu bar. Make sure **Part** is selected. Type *support_bar* as the name of the file > clear the icon of **Use default template > OK**.

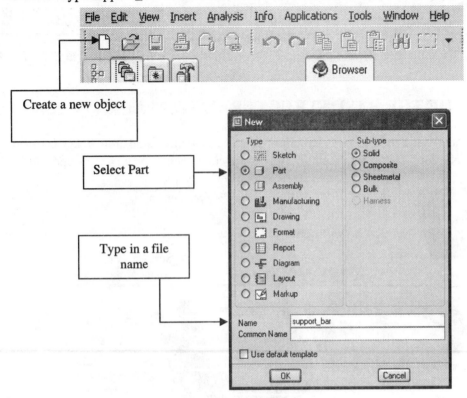

Select mmns_part_solid (units: Millimeter, Newton, Second) and type *support_bar* in **DESCRIPTION**, and *student* in **MODELED_BY**, then **OK**.

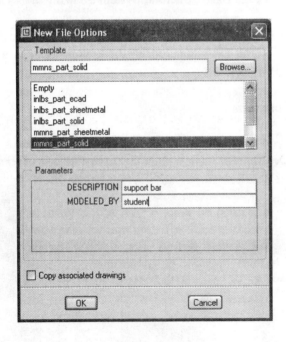

The design screen will be on display, as shown below.

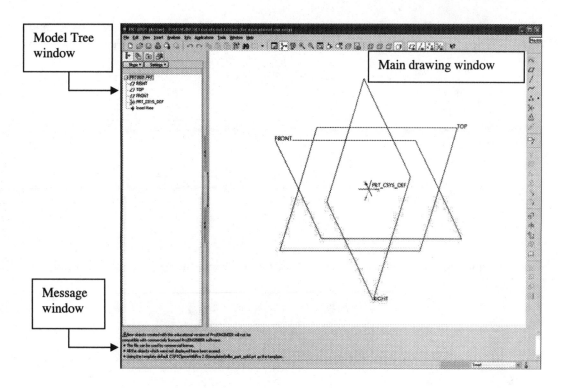

Model Tree window

Main drawing window

Message window

Directly select the icon of **Extrude.** From the dashboard, set the thickness value to 40. Activate **Placement** to define a sketch plane > **Define.**

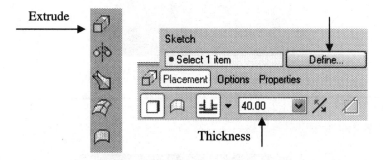

Extrude

Thickness

Select the **FRONT** datum plane as the sketch plane, and accept the default orientation.

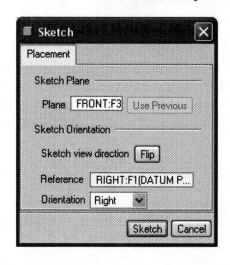

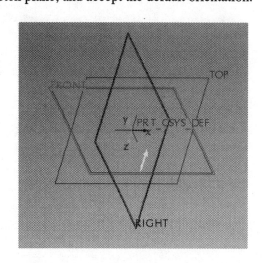

Click the icon of **Rectangle** and sketch a rectangle. The 2 dimensions are 50 and 240, respectively.

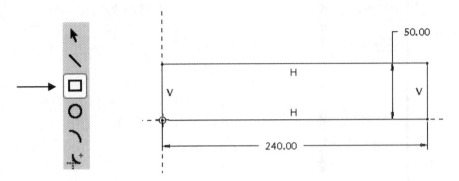

Upon completing this sketch, click the icon of **Done**. From the feature control panel, click the icon of **Complete**.

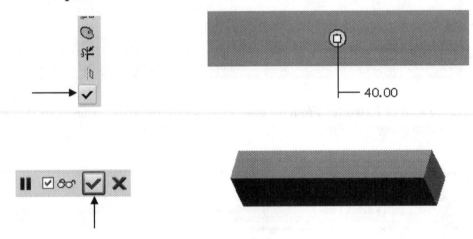

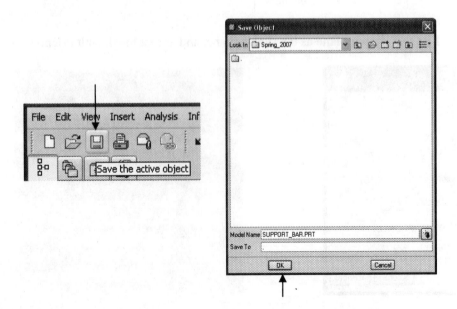

Step 2: Save the created support_bar component.
Click the icon of **Save the active object > OK**.

Creation of the Top_Plate Component

Step 1: Create a 3D solid model for the top plate component.

File > New > Part > type *top_plate* as the file name and clear the icon of **Use default template > OK**.

Select mmns_part_solid (units: Millimeter, Newton, Second) and type *top_plate* in **DESCRIPTION**, and *student* in **MODELED_BY**, then **OK**.

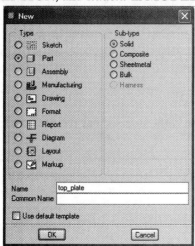 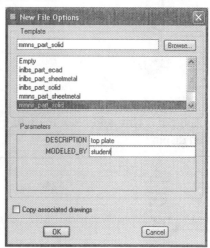

Directly select the icon of **Extrude.** From the dashboard, set the thickness value to 18. Activate **Placement** to define a sketch plane > **Define**.

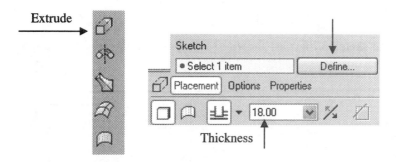

Select the **TOP** datum plane as the sketch plane, and accept the default orientation

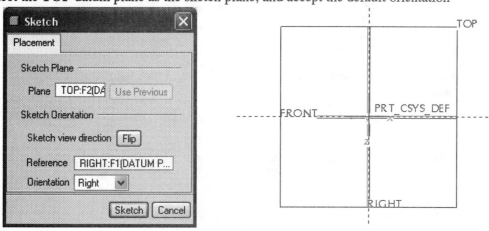

Click the icon of **Circle** and sketch a circle. Note the center of this circle at the origin of the coordinate system. Modify the displayed dimension to 280 by double clicking the displayed dimension.

Icon of Circle →

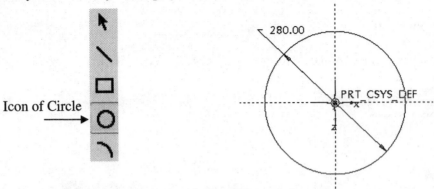

Click the icon of **Trim**, and click the half circle on the right side so that the half circle on the left side remains on display.

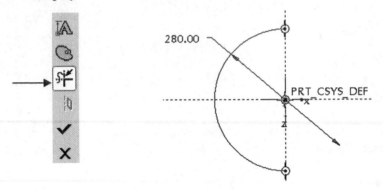

Click the icon of **Line** and sketch 3 lines, as shown. Modify the displayed dimension to 400 by double clicking the displayed dimension.

Icon of Line →

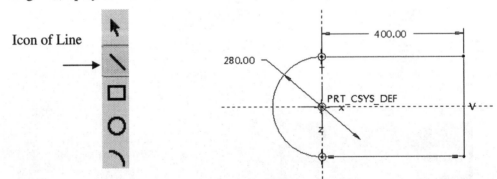

Upon completing this sketch, click the icon of **Done**. From the feature control panel, click the icon of **Complete**.

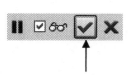

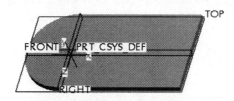

Step 2: Create a circular feature to accommodate the fan blade through a cut operation

Directly select the icon of **Extrude.** From the dashboard, activate the **Cut** operation, set **Thru All** as the depth choice. Activate **Placement** to define a sketch plane > **Define.**

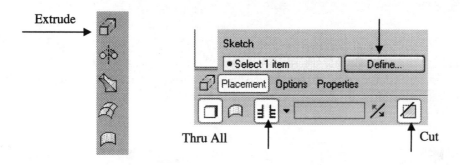

Select the top surface of the plate (do not select the **TOP** datum plane) as the sketch plane, and accept the default orientation.

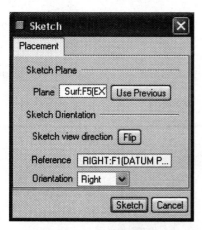

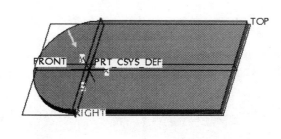

Click the icon of **Circle** and sketch a circle. The diameter value is *170* and the position dimension is *130*, as shown.

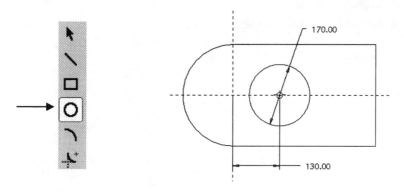

Upon completing this sketch, click the icon of **Done**. From the feature control panel, click the icon of **Complete**.

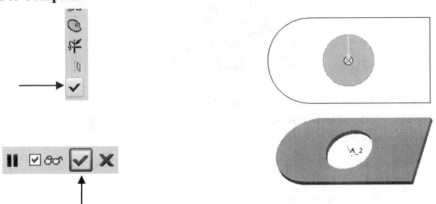

Step 3: Create a slot feature to accommodate the RCX, the size of which is 64 x 94 mm.
Directly select the icon of **Extrude.** From the dashboard, activate the **Cut** operation, specify 2 as the depth of cut. Activate **Placement** to define a sketch plane > **Define.**

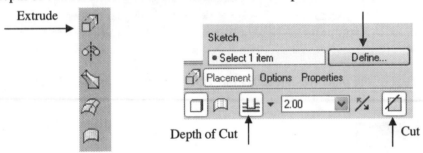

Select the top surface of the plate (do not select the **TOP** datum plane) as the sketch plane, and accept the default orientation.

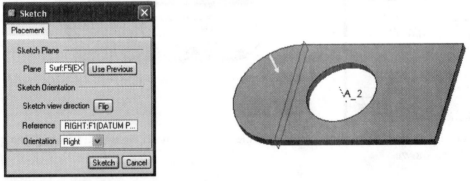

Right-click and hold, select **Centerline**.

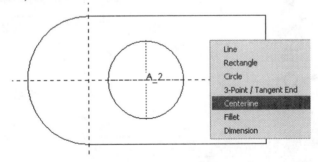

Sketch a horizontal centerline. Click the icon of *Rectangle* and sketch a rectangle, which is symmetric about the horizontal centerline. The 2 dimensions are 64 and 94, respectively.

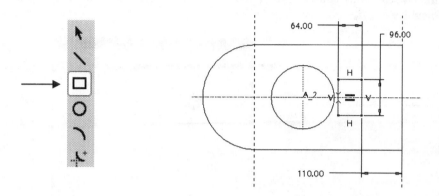

Upon completing this sketch, click the icon of **Done**. From the feature control panel, click the icon of **Complete**.

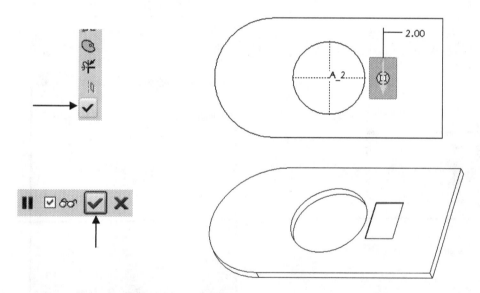

Step 4: Save the created top_plate component.
Click the icon of Save the active object > **OK**.

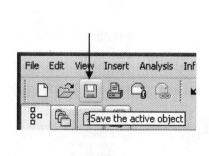
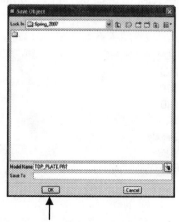

Creation of the Bottom_Plate Component

To create the bottom_plate component, click **File** > we use **Save a Copy** and type *bottom_plate* as the name of the new file > **OK**.

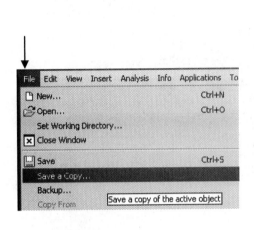

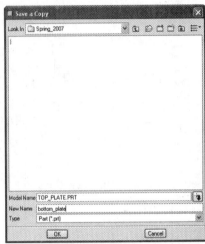

Now let us open the file: bottom_plate.prt. Click the icon of **Open an existing object** > highlight bottom_plate.prt > **Open**.

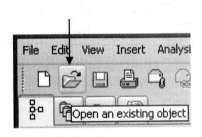

In the model tree, highlight the slot feature or highlight **Extrude 3** > right-click and hold, select **Delete > OK.** The slot feature does not appear on the display any more.

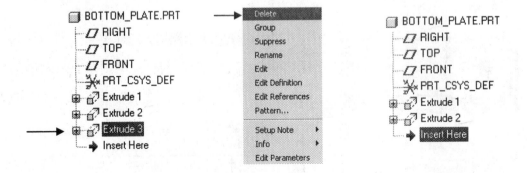

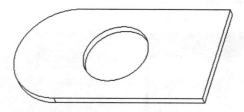

Now let us save the modified bottom_plate component.
Click the icon of **Save the active object > OK**.

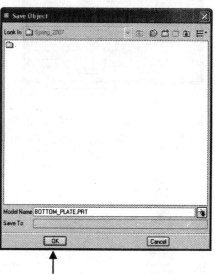

Creation of the Skirt Component

Step 1: Create a 3D solid model for the skirt component.

     File > New > Part > type *skirt* as the file name and clear the icon of **Use default template >**
OK.

     Select mmns_part_solid (units: Millimeter, Newton, Second) and type *skirt* in **DESCRIPTION**,
and *student* in **MODELED_BY**, then **OK.**

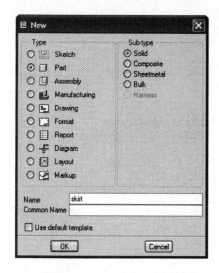

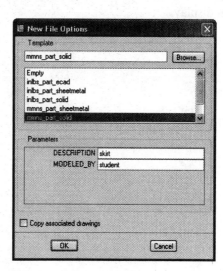

Click the icon of **Sketch Tool**. Pick the **TOP** datum plane as the sketch plane and accept the default setting for orientation > **Sketch.**

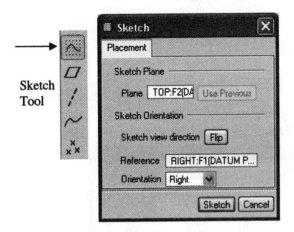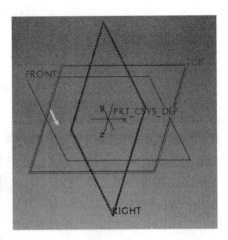

Click the icon of **Circle** and sketch a circle. Note the center of this circle at the origin of the coordinate system. Modify the displayed dimension to 280 by double clicking the displayed dimension.

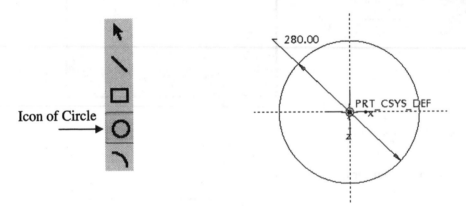

Click the icon of **Trim,** and click the half circle on the right side so that the half circle on the left side remains on display.

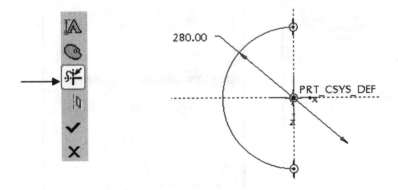

Click the icon of **Line** and sketch 3 lines, as shown. Modify the displayed dimension to 400 by double clicking the displayed dimension.

Icon of Line

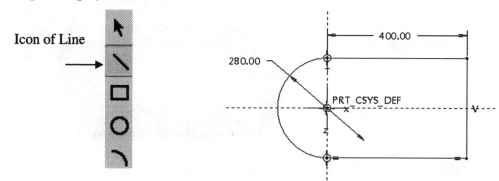

Upon completing this sketch, click the icon of **Done**.

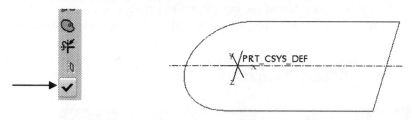

From the main menu, select **Insert > Sweep > Thin Protrusion > Select Trajectory**.
Select **Curve Chain** > pick the sketched trajectory > **Select All > Done**.
Click the icon of **Line** and sketch 3 straight lines, following the centerlines, as shown below. The 2 dimensions are 40 and 80, respectively.

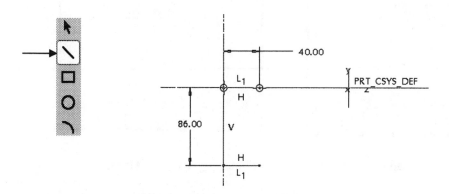

Upon completing, click the icon of **Done**.
For the side of material, **Flip > Okay** to ensure that the material is added to the outside of the skirt, as shown.

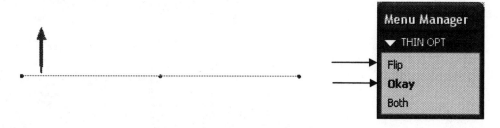

For the thickness of the skirt material, type *0.5* and press the **Enter** key > **OK** to complete the creation.

Creation of the Assembly of the Base Structural System

Step 1: Create a 3D solid model for the assembly of the base structural system.

File > New > Assembly > type *base_structure* as the file name and clear the icon of **Use default template > OK.**

Select mmns_asm_design (units: Millimeter, Newton, Second) and type *base_structure* in **DESCRIPTION**, and *student* in **MODELED_BY**, then **OK.**

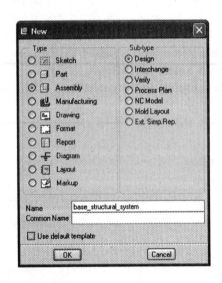
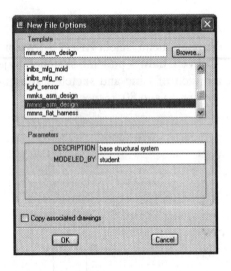

Click the icon of **Add component** > highlight bottom_plate.prt > **Open.**

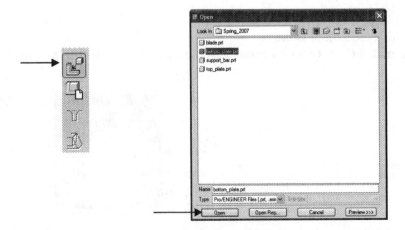

Select **Default** > click the icon of **Complete** from the constraint control panel.

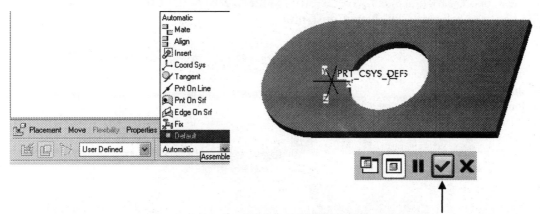

Click the icon of **Add component** > highlight support_bar.prt > **Open**.

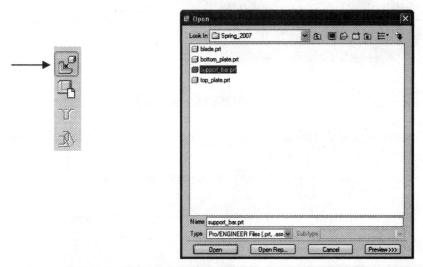

Let us insert the first constraint. Select the top surface from the bottom_plate component and select the bottom surface from the support_bar component. A constraint called **Mate** has been defined. Click **New Constraint**.

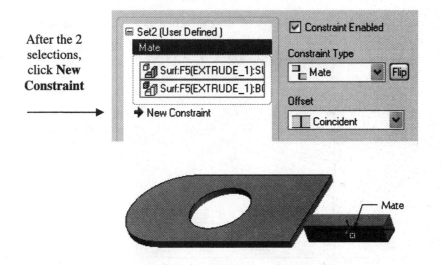

Let us insert the second constraint. Select the **ASM_RIGHT** and select the **RIGHT** datum plane from the support_bar component. Set the offset value to zero or use Coincident. A new constraint called **Align** has been defined. Click **New Constraint**.

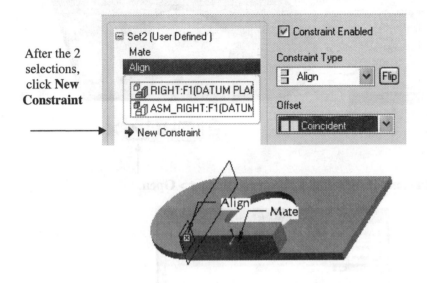

After the 2 selections, click **New Constraint**

Let us insert the third constraint. Select the **ASM_RIGHT** and select the **RIGHT** datum plane from the support_bar component. Set the offset value to 2. A new constraint called **Align** has been defined. At this moment, the support_bar component is fully constrained. Click the icon of **Complete** from the constraint control panel

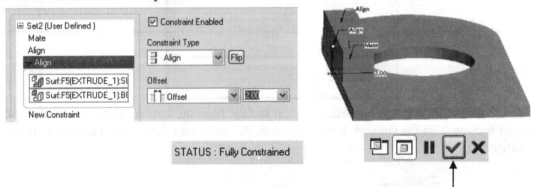

Now let us assemble the support_bar component on the other side. Click the icon of **Add component** > highlight support_bar.prt > **Open**.

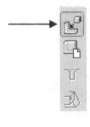

We follow the same procedure to insert 3 constraints as we did in the process of assembling the first support_bar component.

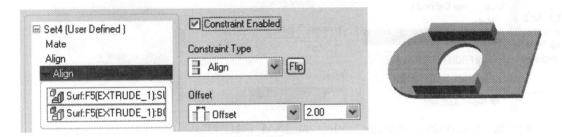

Now let us assemble the support_bar component at the front position. Click the icon of **Add component** > highlight support_bar.prt > **Open**.

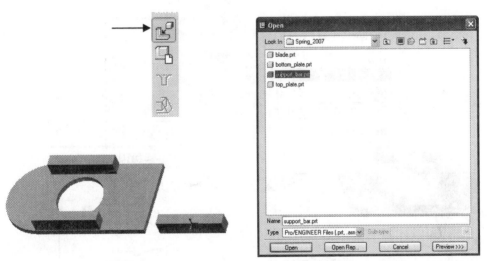

Let us insert the first constraint. Select the top surface from the bottom_plate component and select the bottom surface from the support_bar component. A constraint called **Mate** has been defined. Click **New Constraint**.

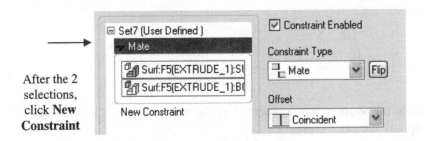

Let us insert the second constraint. Select the **ASM_RIGHT** and select the **FRONT** datum plane from the support_bar component. Set the offset value to zero or use **Coincident**. A new constraint called **Mate** has been defined. Click **New Constraint**.

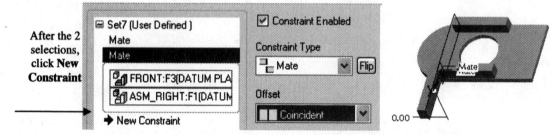

After the 2 selections, click **New Constraint**

Let us insert the third constraint. Select the **ASM_FRONT** and select the end surface from the support_bar component. Set the offset value to 120. A new constraint called **Align** has been defined. At this moment, the support_bar component is fully constrained. Click the icon of **Complete** from the constraint control panel

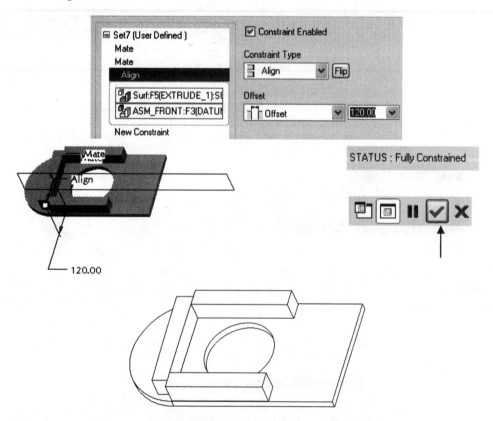

Now let us assemble the support_bar component at the rear position. Click the icon of **Add component** > highlight support_bar.prt > **Open**.

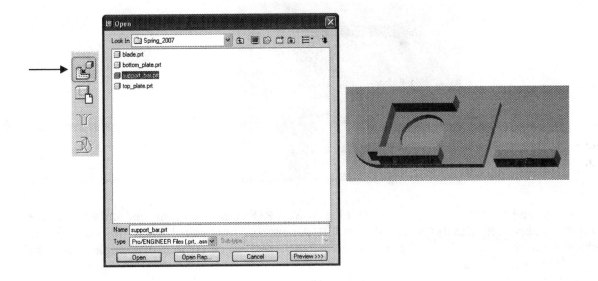

Let us insert the first constraint. Select the top surface from the bottom_plate component and select the bottom surface from the support_bar component. A constraint called **Mate** has been defined. Click **New Constraint**.

After the 2 selections, click **New Constraint**

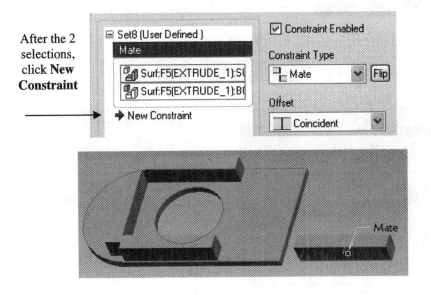

Let us insert the second constraint. Select the **ASM_RIGHT** and select the **FRONT** datum plane from the support_bar component. Set the offset value to 398. A new constraint called **Mate** has been defined. Click **New Constraint**.

After the 2 selections, click **New Constraint**

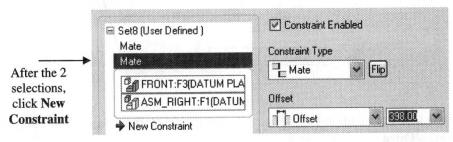

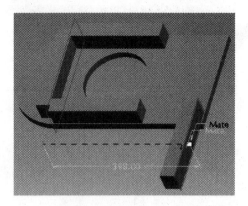

Let us insert the third constraint. Select the **ASM_FRONT** and select the end surface from the support_bar component. Set the offset value to 120. A new constraint called **Align** has been defined. At this moment, the support_bar component is fully constrained. Click the icon of **Complete** from the constraint control panel

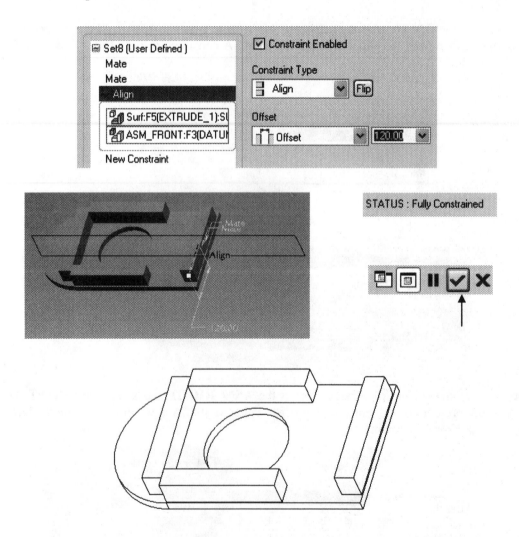

Now let us assemble the top_plate component. Click the icon of **Add component** > highlight top_plate.prt > **Open**.

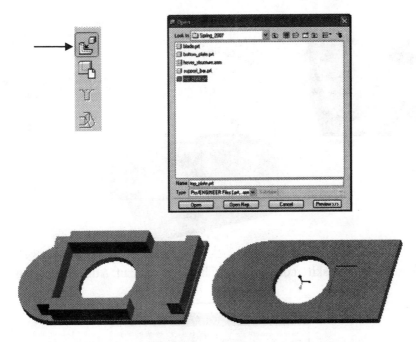

Let us insert the first constraint. Select the top surface from the support_bar component and select the bottom surface from the top_plate component. A constraint called **Mate** has been defined. Click **New Constraint**.

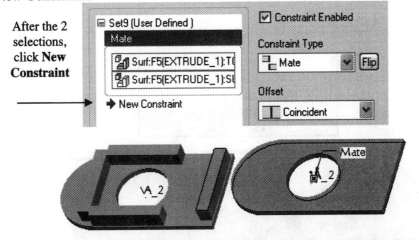

Let us insert the second constraint. Select the axis from the bottom_plate component and select the axis from the top_plate component. A new constraint called **Align** has been defined. At this moment, the top_plate component is fully constrained. Click the icon of **Complete** from the constraint control panel.

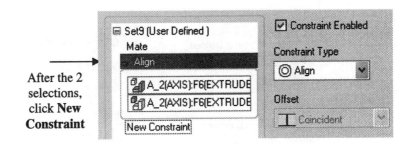

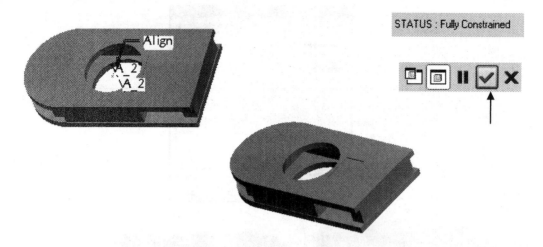

Now let us assemble the skirt component. Click the icon of **Add component** > highlight skirt.prt > **Open**.

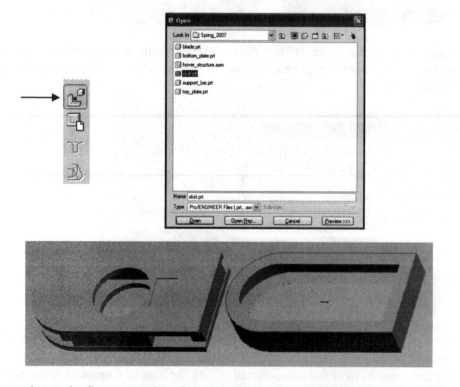

Let us insert the first constraint. Select the top surface from the top_plate component and select the inner surface from the skirt component. A constraint called **Mate** has been defined. Click **New Constraint**.

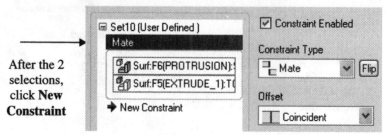

Let us insert the second constraint. Select the cylindrical surface from the top_plate component and select the inner part of the cylindrical surface from the skirt component. A new constraint called **Insert** has been defined. At this moment, the skirt component is fully constrained. Click the icon of **Complete** from the constraint control panel

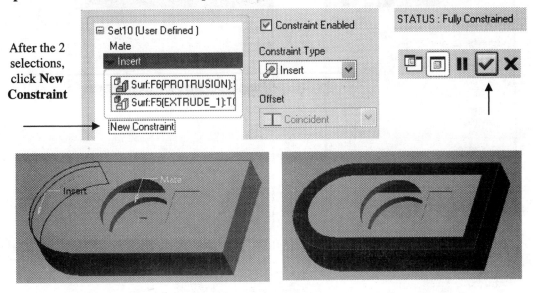

9.3 CREATION OF A LIFT FAN SYSTEM

Figure 9-2 is an exploded view of the lift fan system. As illustrated, this subsystem has a motor, a holding bar, a fan support and a blade for an axial fan or a turning wheel for a centrifugal fan.

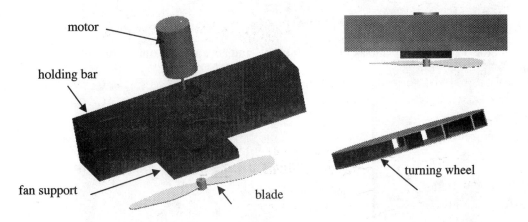

Fig. 9.2 An Exploded View of the Lift Fan System Showing the 4 Components.

We begin with creating the holding_bar component. Afterwards, we construct the motor support component, the motor component, the blade component, and the turning wheel component. Finally, we assemble them together.

Creation of the Holding_Bar Component

Step 1: Create a 3D solid model for the holding bar component.

File > New > Part > type *holding_bar* as the file name and clear the icon of **Use default template > OK.**

Select mmns_part_solid (units: Millimeter, Newton, Second) and type *holding_bar* in **DESCRIPTION,** and *student* in **MODELED_BY,** then **OK.**

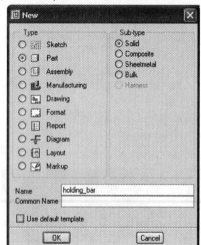 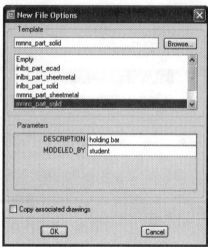

Directly select the icon of **Extrude.** From the dashboard, set the thickness value to 52. Activate **Placement** to define a sketch plane > **Define.**

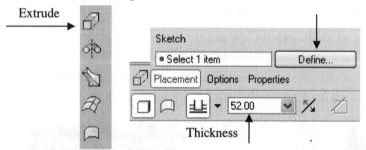

Select the **TOP** datum plane as the sketch plane, and accept the default orientation.

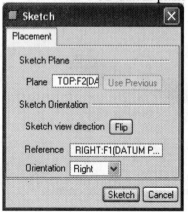 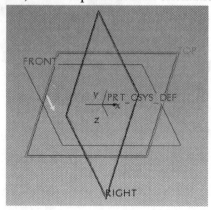

Right-click and hold, select **Centerline**. Sketch a vertical centerline and a horizontal centerline.

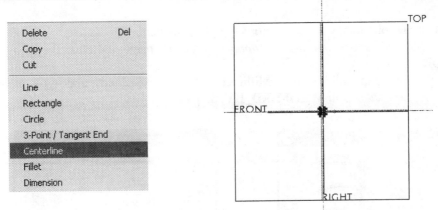

Click the icon of *Rectangle* and sketch a rectangle, which is symmetric about the 2 centerline. The 2 dimensions are 240 and 56, respectively.

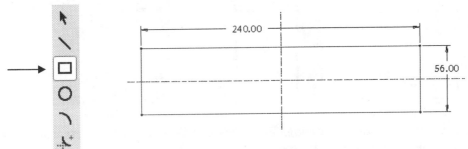

Click the icon of **Circle** to sketch a circle. The diameter value is 36, as shown.

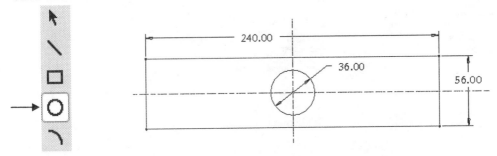

Upon completing this sketch, click the icon of **Done**. From the feature control panel, click the icon of **Complete**.

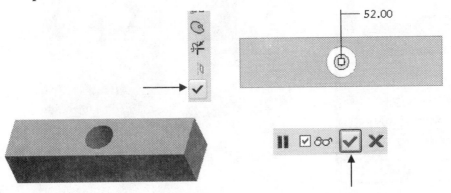

Creation of the Motor Support Component

Step 1: Create a 3D solid model for the motor support component.

File > New > Part > type *motor_support* as the file name and clear the icon of **Use default template > OK**.

Select mmns_part_solid (units: Millimeter, Newton, Second) and type *motor_support* in **DESCRIPTION**, and *student* in **MODELED_BY**, then **OK.**

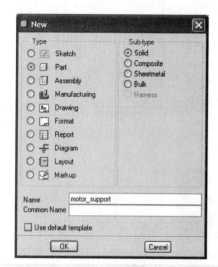
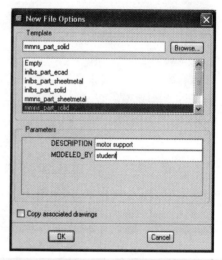

Directly select the icon of **Extrude.** From the dashboard, set the thickness value to 10. Activate **Placement** to define a sketch plane > **Define.**

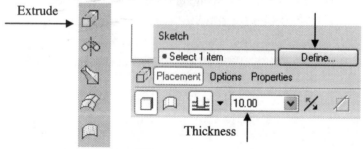

Select the **TOP** datum plane as the sketch plane, and accept the default orientation.

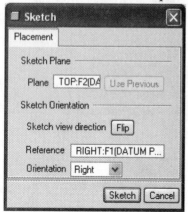
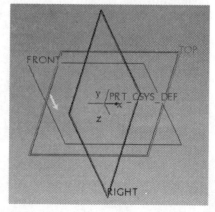

Right-click and hold, select **Centerline**. Sketch a vertical centerline and a horizontal centerline.

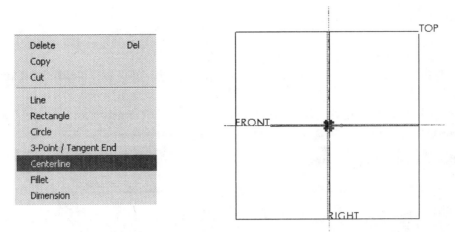

Click the icon of **Rectangle** and sketch a rectangle, which is symmetric about the 2 centerlines. The 2 dimensions are 76 and 56, respectively.

Click the icon of **Circle** to sketch a circle. The diameter value is 12, as shown.

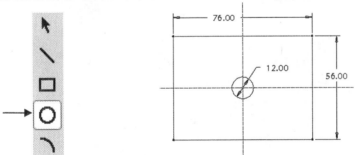

Upon completing this sketch, click the icon of **Done**. From the feature control panel, click the icon of **Complete**.

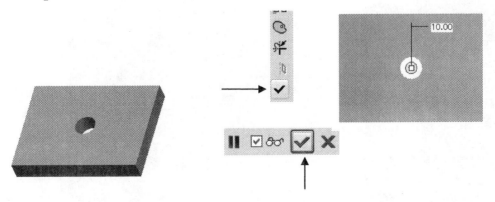

Creation of the Motor Component

Step 1: Create a 3D solid model for the motor component.

File > New > Part > type *motor* as the file name and clear the icon of **Use default template >
OK**.

Select mmns_part_solid (units: Millimeter, Newton, Second) and type *motor* in
DESCRIPTION, and *student* in **MODELED_BY**, then **OK**.

 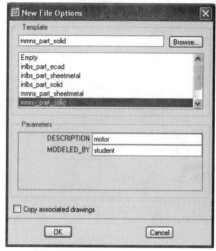

Directly select the icon of **Extrude.** From the dashboard, set the thickness value to 56. **Activate
Placement** to define a sketch plane > **Define.**

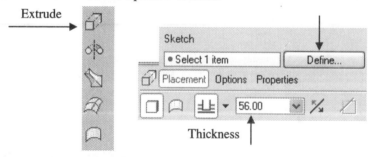

Select the **TOP** datum plane as the sketch plane, and accept the default orientation.

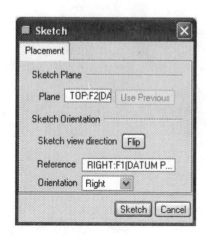 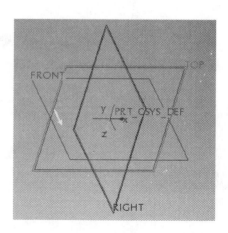

Click the icon of **Circle** to sketch a circle. The diameter value is 36, as shown.

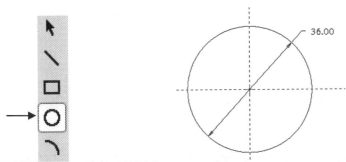

Upon completing this sketch, click the icon of **Done**. From the feature control panel, click the icon of **Complete**.

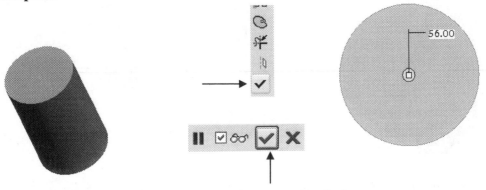

Step 2: Create a shoulder on the bottom surface of the created cylinder.

Select the icon of **Extrude.** From the dashboard, set the thickness value to 6. Activate **Placement** to define a sketch plane > **Define.**

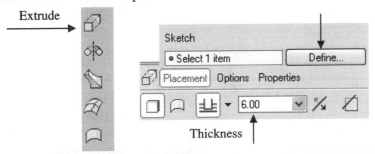

Select the bottom surface of the cylinder as the sketch plane, and accept the default orientation.

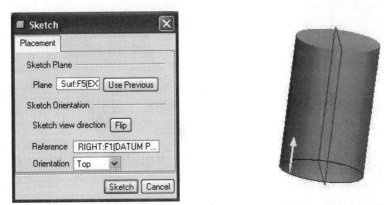

Click the icon of **Circle** to sketch a circle. The diameter value is 12, as shown.

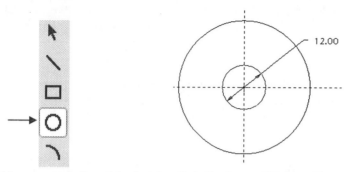

Upon completing this sketch, click the icon of **Done**. From the feature control panel, click the icon of **Complete**.

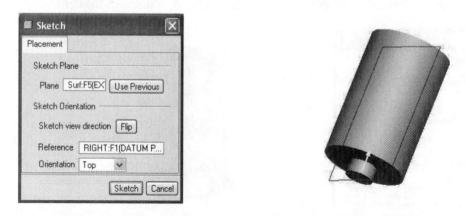

Step 3: Create the shaft of the motor.

Select the icon of **Extrude.** From the dashboard, set the thickness value to 16. Activate **Placement** to define a sketch plane > **Define.**

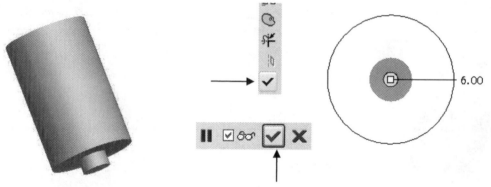

Select the bottom surface of the shoulder as the sketch plane, and accept the default orientation.

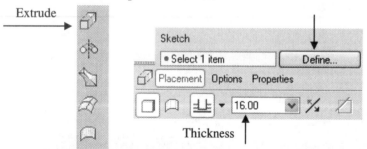

Click the icon of **Circle** to sketch a circle. The diameter value is 3, as shown.

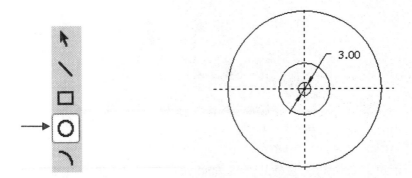

Upon completing this sketch, click the icon of **Done**. From the feature control panel, click the icon of **Complete**.

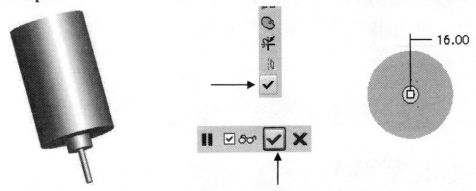

Creation of the Blade Component for an Axial Fan

Step 1: Create a 3D solid model for the blade component.

    File > New > Part > type *blade* as the file name and clear the icon of **Use default template >**
OK.

    Select mmns_part_solid (units: Millimeter, Newton, Second) and type *blade* in
DESCRIPTION, and *student* in **MODELED_BY**, then **OK**.

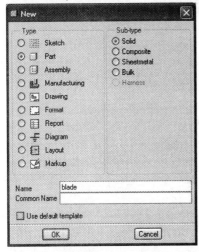

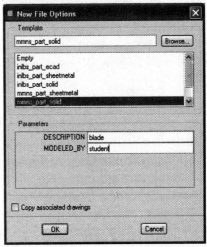

    Directly select the icon of **Extrude.** From the dashboard, use **Symmetry** and set the depth
value to 1. Activate **Placement** to define a sketch plane > **Define**.

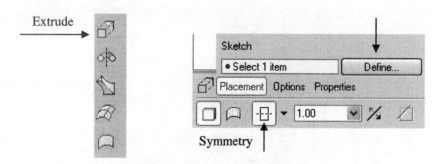

Select the **TOP** datum plane as the sketch plane, and accept the default orientation. Sketch a circle. Its diameter value is 10, as shown below.

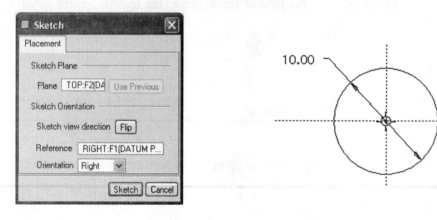

Pick the sketched circle, right-click and hold, select Construction so that the circle in the solid line is converted to a circle in the dashed line.

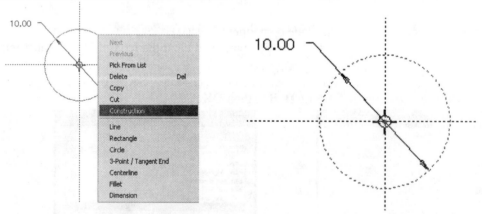

Click the icon of **Point** to define a point on the construction circle, as shown.

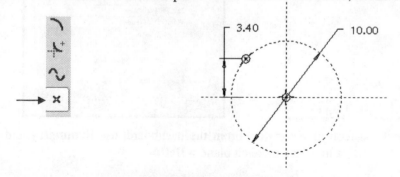

Follow the same procedure to define 4 more points, as shown.

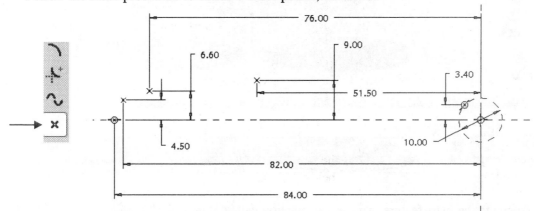

Select the icon of **Spline** to sketch a curve by connecting the 5 points defined.

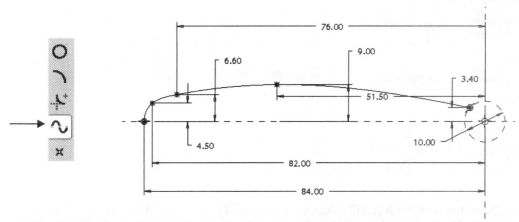

Sketch a horizontal centerline and mirror the sketched curve about the horizontal centerline. Sketch a vertical line to form a closed sketch.

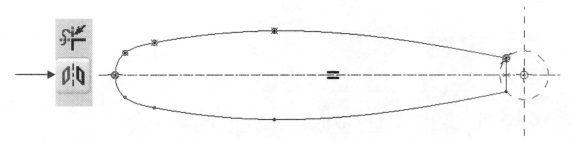

Sketch a vertical centerline and mirror the mapped curve about the vertical centerline.

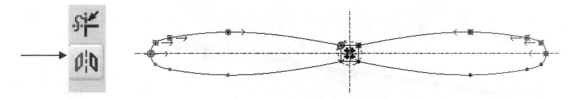

Upon completing this sketch, click the icon of **Done**. From the feature control panel, click the icon of **Complete**.

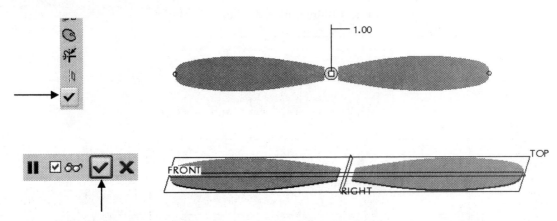

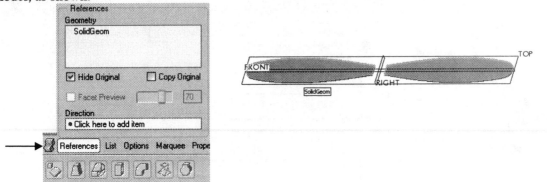

To create the twist feature, we use the function called **Warp**. Let us twist the right side of the blade first. To do so, from the main menu, select **Insert > Warp > References** > pick the created model, as shown.

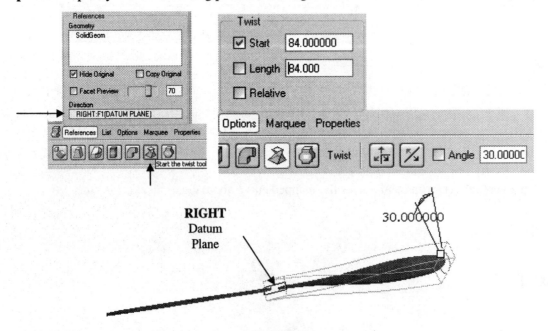

Click **Direction** > select **RIGHT** > click the icon of **Twist** > type *30* as the twist angle. Click **Options** > specify *84* as the starting point > set Length to *84* as well > **Done**.

To twist the left side of the blade, select **Insert > Warp > References** > pick the created model, as shown.

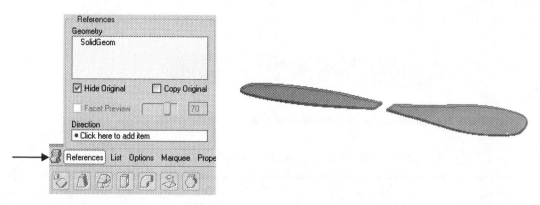

Click **Direction** > select **RIGHT** > click the icon of **Twist** > type *30* as the twist angle. Click **Options** > specify *0* as the starting point > set Length to *84* as well > **Done**.

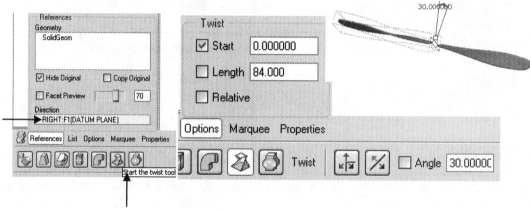

Now let us create the central part. Directly select the icon of **Extrude**. From the dashboard, use **Symmetry** and set the depth value to 10. Activate **Placement** to define a sketch plane > **Define**.

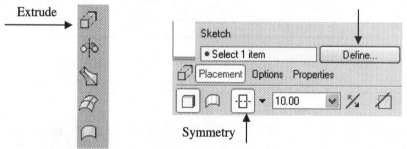

Select the **TOP** datum plane as the sketch plane, and accept the default orientation. Sketch 2 circles. Their diameter values are 10 and 3, respectively, as shown below.

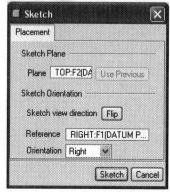

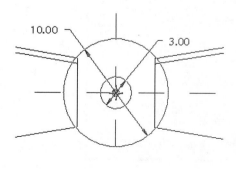

Click the icon of **Done**. From the feature control panel, click the icon of **Complete**.

Creation of a wheel component for a centrifugal fan

Step 1: Create a 3D solid model for the centrifugal blade component.

File > New > Part > type *centrifugal_blade* as the file name and clear the icon of **Use default template > OK**. Select mmns_part_solid (units: Millimeter, Newton, Second) and type *centrifugal blade* in **DESCRIPTION**, and *student* in **MODELED_BY**, then **OK**.

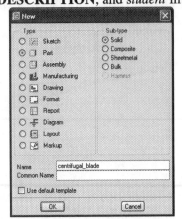

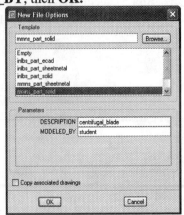

Select the icon of **Extrude**. From the dashboard, set the thickness value to 1. Activate **Placement** to define a sketch plane > **Define.**

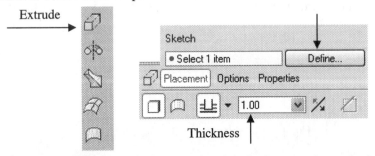

Select the **TOP** datum plane as the sketch plane, and accept the default orientation. Sketch a circle. Its diameter value is 168, as shown below.

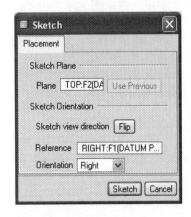

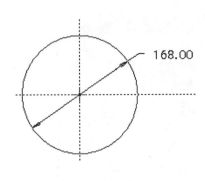

Upon completing this sketch, click the icon of **Done**. From the feature control panel, click the icon of **Complete**.

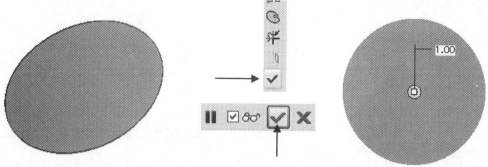

Step 2: Create the first blade.

Select the icon of **Extrude**. From the dashboard, set the thickness value to 15. Activate **Placement** to define a sketch plane > **Define.**

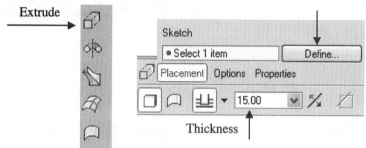

Select the top surface of the cylindrical plate as the sketch plane, and accept the default orientation.

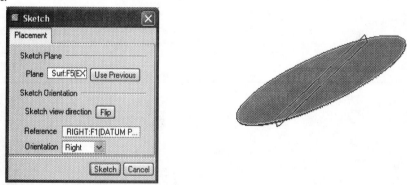

Click the icon of **Line**. Sketch a line inclined at angle equal to 50° with respect to the horizontal axis, as shown.

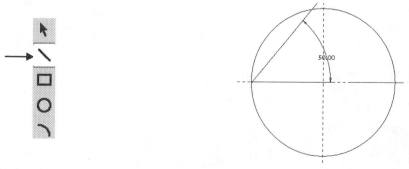

Pick the sketched line > right-click and hold, select Construction, as illustrated below. The solid line is converted to a dashed line.

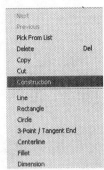

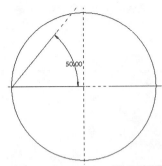

Click the icon of **Point**, define a point at the intersection of the construction line and the circle. Specify the distance of 10 with respect to the horizontal axis.

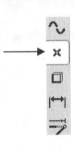

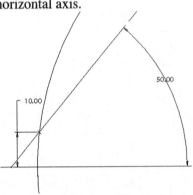

Click the icon of **Line**. Sketch a rectangle, as shown. The 2 dimensions are 30 and 20, as shown.

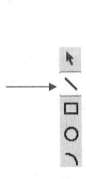

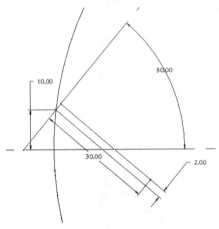

Upon completing this sketch, click the icon of **Done**. From the feature control panel, click the icon of **Complete**.

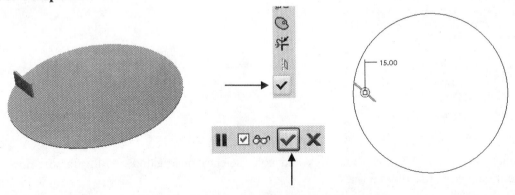

Step 3: Use **Pattern** to create the other 11 blads.

From the model tree, highlight the first blade feature > right click and pick **Pattern** > Axis.

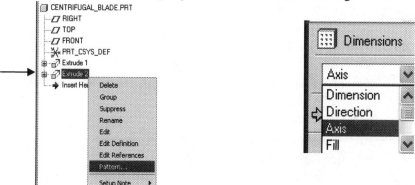

Pick the axis on display > type 12 and 30 > click the icon of **Complete**.

Step 4: Add the top plate.

Select the icon of **Extrude**. From the dashboard, set the thickness value to 1. Activate **Placement** to define a sketch plane > **Define.**

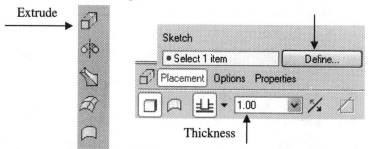

Select the top surface of the blades as the sketch plane, and accept the default orientation.

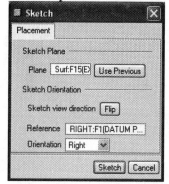

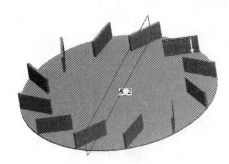

Sketch a circle. Its diameter value is 168, as shown below.

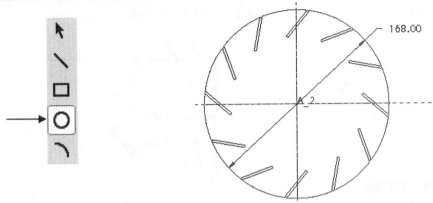

Upon completing this sketch, click the icon of **Done**. From the feature control panel, click the icon of **Complete**.

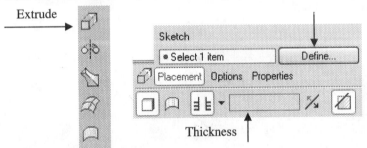

Step 5: Create a through-all hole at the central location.

Select the icon of **Extrude.** From the dashboard, select **Cut** > select **Thru All** as the depth choice. Activate **Placement** to define a sketch plane > **Define.**

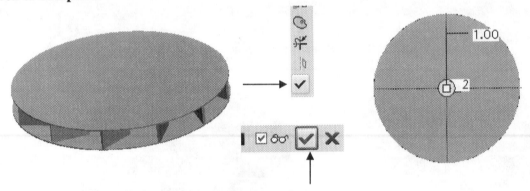

Select the top surface of the second blade as the sketch plane, and accept the default orientation.

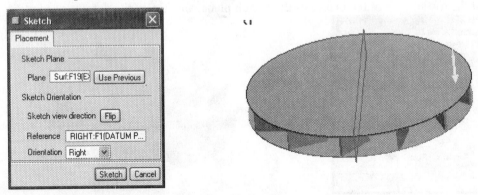

Sketch a circle. Its diameter value is 3, as shown below.

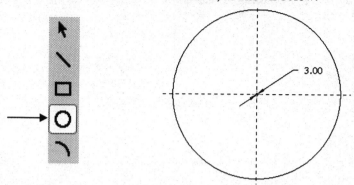

Upon completing this sketch, click the icon of **Done**. From the feature control panel, click the icon of **Complete**.

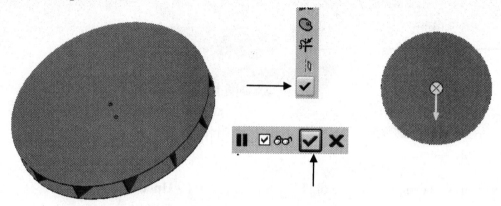

Creation of the Assembly of the Lift Fan System

Step 1: Create a 3D solid model for the assembly of the base structural system.

    File > New > Assembly > type *lift_fan_system* as the file name and clear the icon of **Use default template > OK**.

    Select mmns_asm_design (units: Millimeter, Newton, Second) and type *lift_fan_system* in **DESCRIPTION**, and *student* in **MODELED_BY**, then **OK**.

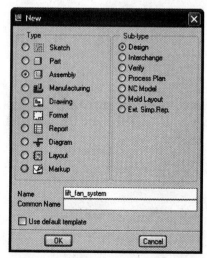
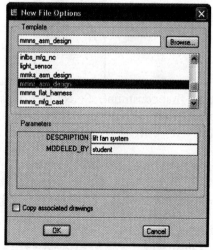

Click the icon of **Add component** > highlight holding_bar.prt > **Open**.

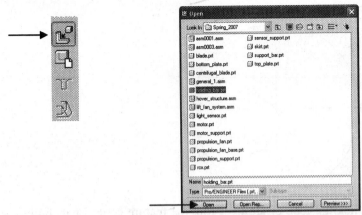

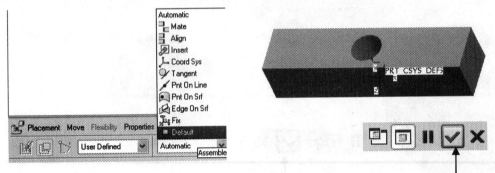

Select **Default** > click the icon of **Complete** from the constraint control panel.

Click the icon of **Add component** > highlight motor_support.prt > **Open**.

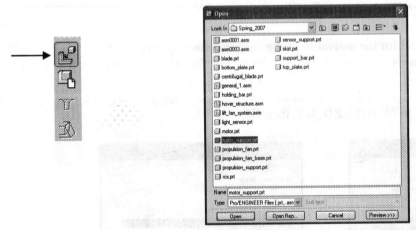

Let us insert the first constraint. Select the top surface from the motor_support component and select the bottom surface from the holding_bar component. A constraint called **Mate** has been defined. Click **New Constraint**.

After the 2 selections, click **New Constraint**

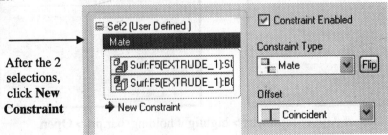

Let us insert the second constraint. Select the axis from the motor_support component and the axis from the holding_bar component. Use **Coincident**. A new constraint called **Align** has been defined. At this moment, the support_bar component is fully constrained. Click the icon of **Complete** from the constraint control panel.

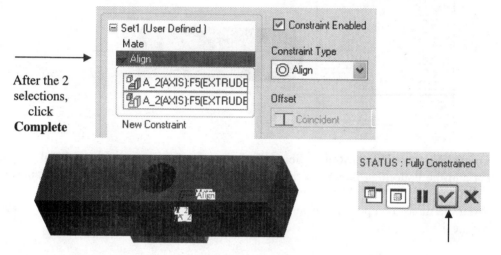

Now let us assemble the motor component on the other side. Click the icon of **Add component** > highlight motor.prt > **Open**.

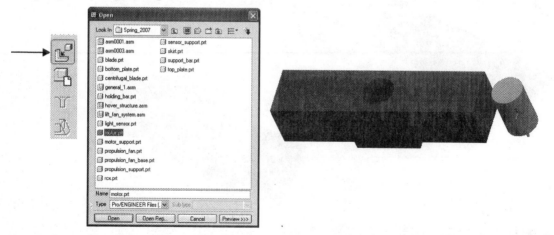

We follow the same procedure to insert 2 constraints as we did in the process of assembling the motor support component.

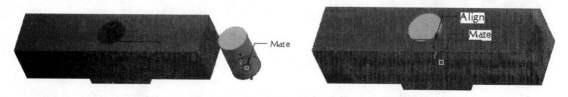

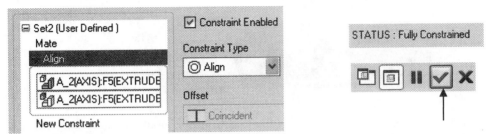

Now let us assemble the blade component. Click the icon of **Add component** > highlight blade.prt > **Open**.

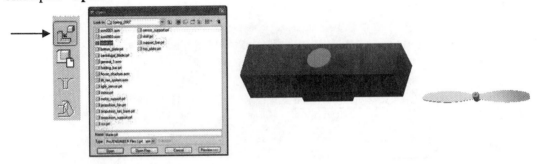

Let us insert the first constraint. Select the bottom surface from the cylindrical part of the blade component and the bottom surface from the shaft of the motor component. A constraint called **Align** has been defined. Click **New Constraint**. Afterwards, select the axis from the blade component and the axis from the motor component. A new **Align** constraint is defined, thus completing the assembly process for the lift fan system.

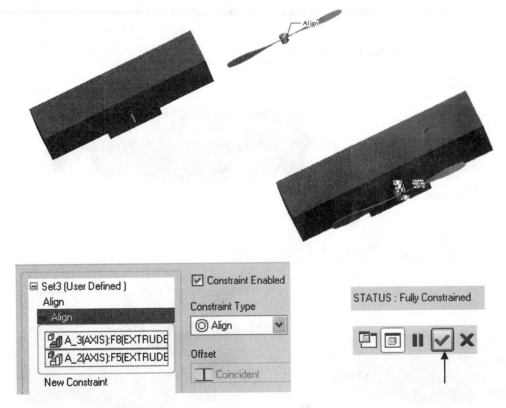

If a user wants to assemble the centrifugal plate, he or she may complete the following:

Open the centrifugal
blade file.

Define an Align
constraint.

Define an Align
constraint.

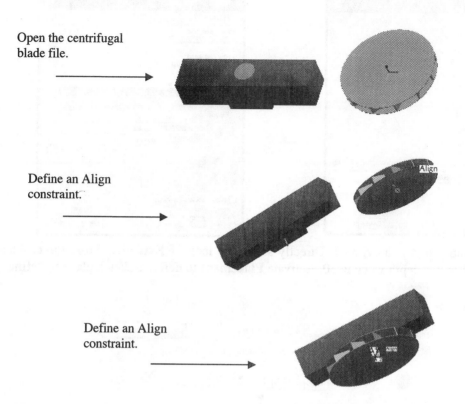

9.4 CREATION OF A PROPULSION FAN SYSTEM

Figure 9-3 is an exploded view of the propulsion fan system. As illustrated, this subsystem has a propulsion fan, a propulsion support component, and a propulsion fan base.

Propulsion support

Propulsion fan

Propulsion fan base

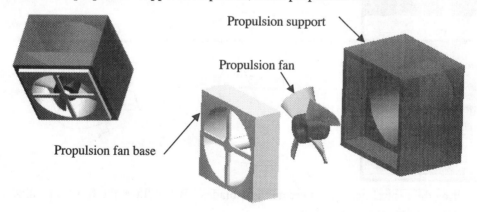

Fig. 9.3 An Exploded View of the Propulsion Fan System Showing the 3 Components.

Creation of the Propulsion Fan Component

Step 1: Create a 3D solid model for the propulsion fan component.
     File > New > Part > type *propulsion_fan* as the file name and clear the icon of **Use default template > OK**. Select mmns_part_solid (units: Millimeter, Newton, Second) and type *propulsion_fan* in **DESCRIPTION**, and *student* in **MODELED_BY**, then **OK**.

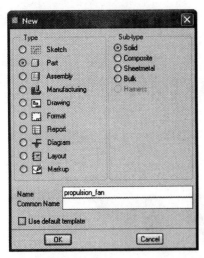
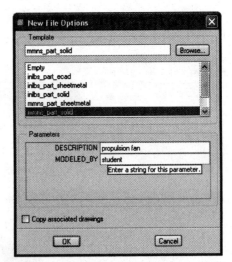

The central part is a cylinder. Directly select the icon of **Extrude.** From the dashboard, use **Symmetry** and set the depth value to 30. Activate **Placement** to define a sketch plane > **Define.**

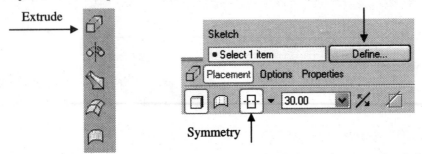

Select the **FRONT** datum plane as the sketch plane, and accept the default orientation. Sketch 2 circles. Their diameter values are 34 and 15, respectively. Click **Done** > **Complete.**

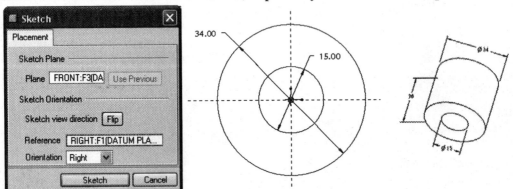

After creating the central part, let us create the 4 blades. We will use the function called surface modeling to create the blades, as shown below.

As illustrated below, a blade consists of 4 curves on the boundaries of the blade. These 4 curves form a closed form. Among the 4 curves, two are straight lines.

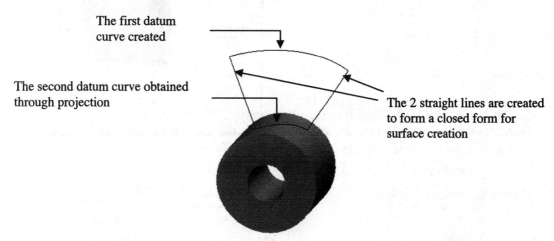

The first datum curve created

The second datum curve obtained through projection

The 2 straight lines are created to form a closed form for surface creation

Step 2: Create 2 datum planes. The offset values are 12 and (-12) with respect to **FRONT** datum plane, respectively.

Click the icon of **Creating Datum Plane** > pick the **FRONT** datum plane > type *12* > **OK**.

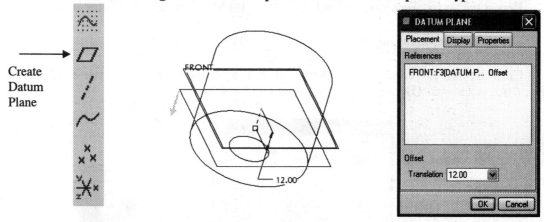

Create Datum Plane

Repeat the above process to create the second datum plane. Make sure that you select the **FRONT** Datum Plane and type *-12*.

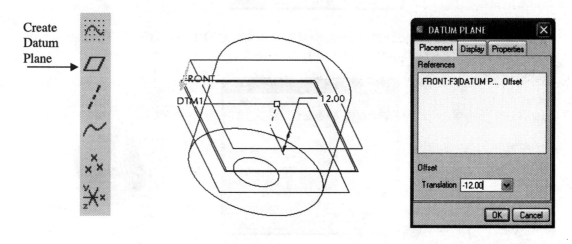

Create Datum Plane

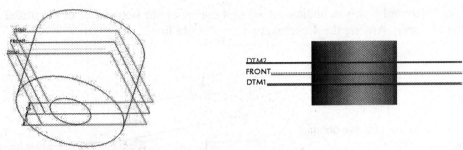

Step3: Sketch 2 circles on **DTM1** and **DTM2**, respectively.

From the toolbar of datum feature creation, select the icon of **Sketch Tool >** pick **DTM1 > Sketch.** Click the icon of **Circle** and sketch a circle and specify *90* as the diameter value.

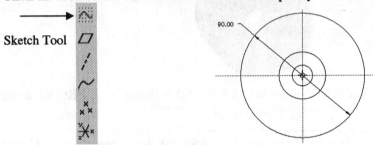

Sketch Tool

Repeat the above procedure to sketch a circle on **DTM2**. The diameter value is *90*.

Step 4: Define a datum point on the circle created on **DTM1** and define a datum point on the circle created on **DTM2**.

Click the icon of **Create Datum Point >** click a location on the sketch circle > change the ratio value to *0.40*, as shown > **OK**, completing the creation of **PNT0**.

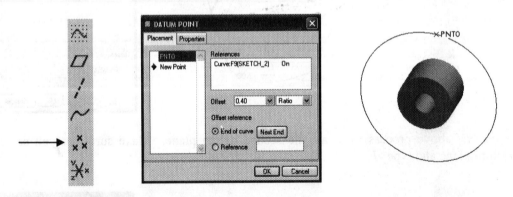

Repeat the above procedure to define a datum point, **PNT1**, on **DTM2**. The datum point is on the sketched circle. The length ratio value is 0.1, as shown.

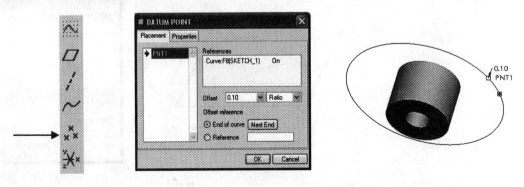

Step 5: Create a datum curve going through PNT0 and PNT1.
Click the icon of **datum curve > Thru Points > Done.**
Spline > Whole Array > Add Point > pick **PNT1** first, then **PNT0 > Done.**

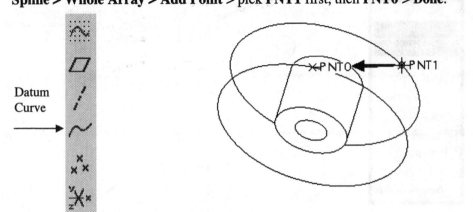

Datum
Curve

From the window called **Tangency > Define.**

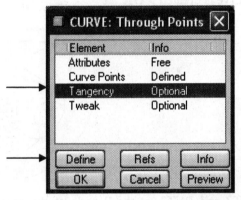

From the pop up window, **Start > Crv/Edge/Axis > Tangent** > pick the datum curve on **DTM2
> Okay.**

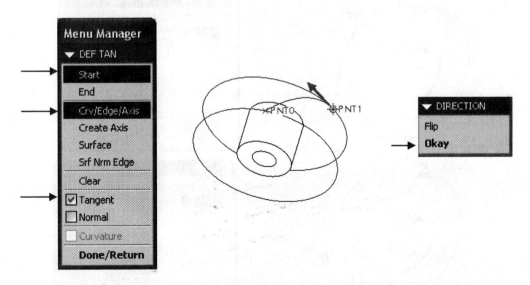

Repeat this process for the end point, **End > Crv/Edge/Axis > Tangent** > pick the datum curve
on **DTM1 > Okay > Done/Return > OK.**

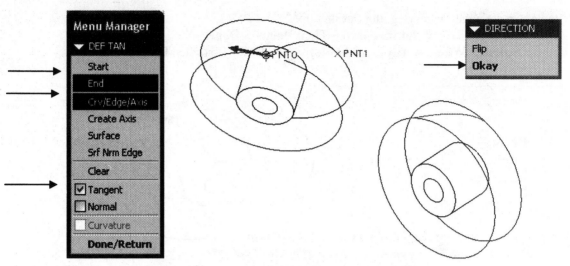

Step 6: To create the second datum curve, we use the function called **Project.**

Highlight the datum curve just created > **Edit** > **Project** > **Normal to surface** > pick the cylindrical surface and a projected curve is created > **OK**.

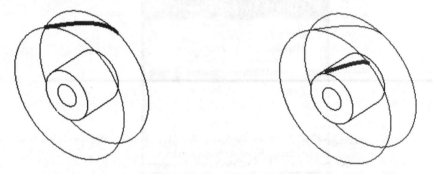

Step 7: Create 2 datum points, PNT2 and PNT3, at the 2 ends of the projected curve so that the 2 straight lines can be created.

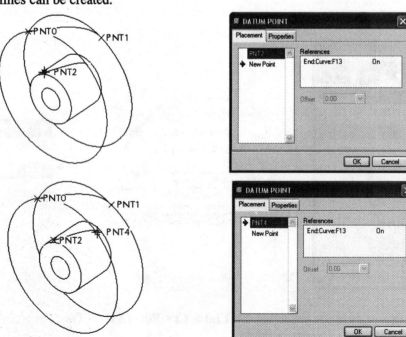

Step 8: Create the 2 straight lines, or 2 datum curves.

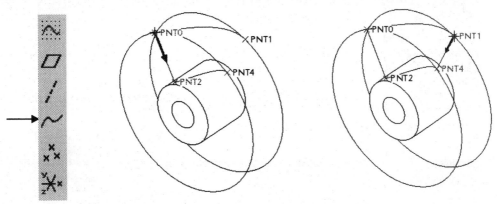

Step 9: Create a surface using the 4 datum curves created.
Click the icon of Boundary blend tool. For the first direction, pick the 2 straight lines.

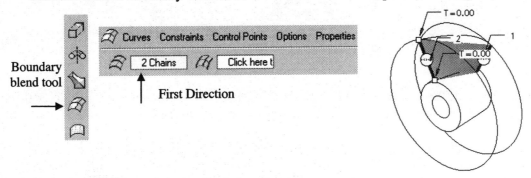

For the second direction, pick the 2 datum curves.

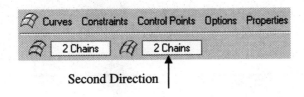

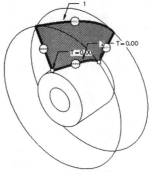

Step 10: Use the Thicken tool to convert the surface feature to a 3D solid feature.
Highlight the created surface > from the main toolbar, select **Edit > Thicken** > specify *1* as the thickness value > **Automatic fit** > click the icon of **Complete**.

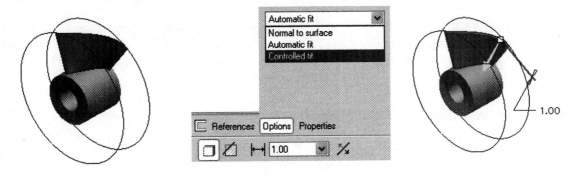

Step 11: Use Pattern and Reference Pattern to create the 3 other fan blades

From the model tree, highlight **Boundary Blend 1** > right click and pick **Pattern** > **Axis** and pick the axis of the cylinder > set the number of copies including the current one to 4 and the increment to 90 > click the icon of **Complete**.

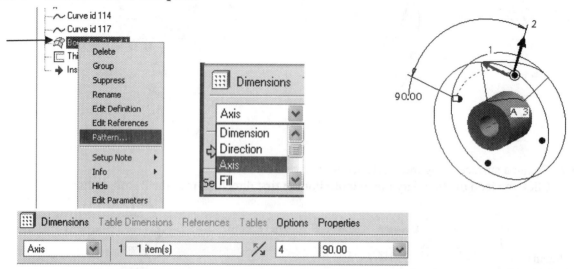

From the model tree, highlight **Thicken** > right click and pick **Pattern** > Axis and pick the axis of the cylinder > set the number of copies including the current one to 4 and the increment to 90 > click the icon of **Complete**. Users may add color to the fan blades.

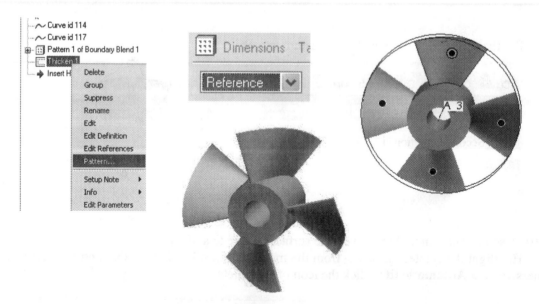

Summary: When a closed form is available, a curved surface can be created in a 3D space. However, a surface is assumed to be weightless. Therefore, the thickness value of a surface is assumed to be zero. To convert this surface feature to a 3D solid feature, we use a function called **Thicken**, thus creating the first blade. Afterwards, we use **Pattern** to obtain the 3 other blades.

Creation of the Propulsion Fan Base Component

Step 1: Create a 3D solid model for the propulsion_fan_base component.

File > New > Part > type *propulsion_fan_base* as the file name and clear the icon of **Use default template > OK**.

Select mmns_part_solid (units: Millimeter, Newton, Second) and type *Propulsion_fan_base* in **DESCRIPTION**, and *student* in **MODELED_BY**, then **OK.**

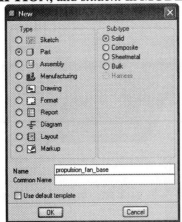
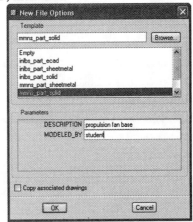

Directly select the icon of **Extrude.** From the dashboard, use **Symmetry** and set the depth value to 50. Activate **Placement** to define a sketch plane > **Define.**

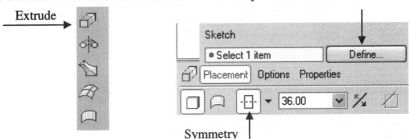

Extrude →

Symmetry |

Select the **FRONT** datum plane as the sketch plane, and accept the default orientation.

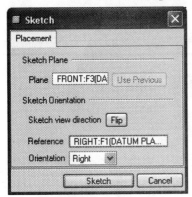
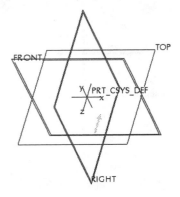

Right-click and hold, select **Centerline.** Sketch a vertical centerline and a horizontal centerline.

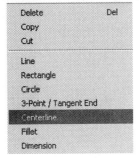
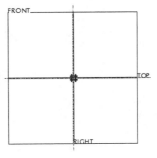

Click the icon of rectangle and sketch a rectangle, which is symmetric about the 2 centerlines. The 2 dimensions are equal with the value equal to 106.   Click **Done > Complete**.

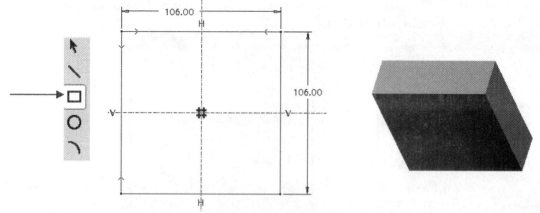

Select the icon of **Extrude.**   From the dashboard, click **Cut** and set the depth value to *33*. Activate **Placement** to define a sketch plane > **Define.**

Select the front surface of the block as the sketch plane, and accept the default orientation.

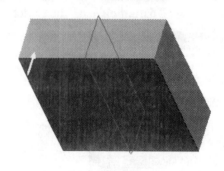

Click the icon of **circle** and sketch 2 circles. The dimensions of the 2 diameters are 94 and 15, respectively.

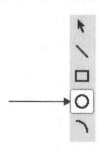

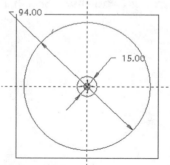

Upon completing the sketch, click the icon of **Done** and the icon of **Complete** from the feature control panel.

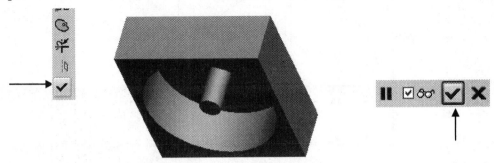

Select the icon of **Extrude.** From the dashboard, click **Cut** and select **Thru All** as the choice of depth. Activate **Placement** to define a sketch plane > **Define.**

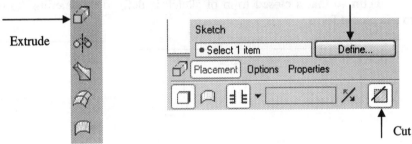

Select the front surface of the block as the sketch plane, and accept the default orientation > **Sketch**.

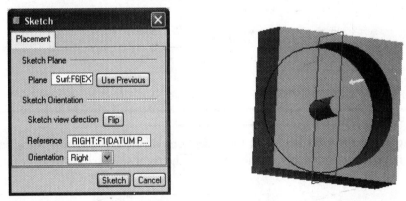

Click the icon of **Use Edge** > pick the 2 sketched circles, as shown below.

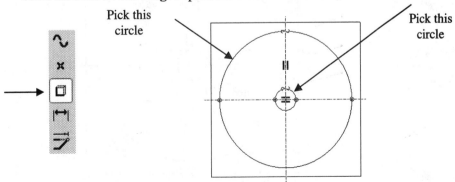

Click the icon of **Line** > sketch 2 lines with the length value equal to 40, as shown below:

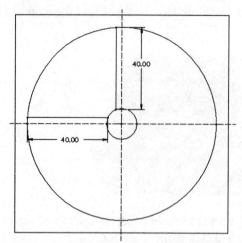

Click the icon of **Trim** so that a closed form of sketch is defined by deleting those sections, which do not belong to this closed form.

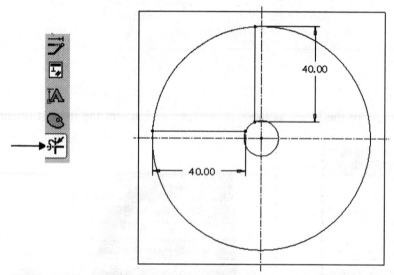

While holding down the **Ctrl** key, pick the 4 sections of the closed form > click the icon of **Mirror** > click the vertical centerline to obtain the image on the right side to the vertical centerline.

Icon of
Mirror
$\longrightarrow$

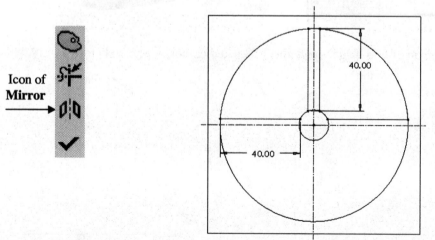

Click the icon of **Mirror** > click the horizontal centerline to obtain a new image.

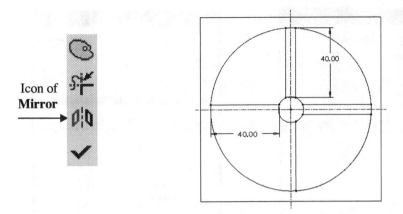

Icon of
Mirror

Click the icon of **Mirror** > click the vertical centerline to obtain another new image.

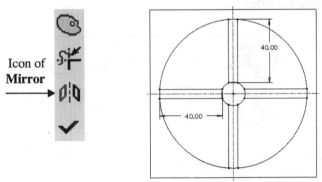

Icon of
Mirror

Upon completing this process, click the icon of **Done** and click the icon of **Complete** from the feature control panel.

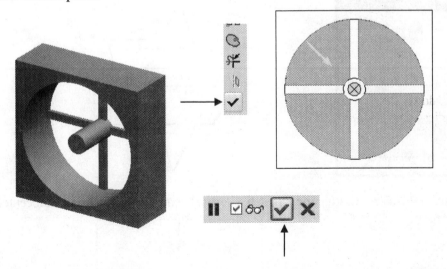

Creation of the Propulsion Support Component

Step 1: Create a 3D solid model for the propulsion_support component.

File > New > Part > type *propulsion_support* as the file name and clear the icon of **Use default template > OK**.

Select mmns_part_solid (units: Millimeter, Newton, Second) and type *propulsion_support* in **DESCRIPTION**, and *student* in **MODELED_BY**, then **OK**.

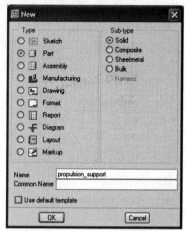

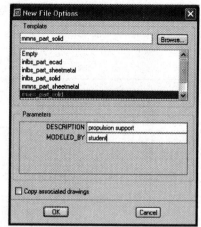

Directly select the icon of **Extrude.** From the dashboard, set the depth value to 90. Activate **Placement** to define a sketch plane > **Define.**

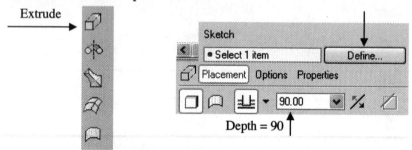

Depth = 90

Select the **FRONT** datum plane as the sketch plane, and accept the default orientation.

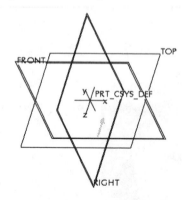

Right-click and hold, select **Centerline.** Sketch a vertical centerline and a horizontal centerline.

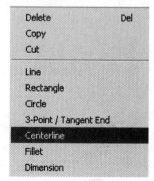

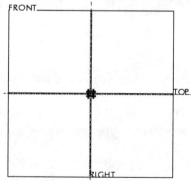

Click the icon of **rectangle** and sketch a rectangle, which is symmetric about the 2 centerlines. The 2 dimensions are equal with the value equal to 120.

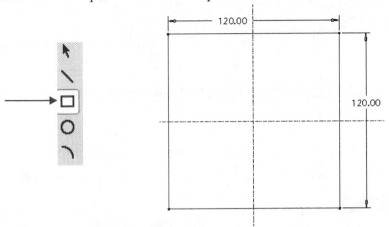

Click the icon of circle and the dimension of diameter is 90.

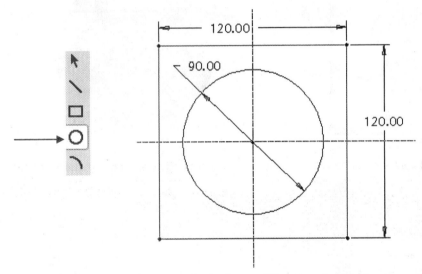

Upon completing this process, click the icon of **Done** and click the icon of **Complete** from the feature control panel.

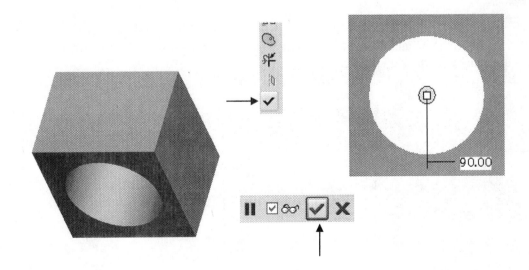

Select the icon of **Extrude.** From the dashboard, click **Cut** and set the depth value to *30.* Activate **Placement** to define a sketch plane > **Define.**

Select the front surface of the block as the sketch plane, and accept the default orientation.

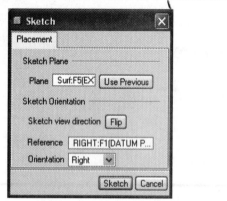

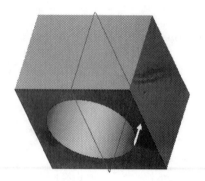

Right-click and hold, select **Centerline**. Sketch a vertical centerline and a horizontal centerline.

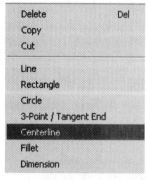

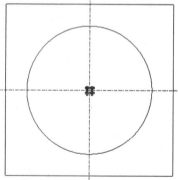

Click the icon of rectangle and sketch a rectangle, which is symmetric about the 2 centerlines. The 2 dimensions are equal with the value equal to 106.

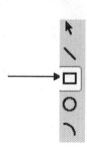

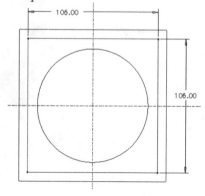

Upon completing this process, click the icon of **Done** and click the icon of **Complete** from the feature control panel.

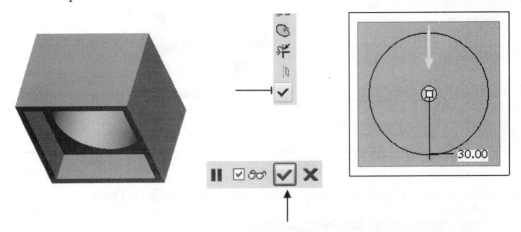

Creation of the Assembly of the Propulsion Fan System

Step 1: Create a 3D solid model for the assembly of the base structural system.

File > New > Assembly > type *propulsion_fan_system* as the file name and clear the icon of **Use default template > OK**.

Select mmns_asm_design (units: Millimeter, Newton, Second) and type *propulsion_fan_system* in **DESCRIPTION**, and *student* in **MODELED_BY**, then **OK**.

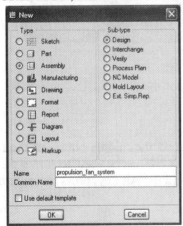
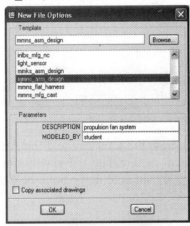

Click the icon of **Add component** > highlight propulsion_support.prt > **Open**.

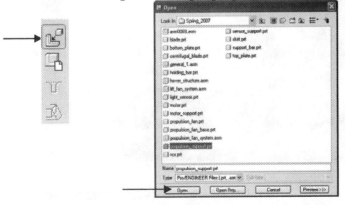

Select **Default** > click the icon of **Complete** from the constraint control panel.

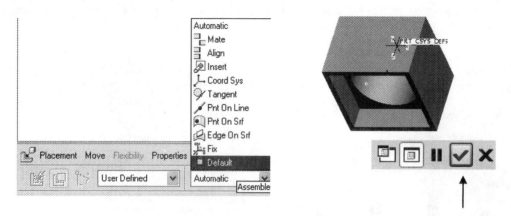

Click the icon of **Add component** > highlight propulsion_fan_base.prt > **Open**.

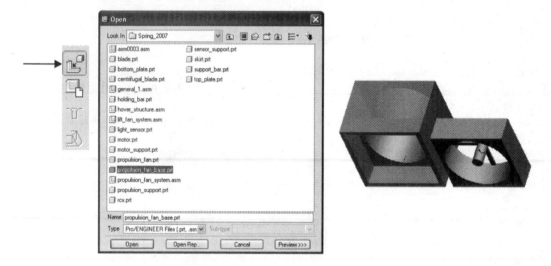

While holding down both the **Ctrl** key and the **Alt** key, press the middle button of the mouse and rotate the base component to facilitate the assembling process. If users want to move the base component, not rotating, they should hold down both the Ctrl key and the Alt key, press the right button of the mouse.

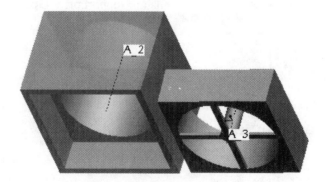

Let us insert the first constraint. Select the axis from the propulsion_support component and the axis from the propulsion_fan_base component. A constraint called **Align** has been defined. Click **New Constraint** to insert 2 Mate constraints, as shown below.

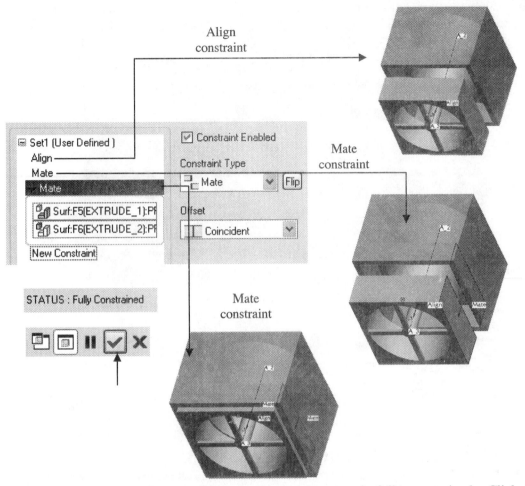

At this moment, the propulsion_fan_base component is fully constrained. Click the icon of **Complete** from the constraint control panel.

Now let us assemble the propulsion fan component on the other side. Click the icon of **Add component** > highlight propulsion_fan.prt > **Open**.

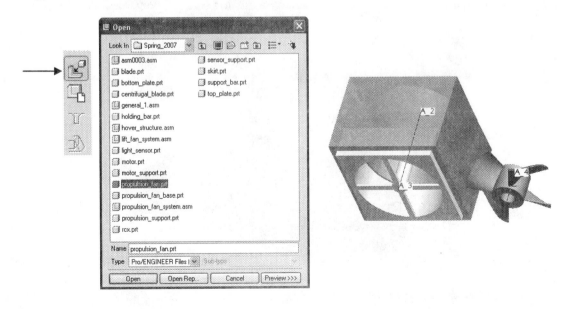

We follow the same procedure to insert 2 constraints as we did in the process of assembling the propulsion_fan_base component.

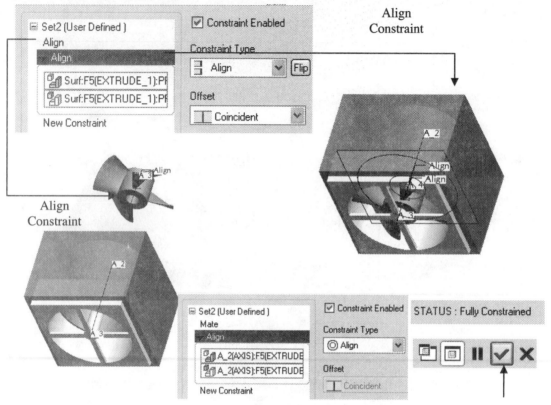

9.5 CREATION OF A HOVERCRAFT ASSEMBLY

Figure 9-4 is an exploded view of the hovercraft assembly. As illustrated, the hovercraft assembly consists of 4 sub-systems. They are the base structure sub-system, the lift fan sub-system, the propulsion fan sub-system, and the sensor-based control system. Note that the sensor-based control system is not presented in this chapter.

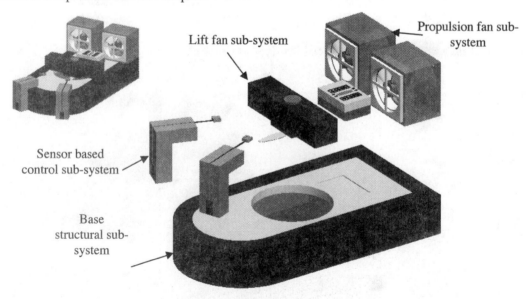

Fig. 9.4 An Exploded View of the Hovercraft Assembly Showing 4 Sub-Systems

Step 1: Create a 3D solid model for the hovercraft assembly.

File > New > Assembly > type *hovercraft_assembly* as the file name and clear the icon of **Use default template > OK**.

Select mmns_asm_design (units: Millimeter, Newton, Second) and type *hovercraft_assembly* in **DESCRIPTION**, and *student* in **MODELED_BY**, then **OK.**

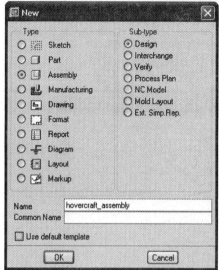 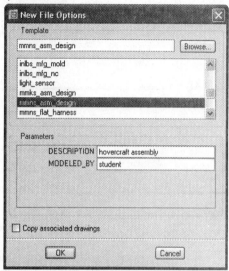

Click the icon of **Add component** > highlight structural_system.asm > **Open**.

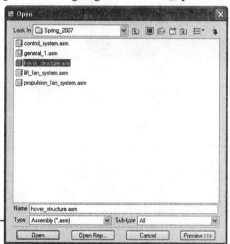

Select **Default** > click the icon of **Complete** from the constraint control panel.

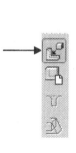 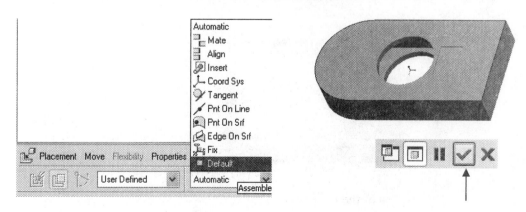

Click the icon of **Add component** > highlight lift_fan_system.asm > **Open**.

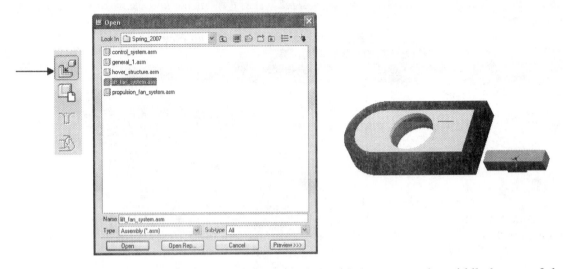

While holding down both the **Ctrl** key and the **Alt** key, press the middle button of the mouse and rotate the lift fan sub-system.

Let us insert the first constraint. Select **ASM_FRONT** from the lift fan assembly and select ASM Right from the structural assembly and set the offset distance to *130*. A constraint called **Align** has been defined. Click **New Constraint** to insert a new constraint.

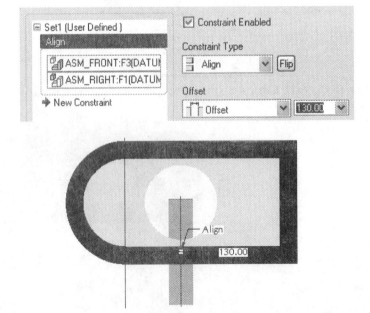

Let us insert the second constraint. Select the bottom surface from the holding bar and the top surface from the skirt. A constraint called **Mate** has been defined. Click **New Constraint** to insert a new constraint.

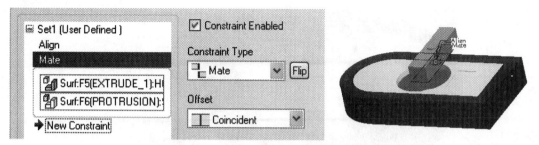

Let us insert the third constraint. Select the axis from the motor and select the axis from the top plate. A constraint called **Align** has been defined. At this moment, the propulsion_fan_base component is fully constrained. Click the icon of **Complete** from the constraint control panel.

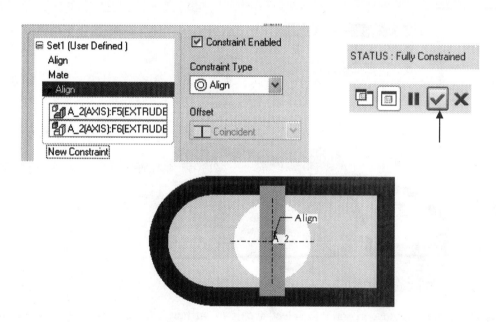

Now let us assemble the propulsion_fan_system. Click the icon of **Add component** > highlight propulsion_fan_system > **Open**.

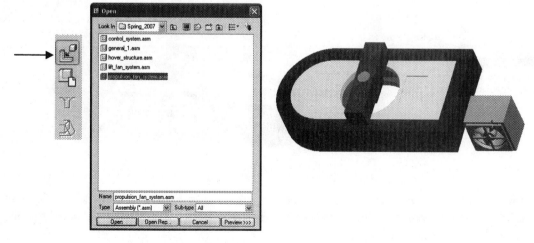

While holding down both the **Ctrl** key and the **Alt** key, press the middle button of the mouse and rotate the propulsion fan system.

Let us insert the first constraint. Select a pair of surfaces, as shown. A constraint called **Align** has been defined. Click **New Constraint** to insert a new constraint.

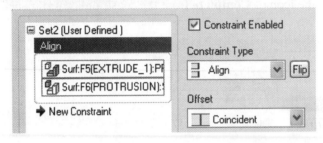

Let us insert the second constraint. Select a pair of surfaces, as shown. A constraint called **Align** has been defined. Click **New Constraint** to insert a new constraint.

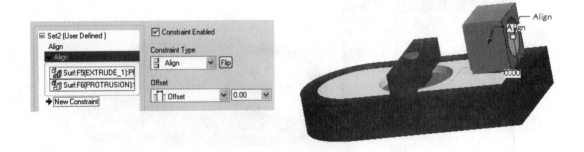

Let us insert the third constraint. Select a pair of surfaces, as shown. A constraint called **Mate** has been defined. At this moment, the propulsion fan system is fully constrained. Click the icon of **Complete** from the constraint control panel.

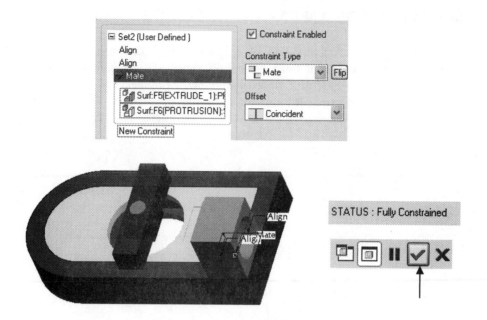

Repeat the above procedure to assemble another propulsion fan system on the other side, as shown.

If a 3D solid model of the sensor_based_control sub-system is available, users may assemble it

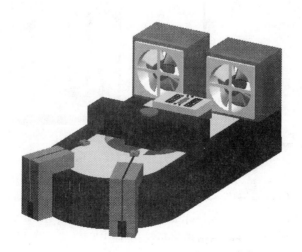

9.6 PREPARATION OF AN ENGINEERING DRAWING

Figure 9-5 presents an engineering drawing of a block with 3 dimensions equal to 10, 6 and 4, respectively. The drawing consists of three projection views and an isometric view. The 3 project views are the front view, top view and right-sided view. Engineering drawings serve as official documents to deliver clear and accurate information in design, manufacturing and assembling.

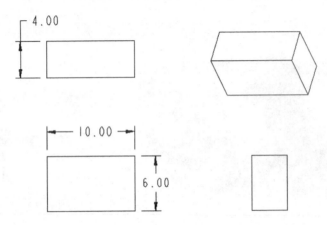

Figure 9-5 Engineering Drawing of a Block

In the following, we prepare an engineering drawing for the propulsion support component created in section 9.4. First, we select the icon, called "**Create a new object**", which is displayed on the menu bar. A **New** window appears, as illustrated below. This is the same step we used to open a new file when creating the 3D solid model of the propulsion support component. However, this time, "**Drawing**" mode, instead of "**Part**" mode, should be selected. Type propulsion_support as the name for the new drawing file. Clear the box of **Use default template** because we do not want to use the default setting for the drawing work. Afterwards, click the button of **OK**.

This brings up a new window called "**New Drawing**", as shown below. Make sure that the file of the 3D solid model of propulsion_support.prt is shown. Otherwise, use the "**Browse**" option to locate it. Select **Empty** under Specify Template, and select the paper size to be **A**. Afterwards, click the button of **OK**.

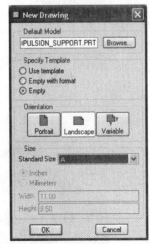

Select the icon of **inserting views** (create a general view), which is displayed on the toolbar. Select a location on the drawing screen as the center point for the Front View (click the left button of mouse). A general view appears on the screen.

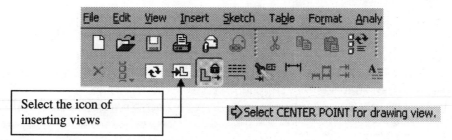

Select the icon of inserting views

Select CENTER POINT for drawing view.

In the pop-up Drawing View window, select **FRONT > Apply > Close**, the construction of the Front View is completed.

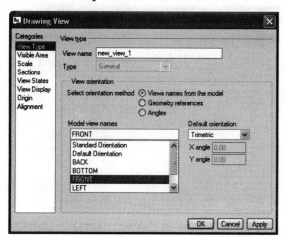

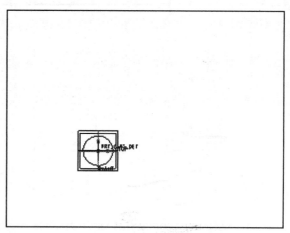

To insert the right side view through projection, first pick the **FRONT** View just created, right-click and hold, and then select **Insert Projection View** > move the curser to the right side and click the left button of mouse, and the construction of the Right-sided View is completed. Follow the same procedure to create the Top View, as shown below.

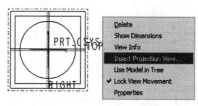

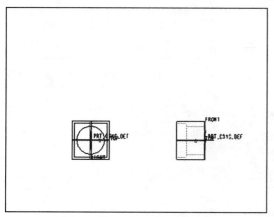

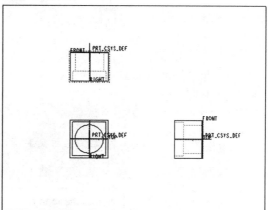

Select the icon of **inserting views** (create a general view) for creating an isometric view through **General** (not through **Projection**). Select a location on the drawing screen as the center point for the Isometric View (click the left button of mouse). A general view appears on the screen.

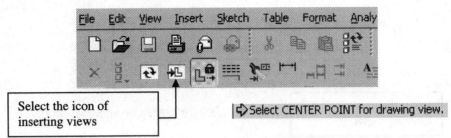

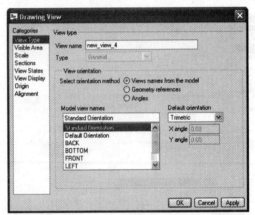

In the pop-up Drawing View window, select **Standard Orientation > Apply > Close**, the construction of the Isometric View is completed.

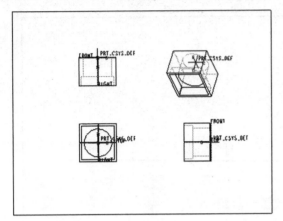

The user may notice that the names of the datum planes, such as FRONT, RIGHT and TOP, appear on the drawing. The name of coordinate system, such as PRT_CSYS_DEF, also appears. To clean the drawing screen, click the icon of **datum planes**, the icon of **datum axes**, the icon of **datum points**, and the icon of **coordinate system** from the main toolbar to disable their displays > click the icon of **repaint** (redraw the current views) from the main toolbar.

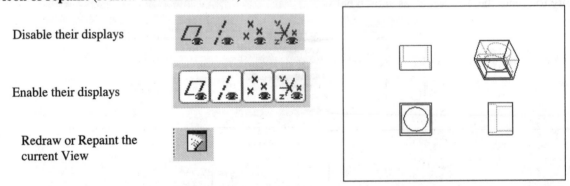

Disable their displays

Enable their displays

Redraw or Repaint the current View

Upon completing the layout, we start adding dimensions. Select the icon of **show/erase**, which is displayed on the toolbar.

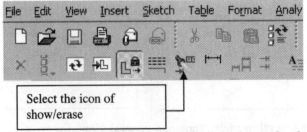

In the pop-up window, select the icon of **dimensions** and the icon of **axis** (centerline), and click the button of **Show All > Yes > Accept All > Close**.

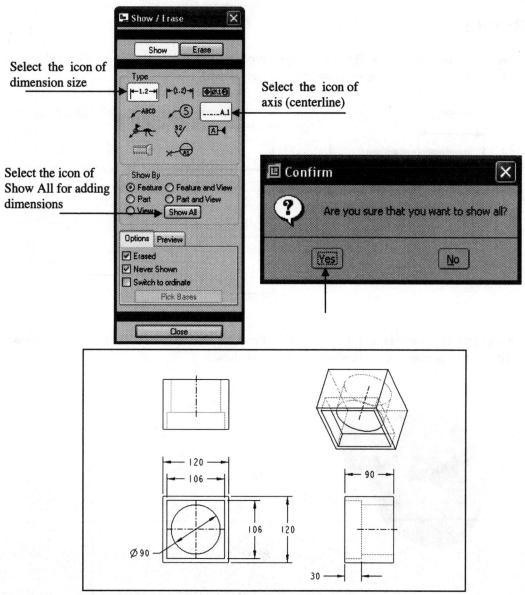

Select the icon of dimension size

Select the icon of axis (centerline)

Select the icon of Show All for adding dimensions

Users may use the icon of pickup to move the dimensions to appropriate locations, thus completing the preparation of an engineering drawing.

REFERENCES

1. Foley, J.D., Dam, A.V., Feiner, S., and Hughes, J., <u>Computer Graphics, Principles and Practice,</u> 2ⁿᵈ edition, McGraw-Hill, 1990.

2. Groover, M.P., and Zimmers, E.W., <u>Computer-aided Design and Manufacturing</u>, Englewood Cliffs, NJ, 1984.

3. Kalameja, A.J., <u>AutoCAD 2006 Tutor for Engineering Graphics</u>, Thomson Delmar Learning, 2006.

4. Qi, G., <u>Engineering Design, Communication and Modeling: Using Unigraphics NX</u>, Thomson, Delmare Learning, 2006.

5. Zhang, G.M., Engineering Design and Pro/ENGINEER Wildfire, Version 3.0, College House Enterprises, LLC., 2006.

EXERCISES

9-1.   An object is shown in a 3D space. Use Pro/ENGINEER to create a 3D solid model of it.
Afterwards, prepare an engineering drawing, as shown.

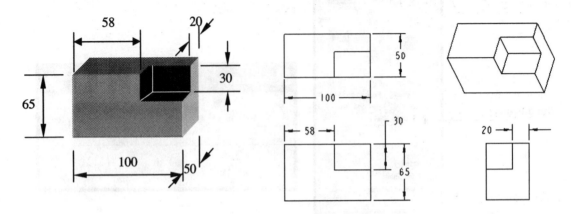

9-2.   An object is shown in a 3D space. Use Pro/ENGINEER to create a 3D solid model of it.
Afterwards, prepare an engineering drawing, as shown.

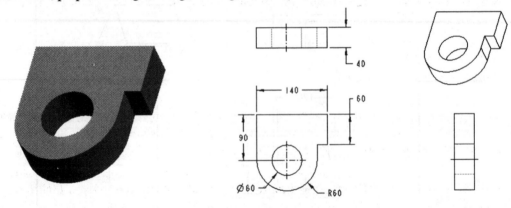

9-3.   An object is shown in a 3D space. Use Pro/ENGINEER to create a 3D solid model of it.
Afterwards, prepare an engineering drawing, as shown.

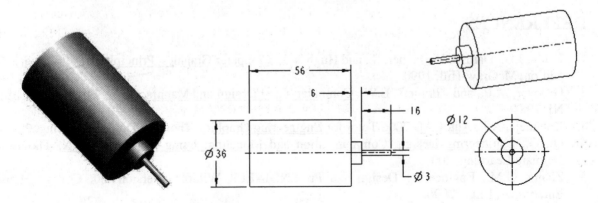

CHAPTER 10

MICROSOFT EXCEL

10.1 INTRODUCTION

Microsoft® Excel, a spreadsheet program, is an extremely powerful tool in engineering because it is useful in many different applications. In addition to providing a basis for using the Excel spreadsheet tool, techniques for three important applications will be described in detail in this chapter. These applications include:

- Preparing tables and graphs
- Making calculations
- Conducting design trade-off studies

While there are several different spreadsheet programs on the market, Excel® has been selected because it is the most popular. Many universities have adopted it and you will likely have access to it in campus computer labs during your tenure in college. The content in this chapter is focused on both developing your entry-level skills in Excel and encouraging you to begin to think like a practicing design engineer. Hopefully, you will find the spreadsheet tool important enough to develop a higher skill level through independent study.

10.2 EXCEL BASICS

The version of Excel described in this chapter is Microsoft Office Excel 2007. However, much of the content in this chapter is applicable to earlier versions of Excel which you may still be running. Microsoft Office 2007 has been given a major facelift as compared to prior versions of Office. The key change is that the menu-based software has been replaced with an all-purpose "ribbon" along the top of the program. The ribbon provides a tabbed browsing environment that contains all of the tools found under the familiar Microsoft menu headers (i.e., File, Edit, View, Insert, etc.), with a minimal user effort required for searching drop down menus. It is advised that you take time to explore each of the tabs along the top ribbon to familiarize yourself with the new configuration for all Microsoft Office applications. The ribbon described is shown in Fig. 10.1.

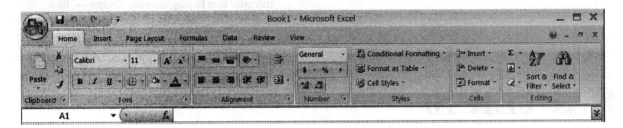

Fig. 10.1 The "ribbon" menu of Microsoft Office Excel 2007

The majority of the screen is occupied by the spreadsheet, which is simply a large table with many columns and rows as shown in Fig. 10.2. The columns are labeled across the top with letters A, B, C,... M and beyond. The rows are labeled down the left side with numbers 1, 2, 3,... 25 and beyond. Each small block, called a cell, is identified with its coordinates (e.g. in Fig. 10.2, the cell A1 is outlined with a border indicating it is active). The letter (column location) is placed before the number (row location). The cells are simply locations in a very large table (spreadsheet) where you can enter numbers, text or formulas.

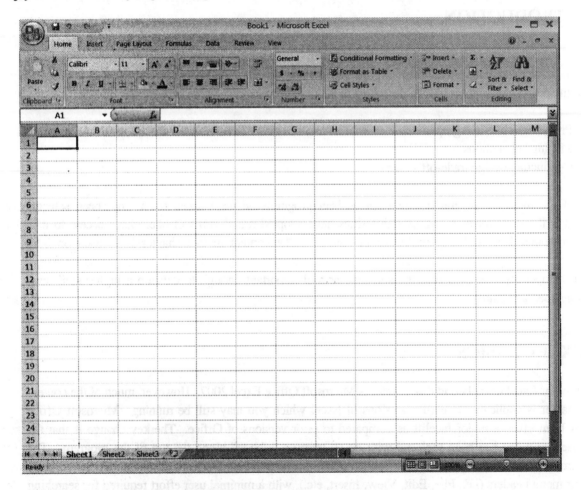

Fig. 10.2 The Excel 2007 spreadsheet showing cell A1 as the active cell

You may move about on the spreadsheet using either the mouse or the keyboard. Because the mouse usually permits the cells to be identified more rapidly, it is the preferred method for placing the cursor over a particular cell. When the pointer is on the spreadsheet, its location is identified with a large plus sign. You may use the mouse to point to any cell on the screen; however, if the cell of interest is not visible on the screen, the scroll bars located below and to the right of the spreadsheet are used to bring this cell into the field of view. When the large plus sign is pointed at the correct cell—click; the cell becomes active and is ready to receive the data you enter. You may also move the active cell by using the arrow keys, the tab key, the shift-tab keys, the enter key, the shift-enter key and the control-arrow keys. Try them and note how the active cell moves about the spreadsheet.

The spreadsheet is much larger than it appears on the screen. In Fig. 10.2, only columns A through M and rows 1 through 25 are shown. Explore the number of rows on the spreadsheet by holding down the control and the arrow-down key. Note that the number of the last row is 1,048,576. Depressing the control and the arrow-right keys moves the active cell to the far right of the spreadsheet. There are a total of 16,384 columns. You have a total of $(1,048,576)(16,384) = 17,179,869,184$ cells on the spreadsheet. Your spreadsheet is huge—let's hope you never have to use all of the available cells!

At the bottom of the spreadsheet, you will find three tabs identifying sheet numbers 1, 2 and 3. When you open Excel and create a new file, you immediately begin with a workbook for a project. The program automatically establishes many sheets (pages) in your workbook, although only three are initially visible. If your project requires more than three worksheets, you can add more very easily. You may enter data and perform calculations on several different worksheets in a file by clicking on the tabs to move from one sheet to another. Now that you are familiar with the layout of Excel 2007, let's turn our attention to using Excel to create tables, graphs, and to perform calculations.

10.3 TABLES AND GRAPHS

On many occasions you will collect numerical data that must be presented at a design briefing or in an engineering report. At other times, you may have extensive results from a mathematical relationship that must be expressed in a meaningful format. You have two choices in presenting large quantities of numerical data—show the numbers in tabular form or on a graph. Both methods of presentation, the table and the graph, have advantages and disadvantages. The method that you select depends on your audience, the message, and the purpose of the presentation.

In this section, the technique for arranging data in a table using a spreadsheet program will be described. Data entry into a spreadsheet is also a convenient method for preparing graphs. Three common methods for representing data in the form of graphs will be described, including:

1. Pie charts
2. Bar charts
3. X-Y graphs

Each type of graph is used for a different purpose and selection of the correct type of presentation is essential in communicating effectively. For instance, the pie chart is used to show distributions, whereas the bar chart is used to compare one set of data with another. X-Y graphs, the most frequently used type of graphic in engineering, shows the variation of some dependant variable Y with some independent variable X. Regardless of the type of table or graphic required, this section will describe how Excel 2007 can assist you in creating high-quality, professional graphics.

Tables

Very early on in the design of a hovercraft, your team will be faced with the requirement of making a special type of table called a parts list. The parts list provides a first estimate of the overall weight of your vehicle, which—as you recall—is needed in order to size your levitation fan. A complete parts list should include all of the components required to construct your team's hovercraft. For each item listed, the parts list should include the quantity of the component required, the vendor, the part number, a weight estimate and the cost. A sample parts list has been created in Excel 2007 and is provided below as Table 10.1.

Table 10.1
Sample hovercraft parts list (albeit incomplete)

	A	B	C	D	E	F
1	**Item Description**	**Vendor**	**Unit Cost ($)**	**Quantity**	**Total Cost ($)**	**Weight (lbs)**
2	Foamular Styrofoam	Lowes	7.15	3	21.45	0.85
3	Liquid Nails (adhesive)	Lowes	1.57	1	1.57	0.10
4	Hobby Plywood (lower deck)	GPA Hobbies	3.99	2	7.98	0.45
5	Balsa Sheet (centrifugal fan)	GPA Hobbies	2.25	1	2.25	0.15
6	APC 4.2x2 Sport Propeller	GPA Hobbies	1.75	2	3.50	0.05
7	Super Speed 9-18V Hobby Motor	GPA Hobbies	5.29	3	15.87	0.30
8	MOSFET N-CH 60V 8A Transistor	Radio Shack	1.89	3	5.67	0.05
9	Lego Light Sensors	Lego Educational	16.50	3	49.50	0.10
10	TENERGY Li-Ion 14.8V 2200 mAh Battery	GPA Hobbies	32.99	1	32.99	1.00
11	RCX Rental	UMD	25.00	1	25.00	0.86
12	**TOTAL**				**165.78**	**3.91**
13						
14	**Sample formulas used to create table above:**					
15	Cell E2: =C2*D2					
16	Cell E12: =sum(E2:E11)					

A number of operations are required in order to make the table shown above. First and foremost, each cell must be formatted. For instance, you may have noticed that each cell along the top row of the table (row 1) is in a bold font type. Cells can be made bold by highlighting the desired cells and clicking the standard Microsoft icon for Bold under the Home tab on the main ribbon. Under this tab, the font style, size, alignment, formatting, color, etc. can all be changed for each cell and each character within a cell. Under the home tab, there is subsection called Alignment. The top right icon in this subsection is called "Wrap Text". This icon can be used to allow the descriptions written in a cell to extend onto multiple rows within the cell instead of being cut off at the start of the cell to the right of the text (see cells A7 and A10). Another important set of commands can be found under the Home ribbon tab in the Number subsection. These icons can be used to set the number of decimal places shown and/or the style of number to be used.

Because the costs were given in US Dollars ($) at the top of the parts list, each of the numbers below were formatted to display two decimal places – this is to a precision of +/- 1 cent. Similarly, the last column showing weight (in lbs) is shown to within +/- 0.01 lbs because the scale used to weigh these components was accurate to within 0.01 lbs. It makes no more sense to provide a weight to a higher level of certainty than can be reasonably known than it does to provide a cost to a higher precision than +/- 1 penny. You should not show numbers in a table with a higher level of precision (number of decimal places) than which you are confident in reporting. There are a number of other formatting operations with which you must become comfortable in order to make professional looking tables. These include adding borders and adjusting the widths of each column/row (which is as simple as left clicking and dragging the line separating the column/row indexes), among others. It is advised that you take some time early in your academic career to better familiarize yourself with the commands required to make highly professional looking tables and graphs.

A second set of operations is required to create the parts list shown in Table 10.1. The use of mathematical formulae simplifies the creation of this table. For instance, because the unit cost is provided in column C and the quantity in column D, the total cost (column E) is found by multiplying the numbers in columns C and D together for each part shown. Excel has a built in mathematical library to make these types of repeated calculations efficiently. From row 2 of the parts list, the unit cost for the Foamular Styrofoam was $7.15 and the quantity was 3, so the total cost is just $7.15 \times 3 = \$21.45$. In Excel, this calculation is made very simply by typing "=C2*D2" into cell E2. The "=" sign indicates to Excel that a formula will follow and the "*" sign indicates the multiplication operation. The basic algebraic operations are called using the following self-explanatory symbols: +, -, *, /. For a much more complete set of formulae available in Microsoft Excel, take some time to explore the Formulas tab on the Excel ribbon.

To enter equations more efficiently, you can select a cell to be used in a formula by simply clicking on it with the mouse instead of typing it in. The really useful aspect of Excel is that once the first equation has been typed in, it can be used to automatically compute any number of similar calculations. For instance, let's automate the computation of the Total Cost column (column E). To do this, copy the formula in cell E2 by pressing CTRL-C and then highlight the cells in column E from rows 2 through 11 where you want to perform a similar calculation. Next, paste this formula into each of these cells by typing CTRL-V. Excel automatically performs all of these calculations using an indexing system in with the cell E# =C#*D#, where the # sign indicates any numbered row you have selected. To check the accuracy of these calculations, double click on any of the boxes in column E to show the formula hidden under that cell.

Excel has hundreds of different mathematical functions that can be used in equations, ranging from trigonometric functions like "=sin(Cell)" to statistical functions like "=average(Cell Range). In creating Table 10.1, one additional computational tool was used in Excel. Instead of adding up all of the cells in the column showing the total cost ("=E2+E3+...+E11), the summation tool has been used. To use this tool, simply type "=sum(" and then highlight the cells of interest – in this case the total cost of each component – and then type ")" and the summation will be performed automatically when you hit enter. If you double click on the cell showing the total cost, you will see "=sum(E2:E11)" as the mathematical operation. Here, "E2:E11" provides the Cell Range over which the operation is performed. Similarly, by copying the formula for the total cost, and pasting it into cell F12, automatically computes the total weight of the entire hovercraft. If you double-click on this cell, it will show "=sum(F2:F11)". Excel indexes to the F column from the E column when the copy/paste function is used across columns in a similar way that it indexed row numbers when calculating the Total Cost as the product of the Unit Cost and the Quantity. As a last note, the user has control over the way in which Excel indexes. For instance, an entire column of numbers (let's say measurements in feet) can be converted into inches by multiplying each cell in feet by a conversion factor (12 in / 1 ft). To do this, the conversion factor of 12 can be placed in a cell, say B2, and can be called as $B$2. The "$" before the B locks in column B and the "$" before the 2 locks in row 2. Essentially, even as the mathematical operations are indexed across columns and rows, the quantity in cell B2 will always be used for the required calculations.

Excel is an extremely powerful computational tool that can be used for much more than making simple tables like the parts list shown in Table 10.1. Again, it is highly recommended that you take the time to further explore the procedure for inserting equations and familiarize yourself with the built-in formulas available in Excel. These can be found under the "Formulas" tab of the main Excel ribbon. If you cannot find the formula you are seeking, do not be discouraged. Excel 2007 allows the user to define custom formulas. With some practice, you will find Excel to be a tool that can help you in all of your engineering classes.

Graphs

Excel is an extremely useful tool for quickly creating high quality graphics. The first graphic that will be created is the pie chart. The pie chart is typically used to show distributions in data. To apply this to the hovercraft project, the data in Table 10.1 has been rearranged to create a table that shows the cost to construct each of the major subsystems. This information is shown below in Table 10.2.

Table 10.2
Hovercraft cost and weight breakdown among subsystems

	Cost ($)	Weight (lbs)
Structure & Levitation	38.54	1.65
Power & Propulsion	52.74	1.3
Sensors & Controls	74.50	0.96
TOTAL	165.78	3.91

Based on Table 10.2, Excel will be used to create a pie chart to illustrate which sub-system requires the largest portion of the $165.78 budget. To do this, click on the Insert tab of the main ribbon. The subsection titled Charts is used to insert any one of the many different built-in chart functions available in Excel. For the time being, let's focus on the task of creating a very simple pie chart by selecting "Pie" as the chart type and choosing "Exploded pie in 3-D" as the specific type. Once this selection has been made, a blank chart should appear on the screen and a new feature on the ribbon should appear called "Chart Tools" with the "Design" subsection selected. The procedure for creating this chart is very similar to the procedure used to create any chart. The data, which you will typically have in tabular format, must be selected for use in the chart. To do this, click on the "Select Data" icon under the "Data" subsection of Chart Tools tab. Having done this, you should be prompted with a pop-up box similar to the one shown in Fig. 10.3 below.

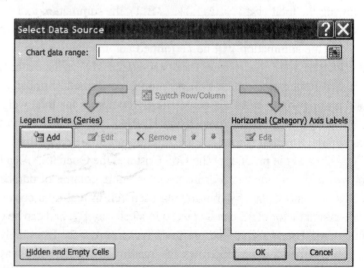

Fig. 10.3 Data source entry pop-up
screen

Creating the chart is as simple as selecting the appropriate set of data to be shown. To do this, click the box to the right of the "Chart data range" entry box of Fig. 10.3. Next, highlight the appropriate set of data. In this case, it is the cell range containing the cost to construct each of the three major sub-systems of the hovercraft design. The next step requires providing the appropriate set of labels to explain the data in the pie chart. To do this, simply click the Edit box underneath the "Horizontal (Category) Axis Labels" box of Fig. 10.3. Similarly to before, highlight the three sets of labels from Table 10.2 (Structure & Levitation, Power & Propulsion, and Sensors & Controls). Though not appropriate in the construction of a pie chart, the "Legend Entries (Series)" box shown in Fig. 10.3 allows the user to add multiple sets of data (or series) to a single graph. This is often used in the construction of engineering X-Y graphs.

The final touches that should be made to every graphic you create include the addition of titles, axis labels, legend labels, etc. To do this, switch to the "Layout" tab under the Chart Tools tab. The Labels subsection allows the user to add the final touches to a graph. Even for the very simple pie chart constructed, it is necessary to add a title. To do so, select the "Chart Title" command and choose "Above Chart" as the type. Next, type in a descriptive title for the chart such as "Hovercraft Cost Breakdown". The only piece of information that is missing from this graphic is the magnitude of each section of the pie chart. In order to add this information to the graph, select the "Data Labels" option and choose "Best Fit" as the type. With only some minor formatting left, the simple procedure described above allows you to create a very professional looking pie chart similar to the one shown in Fig. 10.4.

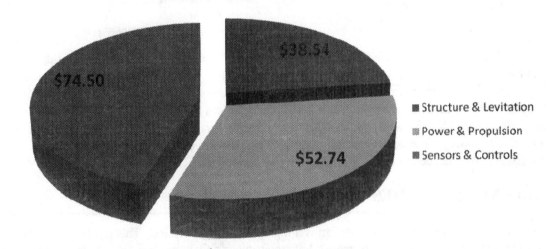

Fig. 10.4 Pie chart made in Excel showing sub-system cost distribution to build a hovercraft

The process described above for inserting a pie chart can be used to insert many different types of charts and graphs. Recall, this chapter will limit the discussion to pie charts, bar charts, and X-Y graphs, but you should explore the other options available through Microsoft Excel. Bar charts can also be used to show distributions, but they are better suited for illustrating comparisons. As an example of a bar chart, consider the results of the hovercraft competition during the 2006-07 academic year at the University of Maryland, as shown in Fig. 10.5. At a glance, the reader can see that the number of hovercraft constructed in the fall was significantly larger than the number of hovercraft built during the spring. At a second glance, the reader can see that the rate of success was significantly higher in the spring than it was in the fall. At a final glance, the reader can see that less than 50% of the hovercrafts built were able to meet the product specifications during either semester. A lot of very valuable information is contained within the very simple bar chart shown as Fig. 10.5.

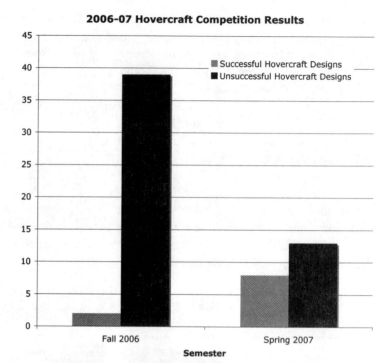

Fig. 10.5 Bar chart illustrating results of hovercraft competition

The final type of graphic to be discussed is the X-Y graph. It is inserted in a similar way to the pie and bar charts. First, create a table of data. Next, insert a chart and select X-Y or X-Y scatter as the type. In engineering, X-Y scatter plots are typically used because they show data points without lines drawn through the points. If the data has a story to tell, it is customary to fit a trend line to the points. Finally, make sure to label all axes (with units when applicable), to provide the graph with a descriptive title, and to provide a legend if necessary. The actual creation of an X-Y graph will be shown in the following section on how to use Excel to conduct design trade-off studies for your hovercraft.

10.4 EXCEL AS A DESIGN TOOL

As you may have realized by now, design is a highly iterative discipline. This can be understood when you consider the number of times your team's hovercraft weight estimate, size and shape has changed over the first few weeks of the semester and how often you were required to solve the static force balance equation $P_{REQ} = W/A$ to determine the required pressure (P_{REQ}) your levitation fan must create in order to ensure that your vehicle can hover. Similarly, how many times has your team solved $Q = h_{gap} \cdot l \cdot \sqrt{2P_{REQ}/\rho_{SL}}$ to determine the required air flow rate (Q) your levitation fan must output to hover with the desired gap height (h_{gap})? A tool like Excel can make calculations like these much less tedious, as well as provide guidance on the ideal size and shape for your vehicle. Let's consider the design of a hovercraft with a square planform of dimension L × L, as shown in Fig. 10.6.

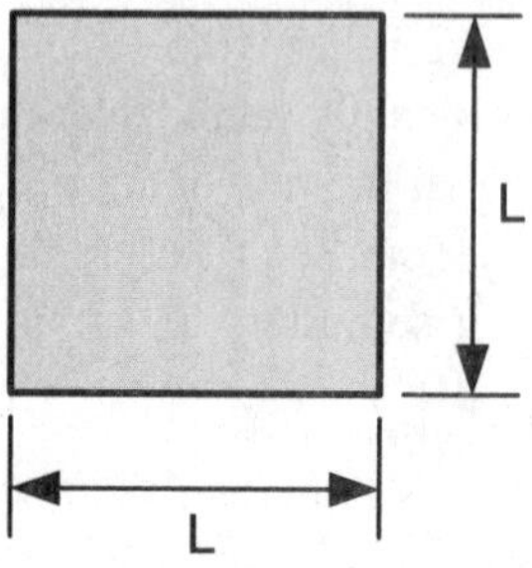

Fig. 10.6 Generic hovercraft configuration with a square planform section

For a simple shape like a square, it is very easy to determine the plenum area $\left(A = L^2\right)$ and leakage perimeter $\left(l = 4L\right)$. Using Excel, a levitation-system sizing calculator can be created to solve the pressure and flow rate fan requirements for a range of vehicle sizes. As a starting point, assume a constant vehicle weight of 3.91 lbs (17.4 N), as this is the weight estimate provided in Table 10.1. While the assumption that the weight of the vehicle is constant regardless of vehicle size (L) is certainly a flaw in this analysis, it is a reasonable assumption because the hull material comprises a relatively small proportion of the overall vehicle weight (~20%). An improved sizing calculator can be created with size-specific weight estimates by considering both the density of the hull material selected and the volume of material required. Table 10.3 illustrates a hovercraft levitation-system sizing calculator created in Microsoft Excel. This table calculates the levitation fan pressure and flow rate requirements over a range of vehicle sizes from L = 0.23 m (~9.0 in) to L = 0.65 m (~25.6 in) for a hovercraft with a fixed weight and a constant hover height requirement. It is recommended that your team create a calculator similar to this one for the configuration (size and geometry) and weight estimate for your particular vehicle.

Table 10.3
Hovercraft sizing calculator

	A	B	C	D	E	F	G	H	I
1	**Hovercraft Sizing**								
2									
3	**Weight Estimate (N):**			17.4			**Conversion Factors:**		
4	**h_{gap} Req. (m)**			0.002			0.00401	**inH$_2$0 / Pa**	
5	**Density (kg/m³):**			1.225			2119	**CFM / m³/s**	
6									
7	**L (m)**	**l (m)**	**A (m²)**	**P$_{req}$ (Pa)**	**P$_{req}$ (inH$_2$0)**	**Gap Area (m²)**	**V$_e$ (m/s)**	**Q$_{req}$ (m³/s)**	**Q$_{req}$ (CFM)**
8	0.23	0.92	0.053	328.9	1.32	0.00184	23.2	0.0426	90.4
9	0.26	1.04	0.068	257.4	1.03	0.00208	20.5	0.0426	90.4
10	0.29	1.16	0.084	206.9	0.83	0.00232	18.4	0.0426	90.4
11	0.32	1.28	0.102	169.9	0.68	0.00256	16.7	0.0426	90.4
12	0.35	1.40	0.123	142.0	0.57	0.00280	15.2	0.0426	90.4
13	0.38	1.52	0.144	120.5	0.48	0.00304	14.0	0.0426	90.4
14	0.41	1.64	0.168	103.5	0.42	0.00328	13.0	0.0426	90.4
15	0.44	1.76	0.194	89.9	0.36	0.00352	12.1	0.0426	90.4
16	0.47	1.88	0.221	78.8	0.32	0.00376	11.3	0.0426	90.4
17	0.50	2.00	0.250	69.6	0.28	0.00400	10.7	0.0426	90.4
18	0.53	2.12	0.281	61.9	0.25	0.00424	10.1	0.0426	90.4
19	0.56	2.24	0.314	55.5	0.22	0.00448	9.5	0.0426	90.4
20	0.59	2.36	0.348	50.0	0.20	0.00472	9.0	0.0426	90.4
21	0.62	2.48	0.384	45.3	0.18	0.00496	8.6	0.0426	90.4
22	0.65	2.60	0.423	41.2	0.17	0.00520	8.2	0.0426	90.4
23									
24	**Sample formulas used to create table above:**								
25	Cell B8 =4*A8				Cell F8 =$D$4*B8				
26	Cell C8 =A8^2				Cell G8 =SQRT(2*D8/$D$5)				
27	Cell D8 =$D$3/C8				Cell H8 =F8*G8				
28	Cell E8 =D8*$G$4				Cell I8 =H8*$G$5				

To truly appreciate the power of the Excel spreadsheet tool, consider the impact that adding a second battery would have on the size and type of levitation fan required. By hand, a small addition to your team's part list (and weight estimate) would require a timely and tedious recalculation effort. However, using Excel, a modified weight estimate can easily be considered by simply updating the value in cell D3 above. Excel allows the iterative nature of design to be managed in a swift and efficient manner. In addition, it provides the user with the ability to conduct design trade off studies instantaneously. How would the size of the levitation fan required change if you wanted to hover at a gap height of 3 mm (0.003 m) instead of 2 mm (0.002 m)? Simply change the value of cell D4 and note how the rest of the table automatically updates.

The sizing calculator created is also valuable because it allows its creator to quickly perform parametric design studies. For instance, look at column I of Table 10.3. For a constant vehicle weight and hover height, how does the required flow rate change with an increase in vehicle size (L)? Clearly, the required flow rate is independent of the vehicle size. Similarly, how does the required pressure change with an increase in vehicle size? By examining column D, it is clear that the pressure a fan must create decreases as the vehicle size (L) is increased. But how much does it decrease? To establish a better idea of the impact of this change, let's create an X-Y scatter plot of vehicle defining length L (column A) versus required plenum pressure (column D), as shown in Fig. 10.7 To do this, follow the procedure outlined in Section 10.3 for inserting a chart. The X-data series consists of cells A8:A22 and the Y-data series consists of cells D8:D22. This data can be inserted into the chart by choosing "Add" under the "Legend entries (Series)" section shown in Fig. 10.3.

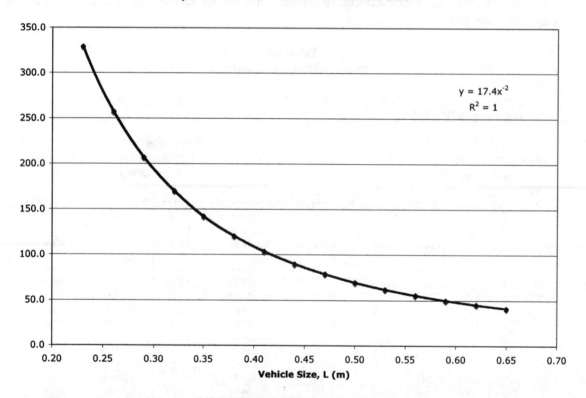

Fig. 10.7 X-Y scatter plot created in Microsoft Excel with a power series trendline added.
The Y axis (the ordinate) shows the required pressure in Pa.

Having created an X-Y scatter plot, you may be wondering why the figure above has a curve drawn through the data points. That is an excellent question. The data points shown in Fig. 10.7 are certainly trying to tell a story. To better understand that story, a trendline has been fit to the data points. To do this, select the "Layout" tab underneath the "Chart Tools" section of the ribbon. There is an icon for Trendline in the Analysis subsection. Click on the Trendline icon and select "More trendline options..." For the trend type, select Power. Also, put an X in the boxes next to "Display Equation on chart" and "Display R-Squared value on chart". Select "Close" and note how the chart has been updated with a trendline through each point on the graph.

By choosing to display the equation of the power series and the R^2 value on the plot, valuable information is provided about the nature and accuracy of the curve fit. The R^2 value is a measure of how well a curve fits a set of data points and typically ranges from 0 to 1. An R^2 value of 1 represents a perfect fit. Excellent experimental data may approach an R^2 value of 1, but rarely ever reaches this level of perfection. As a design engineer, you will need to look at charts like the one shown in Fig. 10.7 to make informed design decisions. What pressure is required for an extremely small vehicle ($L \Rightarrow 0$)? What pressure is required for an extremely large vehicle ($L \Rightarrow \infty$)? What is the benefit in designing a vehicle with a defining length L of 0.35 m instead of 0.25 m? What happens when you define a length of 0.65 m instead of 0.55 m?

Hopefully, you are beginning to see the power and utility of spreadsheet tools. You are encouraged to use Excel to aid your team's hovercraft design. With design being an inherently iterative discipline, you will not regret the initial investment of time required to make a robust hovercraft sizing calculator for your team.

10.5 SUMMARY

Microsoft Excel has been introduced and the procedure for creating tables and graphs has been described. Tables represent an excellent method for reporting data. They are well suited for the presentation of precise numerical data, such as the data required in a parts list. If you need six-figure accuracy, it is easy to obtain that accuracy in a tabular representation of the data.

Many different types of graphs are used to aid in visualizing data. Graphs are much less precise than tables, but they show trends, comparisons, and distributions much more effectively. The pie chart is used to indicate distribution of some quantity. The procedure for creating a pie chart using Excel was described in detail. The bar chart, used to compare the magnitudes of two or more quantities, was introduced. The bar chart was demonstrated and the methods used in its construction were discussed.

Trends are best-illustrated using X-Y graphs. X-Y graphs effectively show the trend of one variable with respect to another. The procedure for constructing this type of graphic has been provided, along with the steps necessary to add a trendline using Microsoft Excel.

Finally, the use of Excel as a design tool has been introduced. A sizing example has been given for the levitation system, but a similar analysis can be made for many of the other hovercraft sub-systems. The Excel spreadsheet greatly aids the iterative design process by providing a means for conducting automated and immediate design trade-off studies.

EXERCISES

10.1 Using Microsoft Excel, create a parts list for the hovercraft your team is developing. Remember to include the manufacturer, model number, description, cost, and weight for each component on your team's hovercraft.

10.2 Using Microsoft Excel, prepare a table similar to the one presented as Table 10.2 showing the breakdown of weight and cost between the main sub-systems of your hovercraft design. This table should be based on the parts list created in Exercise 10.1.

10.3 Use the data from Exercise 10.2 to construct a pie chart showing the distribution of your team's budget between each of the main sub-systems.

10.4 Use the data from Exercise 10.2 to construct a pie chart showing the distribution of your hovercraft's weight between each of the main sub-systems.

10.5 Use the data from Exercise 10.2 to construct a bar chart showing the distribution of your team's budget between each of the main sub-systems.

10.6 Use the data from Exercise 10.2 to construct a bar chart showing the distribution of your hovercraft's weight between each of the main sub-systems.

10.7 Using Microsoft Excel, create a table to calibrate an ultrasonic proximity sensor. The data should include distance in inches from an object as well as the associated NXT reading.

10.8 Prepare an X-Y scatter graph showing the data points found in Exercise 10.7. What trend can be seen in the data points plotted?

10.9 Fit an appropriate trend line to the X-Y scatter graph created in Exercise 10.8. Include the R^2 value on the plot. Was the trend line you have chosen a good fit?

10.10 Create a levitation sizing calculator, similar to the one described in Section 10.4, using a bullet shape configuration as shown below. Note that the bullet shape consists of an aft L × L square with a semi-circle nose.

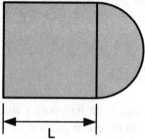

Fig. Ex 10.10

10.11 Based on the results of Exercise 10.10, create an X-Y scatter plot of the plenum area (A) versus the required plenum pressure (P_{REQ}).

10.12 Create a sizing calculator for your team's power sub-system that computes the required battery capacity based on the current draw of each electrical component on your hovercraft.

PART III

COMMUNICATION

CHAPTER 11

TECHNICAL REPORTS

11.1 INTRODUCTION

It is important that you enjoy writing, because engineers often have to prepare several hundred pages of reports, theoretical analyses, memos, technical briefs and letters during a typical year on the job. It is essential that you learn how to effectively communicate—by writing, speaking, listening and employing superb graphics. Communication, particularly good writing, is an extremely important skill. Advancement in your career will depend on your ability to write well. It is recognized that you will be taking several courses in the English Department and other departments in Social Sciences, Arts and the Humanities that will require many writing assignments. These courses should help you immensely with the structure of your composition and the development of good writing skills. Most of the assignments will be essays, term papers, or studies of selected works of literature. However, there are several differences between writing for an engineering company and writing to satisfy the requirements of courses such as English 101 or History 102.

In college, you write for a single reader—your instructor. He or she must read the paper to grade it, subsequently determining how well you are doing in class. In industry, many people within and outside the company, and with different backgrounds and experience, may read your report. In class, the teacher is the expert, but in industry the writer of the report is supposed to be the individual with the knowledge. Many of those working in industry do not want to read your report because they are busy: their phone is ringing continuously, meetings are scheduled back to back, and many important tasks must be completed before the end of the day. They read—if not skim—the report only because they must be aware of the information that it contains. They want to know the key issues, why these issues are important and who is going to take the actions necessary to resolve them. In writing reports in industry, you cut to the chase. There is no sense in writing a 200-word essay when the facts can be provided in a 40 to 50-word paragraph that is brief and cogent. An elegant writing style, so valued by instructors in the arts and humanities, is usually avoided in engineering documentation. In writing an engineering document, do not be subtle; instead be obvious, direct and factual.

In the following sections, some of the key elements of technical writing, including an overall approach, report organization, audience awareness and objective writing techniques will be described. Then a process for technical writing, which includes four phases: composing, revising, editing and proofreading is discussed.

11.2 APPROACH AND ORGANIZATION

The first step in writing a technical report is to be humble. Realize that only a very few of the many people who may receive a copy of your report will read it in its entirety. Busy managers and even your peers read selectively. To adapt to this attitude, organize your report into short, stand-alone sections that attract the selective reader. Three very important sections—the title page, summary and introduction are located at the front of the report so they are easy to find. At the front of your report, they attract attention and are more likely to be read. In college, you call the page summarizing your essay an abstract, but in industry it is often called an executive summary. If it is prepared for an executive, perhaps a manager will consider it sufficiently important to take time to read it. Follow the executive summary with an introduction and then the body of the report. A common outline to follow in organizing your report is provided below:

- Title page
- Executive summary
- Table of contents, list of figures and tables
- Nomenclature—only if necessary
- Introduction
- Technical issue sections
- Conclusions
- Appendices

The title page, the executive summary and the introduction are the most widely read parts of your report. Allow more time polishing these three parts, because they offer the best opportunity to convey the most important results of your investigation.

Title Page

The title page provides a concise title of the report. Keep the title short—usually less than ten words; you will have an opportunity to provide more detail in the body of the report. The authors list their names, affiliations, addresses and often their telephone and fax numbers and e-mail addresses. The reader, who may be anywhere in the world, should be provided with a means to contact the authors to ask questions. For large firms or government agencies, a report number is formally assigned and is listed with the date of release of the report on the title page. A title page of a report on a computer code for designing combustors is shown in Fig. 11.1. This report was prepared by several authors from different organizations and was sponsored by NASA (National Aeronautics and Space Administration) and ARL (Army Research Laboratory).

Executive Summary

The executive summary should, with rare exceptions, be less than a page in length (about 200 words). As the name implies, the executive summary provides, in a concise and cogent style, all the information that the busy executive needs to know about the content in the report. What does the big boss want to know? Three paragraphs usually satisfy the chief. First, briefly describe the objective of the study and the problems or issues that it addresses. Do not include the history leading to these issues. Although there is always history, the introduction is a much more suitable section for historical development and other background information related to the problem. In the second paragraph, describe your solution and/or the resolution of the issues. Give a very brief statement of the approach that you employed, but reserve the details for the body of the report. If actions are to be taken by others to resolve the issues, list the actions, those responsible and the dates of implementation. The final paragraph indicates the importance of the study to the business. Cost savings, improvements in the quality of the product, gains in the market share, enhanced reliability, etc. can be cited. You want to convince the executive that your work was worth its cost, valuable to the corporation and that your group performed admirably.

NASA
Technical Memorandum 107204

Army Research Laboratory
Memorandum Report ARL–MR–307

ALLSPD-3D
Version 1.0a

K.-H. Chen
Ohio Aerospace Institute
Brook Park, Ohio

B. Duncan
NYMA, Inc.
Brook Park, Ohio

D. Fricker
Vehicle Propulsion Directorate
U.S. Army Research Laboratory
Lewis Research Center
Cleveland, Ohio

J. Lee and A. Quealy
NYMA, Inc.
Brook Park, Ohio

April 1996

National Aeronautics and
Space Administration

U.S. ARMY

RESEARCH LABORATORY

Fig. 11.1 Illustration of a title page of a professional report.

Introduction

In the introduction, you restate the problem or issues. Although the problem was already defined in the executive summary, **redundancy is permitted in technical reports**. Recognize that you have many readers, so important information, such as a clear definition of the problem is placed in different sections of the report. In the introduction, the problem statement can be much more complete by expanding the amount of information describing the problem. A paragraph or two on the history of the problem is in order. People reading the introduction are more interested than those reading only the executive summary; therefore, more detail is appropriate.

Following the problem statement, establish the importance of the problem to both the company and the industry. Again, this is a repeat of what was included in the executive summary, but your arguments are expanded. You can cite statistics, briefly describe the analyses that lead to the alleged cost savings, give a sketch of customer comments indicating improved quality, etc. In the executive summary, you simply stated why the study was important. In the introduction, arguments are presented to convince the reader that the investigation was important and worth the time, effort and cost.

In the next segment of the introduction, you briefly describe the approach followed in addressing the problem. You begin with a literature search, followed by interviews with select customers, a study of the products that failed in service, interviews with manufacturing engineers and the design of a modification to decrease the failure rate and improve the reliability of the product. In other words, tell the reader what actions were required to solve the problem.

The final paragraph in the introduction outlines the remaining sections of the report. True, that information is already covered in the table of contents, but again, you repeat to accommodate the selective reader. Also, placing the report contents in the introduction gives you the opportunity to add a sentence or two describing the content of each section. Perhaps you can attract additional reader interest in one or more sections of the report and convince a few people to study the report more completely.

Organization

An organization for the report was suggested by the bullet list shown at the beginning of this section. However, you should recognize that an organization exists for every section and every paragraph in the report. In writing the opening paragraph of the section, you must convey the reason for including it in the report and the importance of its content. You certainly have a reason for including this section in the report, or you would not have wasted the time required to write it. By sharing this reason with your readers, you convince them that the section is important.

A paragraph is written to convey an idea, a thought or a concept. The first sentence in the paragraph must convey that idea, clearly and concisely. The second sentence should describe the significance or importance of the idea. The following sentences in the paragraph support the idea, expand its scope and give cogent arguments for its importance. State the idea in the first sentence and then add more substance in the remaining sentences. If the report is written properly, a speed-reader should be able to read the first two sentences of every paragraph and glean 80 to 90% of the important information conveyed in the report.

As you add detail to a section or a paragraph, develop a pattern of presentation that the reader will find easy to read and understand. First, tell the reader what he or she will be told, then provide the details of the story and finally summarize with the action to be taken to implement the solution. Do not try this approach in the theme that you have to write for English class, because the instructor will not appreciate your redundant style. Technical writing is not the same as theme writing. In technical writing, you will often reiterate facts to insure that the message is conveyed to the readers. Often one or more of the readers must take corrective actions for the benefit of the company and the industry.

You will frequently include analysis sections in engineering reports. In these sections, the problem is stated and then the solution is described. After the problem statement and the solution, you introduce the details. The reader is better prepared to follow the details (the difficult part of the analysis) after they know the solution to the problem.

11.3 KNOW YOUR READERS AND OBJECTIVE

Another important aspect to technical writing is to know something about the readers of your report. The language that you use in writing the report depends on the knowledge of the audience. For example, I am writing this book for engineering students. I know your math and verbal skills from the SAT scores required for admission. I recognize that you have very good language skills and even better math skills. You are bright and articulate. I believe that the content in this book closely matches your current abilities. However, there are a few sections in the book that deal with topics usually found in the business school. Profits and costs are important in product development, and although they may not be as interesting as a computer programming or circuit analysis, they need to be clearly understood.

When you write a report for a class assignment, you understand that the instructor is the only reader with which you need to communicate. You also know that he or she is knowledgeable. In that sense, the writing assignments in a typical college class are not realistic. For example, suppose your manager asks you to write an assembly routing (a step by step set of instructions for the assembly of a certain product) for the production of your new widget in a typical factory in the U. S. What language would you use as you write this routing? If the plant is in Florida or the Southwest for example, Spanish may be the prevalent language. You should also be aware that a significant fraction of the factory workers in the U. S. are functional illiterates and many others detest reading. In this case, it might be a good approach to use fewer words and many cartoons, photographs and drawings.

Know the audience **before** you begin to write, and adapt your language to their characteristics. Consider four different categories of readers:

> Specialists with language skills comparable to yours.
> Technical readers with mixed disciplines.
> Skilled readers, but not technically oriented.
> Poorly prepared readers who may be functional illiterates.

Category 1 is the audience that is the least difficult to address in your writings; the audience in the last category is the most difficult. Indeed, with poorly prepared readers, it is probably better to address them with visual presentations conveyed with television monitors and videotapes.

A final topic, to be considered in planning, is the classification of the technical document that you intend to write. What is the purpose of the task? Several classifications of technical writing are listed below:

- Reports: trip, progress, design, research, status, etc.
- Instructions: assembly manuals, training manuals and safety procedures.
- Proposals: new equipment, research funding, construction and development budget.
- Documents: engineering specifications, test procedures and laboratory results.

Each classification of writing has a different objective and requires a different writing style. If you are writing a proposal for development funding, you need persuasive arguments to justify the costs of the proposed program. On the other hand, if you are writing an instruction manual to assemble a product, arguments and reasons for funding are not an issue. Instead, you would prepare complete and simple descriptions with many illustrations to precisely explain how to accomplish a sequence of tasks involved in the assembly.

11.4   THE TECHNICAL WRITING PROCESS

Whether you are writing a report as part of your engineering responsibilities in industry or as a student in college, you will face a deadline. In college, the deadlines are imposed well in advance and you have a reasonable amount of time to prepare your report. In industry, you will have less time, and an allowance must be made for the delay in obtaining approvals from management before you can release the report. In both situations, you have some limited period of time to write the report. The idea is to start as soon as possible. Waiting until the day before the deadline is a recipe for disaster.

Many professionals do not like to write; consequently, they suffer from writer's block. They sit at their keyboard hoping for ideas to occur. Clearly, you do not want to join this group, and there is no need for you to do so. A technique is described in this section that should help you to avoid writer's block.

Define Your Task

Start your report by initially following the procedures described in the previous section. Understand the task at hand, and classify the type of technical document that you have to write. Define your audience; before starting to write, establish the level of detail and the language to employ. Prepare a skeletal outline of the report. Start with the outline presented previously and add the section headings for the body of the report. This outline is very brief, but it provides the structure for the report. This structure is important because it has divided the big task (writing the entire report) into several smaller tasks (writing one section of the report at a time).

Gather Information

It is impossible to write a report without information. You must generate the information to be presented in the report. You can use a variety of sources: interviews with peers, instructors or other knowledgeable persons who are willing to help; complete literature searches at the library; read the most suitable references; conduct additional Internet searches for information. Take notes as you read, gleaning the information applicable for your report. Be careful not to plagiarize. While you can use material from published works, you cannot copy the exact wording; the statements must be in your own words. If the report has an analysis, go to work and prepare a statement of the problem, execute its solution and make notes about interpreting the solution.

Organize Data

When you have collected most of the information that is to be discussed in your report, organize your notes into different topics that correspond to the section headings. Then incorporate your topics in an initial outline that will grow from a fraction of a page to several pages as you continue to incorporate notes in an organized format.

Compose Document

You are now ready to write. Writing is a tough task that requires a great deal of discipline and concentration. It is suggested that you schedule several blocks of time and reserve them exclusively for writing. The number of hours that should be scheduled will depend on the rate at which you compose, revise, edit and proofread. Some authors can compose about a page or so per hour, but most students need more time.

In scheduling a block of time for composing, you should be aware of your productive interval. Most writers take about a half an hour to come up to speed, and then they compose well for an hour or two before their attention and/or concentration begins to deteriorate. The quality of the composition begins to suffer at this time, and it is advisable to discontinue writing if you want to maintain the quality of your text. This fact alone should convince you to start your writing assignments well before the deadline date.

While you are writing, avoid distractions. Writing requires deep concentration—you must remember your message, the supporting arguments, the paragraph and sentence structure, grammar, vocabulary and spelling. Find a quiet, comfortable place and focus your entire concentration on the message and the manner in which you will present it in your report.

Naturally, some sections of a report are easier to write than others. The easiest are the appendices, because they carry factual details that are nearly effortless to report. The interior sections of the body of the report carry the technical details that are also easy to prepare, because describing detail is less concise and less cogent. Do not become careless on these sections because they are important; however, each sentence does not have to carry a knockout punch.

The most difficult sections are the introduction and the executive summary. It is advisable to write these two sections after the remainder of the first draft of the report is completed. Usually the introduction is written before the executive summary. Although the introduction contains much of the same information as the executive summary, it is more expansive. The introduction can be used as a guide in preparing the executive summary.

11.5   REVISING, EDITING AND PROOFREADING

Writing is a difficult assignment. Do not expect to be perfect in the beginning. Practice will help, but for most of us, it takes a very long time to improve our skills because writing is such a complex task. Expect to prepare several drafts of a paper or report, before it is ready to be released. In industry, several drafts are essential because you often will seek peer reviews and manager reviews are mandatory. In college, you have fewer formal requirements for multiple drafts, but preparing several drafts is a good idea if you want to improve the report and your grade.

First and Subsequent Drafts

The first draft is focused on composition, and the second is devoted to revising the initial composition. Use the first draft for expressing your ideas in reasonable form and in the correct sections of the report. In the second draft, focus on revising the composition. Defer the editing process, and concentrate on the ideas and their organization. Make sure the message is in the report and that it is clear to all of the readers. Polish the message later in the process.

Several hours should elapse between the first and second drafts of a given section of the report. If you read a section over and over again, you soon lose your ability to judge its quality. You need a fresh, rested brain for a critical review. In preparing this textbook, the author composes on one day and revises on the next day. Revising is always scheduled for an early morning block of time. When you are rested, concentration is usually at its highest level.

Let's make a clear distinction between composing, revising, editing and proofreading. Composition is writing the first draft where you formulate your ideas and organize the report into sections, subsections and paragraphs. Unless you are a super talented writer, your first draft is far from perfect. The second draft is for revisions where you focus on improving the composition. The third draft is for editing where errors in grammar, spelling, style, and usage are corrected. The fourth draft is for proofreading where typographical errors are eliminated.

Revising

When you revise your initial composition, be concerned with the ideas and the organization of the report. Is the report organized so that the reader will quickly ascertain the principle conclusions? Are the section headings descriptive? Sections and their headings are helpful to the reader because they help organize his or her thoughts. Additionally, they aid the writer in subdividing the task and keeping the subject of the section in focus. Are the sections the correct length? Sections that are too long tend to be ignored or the reader becomes tired and loses concentration before completely reading them.

Question the premise of every paragraph. Are the key ideas presented together with their importance before the details are included? Is enough detail given or have you included too many trivial items? Does the paragraph contain a single idea or have you tried to include two or even three ideas in the same paragraph? It is better to use a paragraph for every idea even if the paragraphs are short.

Have you added transition sentences or transitional phrases? The transition sentences, usually placed at the end of a paragraph, are designed to lead the reader from one idea to the next one. Transitional phrases, embedded in the paragraph, are to help the reader place the supporting facts in proper perspective. You contrast one fact with another, using words like **however** and **although**. You indicate additional facts with words such as **also** and **moreover**.

Editing

When the ideas flow smoothly and you are convinced that the reader will follow your concepts and agree with your arguments, begin to edit. Run the spell checker, and eliminate most of the typographical errors and the misspelled words in the report. Search for additional misspelled words, because the spell checker does not detect the difference between certain words like **grate** or **great**, or say **like** and **lime** and between **from** and **form**.

Look for excessively long sentences. When sentences become 30 to 40 words in length, they begin to tax the reader. It is better to use shorter sentences where the subject and the verb are close together. Make sure that the sentences are actually sentences with a subject and a verb of the same tense. Have you used any comma splices (attaching two sentences together with a comma)? Examine each sentence and eliminate unnecessary words or phrases. Find the subject and the verb, and attempt to strengthen them. Look for redundant words in a sentence, and substitute different words with similar meaning to eliminate redundancy. Be certain to employ the grammar check incorporated in the word processing program. It is not perfect, but it does locate many grammatical errors.

Proofreading

The final step is proofreading the paper to eliminate errors. Start by running the spell checker for the final time. Then print out a clean, hard copy to use for proofreading. Check all the numbers and equations in the text, tables and figures for accuracy. Then read the text for correctness. Most of us have trouble reading for accuracy because we read for content. We have been trained since first grade to read for content, but we rarely read for correctness.

To proofread, read each word separately. You are not trying to glean the idea from the sentence, so do not read the sentence as a whole. Instead, read the words as individual entities. If it is possible, arrange for some help from a friend—one person reading aloud to the other with both having a copy of the manuscript. The listener concentrates on the appearance of each word and then checks the text against the spoken word to verify its correctness.

11.6   WORD PROCESSING SOFTWARE

Word processing software is a great tool to use in preparing your technical documents. The significant advantage of using word processing software is the ability to revise, edit and make the necessary changes without excessive retyping. You can mark, delete, cut, copy and move text. You can easily insert words, phrases, sentences and/or paragraphs.

While working with word processing software, it is difficult to revise, edit, and proofread on the screen because only a portion of the page is visible. Better results are obtained if you print a hard copy and view the entire page. With the entire page, you can see several paragraphs at once and can check that they are in the correct sequence. Use double spacing when printing the first few drafts to give adequate space between the lines for your modifications.

After you have completed the revisions on the hard copy, make the required modifications to the text using word processing software and save the results to a floppy disk. It is recommended that you keep only the most recent version (draft) of the document. If you save several versions of the document, it is necessary to keep a logbook of the changes to each version. If the writing takes place over several weeks, it is easy to lose track of what changes you made and which version of the hard copy goes with which electronic copy. You will find it easier to keep one electronic file (on your floppy disk) of the most recent draft with a hard (paper) copy to serve as a back up.

Word processing software has several features that are helpful in editing. The spellchecker finds most of your typographical errors and provides suggestions to correct many misspelled words. The search or find command permits you to systematically examine the entire document so you may replace a specific word with a better substitute. A thesaurus is available to help you with word selection, but you must be careful when using it. Make sure you understand the meaning of the word that you select; do not try to impress the reader by using long or unusual words. Short words that are easily understood by the reader are preferred in technical writing.

Formatting the report is another significant advantage of word processing software. You can easily produce a document with a professional appearance. The formatting bar permits you to select the type of font, the point size (the height of the characters) and emphasis such as bold, underline or italic.

You can also format the page with four different types of line justification, which are commonly available. I am using word processing software to prepare this textbook, and applying "justify" alignment for both the right and left margins. The word processing software automatically added the spaces required in each line and aligned both margins. In designing your page, use generous margins. One-inch margins all around are standard unless the document is to be bound; then the left margin is usually increased from 1.00 to 1.25 inch.

Tables and graphs can be inserted in the text. Take advantage of the ability to introduce clip art into the text or to transfer spreadsheets and drawings produced in other software programs into your report. Position the tables and figures as close as possible to the location in the text where they are introduced. Try to avoid splitting a table between two pages. If the table is longer than a page or two, consider placing it in an appendix. Identify figures and photographs with a suitable caption. The caption, placed directly beneath the figure, is to reinforce the message conveyed by the illustration.

11.7 SUMMARY

Writing is a difficult skill to master; most engineers experience significant problems when they prepare reports early in their careers. Unfortunately, the writing experiences in college do not correspond well with the writing requirements in industry. In college, you write for a knowledgeable instructor. In industry, you write for a wide range of people with different reading abilities. Moreover, the audience often varies from assignment to assignment. In both college and industry, you write to meet deadlines imposed by others. While writing may not always seem to be fun, there are many techniques that you can employ to make writing much easier and more enjoyable.

The first technique is organization; an outline for a typical report was suggested. Gather information for the report from a wide variety of sources, and generate notes that you can use to refresh your memory as you write. Sort the notes and transpose the information to expand the outline for your report.

When you organize the outline, but before you begin to write, determine as much as possible about your audience. They may be technically knowledgeable regarding the subject or they may be functionally illiterate. The language that you use in the report will depend on the reader's ability to understand. Also understand the objective of your document. Is it a report, an extended memo, a proposal or an instructional manual? Styles differ depending on the objective, and you must be prepared to change accordingly.

There is a process to facilitate the preparation of any document. It begins with starting early and working systematically to produce a very professional document. Divide the report into sections and write the least difficult sections (appendices and technical detail portions) first. Defer the more difficult sections, such as the executive summary and the introduction until the other sections have been completed.

The actual writing is divided into four different tasks—composing, revising, editing and proofreading. Keep these tasks separate:

- Compose before revising
- Revise before editing

- Edit before proofreading
- Proofread with great care.

Multiple drafts are necessary with this approach, but the results are worth the effort.

Word processing software does not substitute for clear thinking, but it is extremely helpful in preparing professional documents. Word processing saves enormous amounts of time in a systematic editing process. It enables a mix of art, graphics and text neatly integrated into a single document. Word processing software has a thesaurus and word search features that are helpful in editing. It essentially turns a computer and printer into a print shop so that you have wide latitude in the style and appearance of your professional documents.

REFERENCES

1. Eisenberg, A., Effective Technical Communication, 2nd Edition, McGraw Hill, New York, NY, 1992.
2. Goldberg, D. E., Life Skills and Leadership for Engineers, McGraw Hill, New York, NY, 1995.
3. Elbow, P. Writing with Power, 2nd Edition Oxford University Press, New York, NY, 1998.
4. Alreb, G. T., C. T. Brusaw and W. E. Oliu, The Handbook of Technical Writing, 7th Edition, St. Martins Press, New York, NY, 2003
5. Struck, W. C. Osgood and R. Angell, Elements of Style, 4th Edition, Allyn & Bacon, Needham Heights, MA, 2000.
6. Pickett, N. A. et al, Technical English: Writing, Reading and Speaking, Longman, Reading MA, 2000.

EXERCISES

11.1 Prepare a brief outline of the organization for the final report describing the development of the product that your team is designing.

11.2 Prepare an extended outline of the final report for the development of the product that your team is designing.

11.3 Write a section describing one of the subsystems for your team's project.

11.4 Write a section covering the testing of a subsystem of the product that your team is designing.

11.5 Write the Introduction for the final report on the product that your team is developing.

11.6 Write the Executive Summary for the final report on the product that your team is developing.

11.7 Revise the Introduction that another team member wrote.

11.8 Edit the Introduction after it has been composed and revised by others.

11.9 Proofread the Introduction after it has been composed, revised, and edited by others.

11.10 Edit the final report after other team members have completed it.

11.11 Proofread the final report after other team members have completed it.

CHAPTER 12

DESIGN BRIEFINGS

12.1   INTRODUCTION

You communicate by writing, speaking and with a number of different visual methods. All these modes of communication play a role as you attempt to convey your thoughts, ideas and concerns to others, ands they all are important. Let's focus your attention on speaking in this chapter. You use speech almost continuously in your daily life. Why do you need to study about design briefings? There are several reasons. You usually speak with your friends and family in an informal style. You know them and feel comfortable talking with them. They know you. They are concerned with your well being, genuinely like you and are usually interested in what you have to say. A professional presentation is different. It is a formal event that is usually scheduled well in advance. The audience may include a few friends, but usually it consists of peers (coworkers) and strangers. The time you have in which to convey your message is limited, and the audience may not be interested in your message. A person or two in the audience may be managers who control your advancement in the company. There are many reasons for tension headaches as you prepare for a professional presentation.

The design briefing is extremely important to both the product development process and to your career. Information must be effectively transmitted to your peers, management and any external parties involved in the project. Clear messages accurately defining problems that the development team is effectively addressing are imperative. On the other hand, ambiguous messages are often misunderstood, hinder the definition of the problem and lead to delays in implementing timely solutions. Clearly, such messages should be avoided. The design briefing provides an opportunity to review the status of a specific product development. It also permits peers to share their ideas with you and affords management an open forum for assessing the quality of your work and the progress made by your development team. Because the professional presentation is critically important, you should become knowledgeable about some effective techniques for accurately delivering your messages to a mixed group of strangers, peers and acquaintances.

12.2   SPEECHES, PRESENTATIONS AND DISCUSSIONS

To begin, let's distinguish between three types of formal methods of oral communication—speeches, presentations and group discussions.

Speeches

The speech is the most formal of the three modes of communication. In the year of a presidential or senatorial election, there are always many examples of speeches. Everyone is besieged with political addresses that clearly illustrate the key features of speeches. These speeches are given to large audiences in huge rooms, stadiums or coliseums. The setting is usually not appropriate for visual aids although the speaker usually uses a teleprompter. The audience is diverse with a wide range of concerns. Those listening to the speech are of widely different ages with different interests and persuasions. The speech is carefully scripted with the speaker usually reading or closely following the script. Ad-libbing is carefully avoided. Communication is one way—from the speaker to the audience. Questions are usually not appropriate and generally not appreciated. Time is strictly enforced unless you are the President of the United States giving the State of the Union address. Fortunately, engineers are rarely called upon to make speeches; hence, this topic will not be developed further in this book.

Presentations

Professional presentations differ from speeches. Presentations are made to smaller audiences in smaller rooms that are usually equipped with electric power, light controls and projection devices. You depend on visual aids, demonstrations, simulations and other props to help convey your message during a typical presentation. The audience is knowledgeable (about your topic) and usually has many common characteristics. The presentation is very carefully prepared because of its importance. It has order and structure, but it is not scripted. The flow of information is largely from the speaker to the audience; however, questions are permitted and often encouraged. The speaker is considered the expert, but discussion, comments and questions give the audience an opportunity to share their knowledge of the topic. Time is carefully controlled and is often insufficient from the speaker's viewpoint. The professional presentation is a vitally important mode of oral communication that you must quickly learn to master.

Group Discussions

Group discussions are also very important to the design engineer. The audience is smaller with much more in common. The group may all be members of a development team. The topic being discussed is narrowly focused. The speaker (central person) serves as a moderator, and he or she is an expert on the topic being considered. However, members of the discussion group (audience) may be as knowledgeable as the speaker (moderator). The moderator works in two modes. He or she may act as a presenter giving brief background information to frame an issue that prompts any member of the group to discuss the topic. The moderator may also direct questions to a group member known to be the most knowledgeable concerning the issue being addressed. The speaker (moderator) controls the flow of the information, but the flow is clearly multi-directional—from speaker to the group and from one group member to another. When a group (team) discusses a design issue, time is difficult to control and the content and range of coverage are strongly dependent on the skill of the moderator. Group discussion is very important in industry because the leader of a development team will often use this method of communication to identify problems and to initiate efforts directed toward their solution. Group discussion methods will not be described in this textbook; however, it is recommended that you watch *Washington Week in Review* on the PBS television network to gain some insights. While the participants are journalists, you can adopt many of moderator's clever techniques to guide engineering discussions.

Design Briefing

A design briefing is a type of professional presentation. It is the method you will use in speaking before the class and will be an educational objective emphasized in this course. Indeed, you probably will be required to make presentations describing your team's product development on at least two occasions—the preliminary and the final design reviews. In these two design reviews, you will employ visual aids to assist in delivering the information required to describe the status of your project with conciseness and clarity.

12.3   PREPARING FOR THE BRIEFING

A design briefing is too important to take casually. You should prepare very carefully to insure that you will accurately report the status of your team's development and to identify problems or unresolved issues. There are three aspects that you should consider in your preparations:

- Identify the audience
- Plan the organization of the presentation and your coverage of each topic
- Prepare interesting and attractive visual aids.

Identifying the Audience

In a typical college classroom, it is much too easy to identify the audience. You have your peers, the instructor and perhaps a visitor or two from industry. In an industrial setting, the audience will be more difficult to identify because it will be much more diverse. The size and diversity of the audience depends on the magnitude of the development project. A small briefing will have 10 to 20 people in attendance with most participants from within the company. A large briefing may include 50 to 100 representatives who are both internal and external to the company. The characteristics of the audience are important because you must adapt the content of the presentation to the audience. Classify your audience with regard to their status, interest and knowledge before you begin to plan the style and content for your presentation.

The status of the audience refers to their position in the various organizations they represent. Are they peers, managers or executives? If they are a mix, which is likely, who is the most important? If you are preparing a design briefing for high-level management, it must be concise, cogent and void of minor detail. Executives are busy, stressed and always short on time. High-level executives and managers are impatient and rarely interested in the technical details that engineers love to discuss in their presentations. Recognize these characteristics and adjust the content and the timing of your presentation accordingly. The executives are interested in costs, schedule, market factors, performance, risk or any other critical issue that will delay the development or increase projected costs of either the development or the product. Usually you will have only 5 to 10 minutes to convey this information at an executive briefing.

First and even second level managers have more time and are more interested in you and the designs that you are creating. If you prepare the presentation for this lower level of management, you can safely plan for more time (10 to 20 minutes). These managers are likely to be engineers, and they will share your knowledge of the subject. In fact, they probably will be more experienced, expert and knowledgeable than you. They will want to know about the schedule and costs because they share responsibilities with the higher-level management for meeting these goals. However, they are also interested in the important details, and you can engage them in a discussion of subsystem performance. The first and second level managers usually control the resources for the development team. If you need help, make sure that they receive the message in enough detail to provide the assistance required. Do not hide the problems that your team is encountering. Managers do not like surprises. If you have a problem, make sure they understand it and are able to participate in its solution. If you hide a problem from management and it causes a delay or escalation in costs, the manager will take the heat and you will be in a very cold doghouse.

Peer Reviews

Design briefings for peers are often less tense, and you will have more time (20 to 30 minutes). You will address schedule and cost issues because everyone needs to know if your team is meeting the milestones on the Gantt charts. However, the main thrust of your presentation will pertain to design details. If you are working on a subsystem that interfaces with other subsystems, you will cover your subsystem in sufficient detail for all the team members to understand its geometry, interfaces, performance, etc. It is particularly important that the interface issue be fully addressed in the presentation. If four subsystems are to be integrated in a given product, many details about all four subsystems must be addressed to ensure that the integration goes smoothly. Suppose that your team is developing a power tool with a motor that draws a current of 10 amps, and some team member plans to employ a switch, rated at 6 amps, to turn the power on/off. The switch will probably function satisfactorily in the short-term tests with the prototype, but it will malfunction in the field with extended usage sometime after the product has been released to the market. Clearly, the two persons responsible for the interface between the motor and the switch did not communicate properly. The integration of the two subsystems failed.

The most important purpose of design briefings with peers is to make certain that all of the details have been addressed. You strive to integrate the subsystems without problems arising when the prototype is assembled and tested.

Peer reviews in the absence of management are very beneficial in the development process. The knowledge of most of the team members is about the same although some members are more experienced than others. The topic is usually a detailed review of one subsystem or another. The audience is fresh and capable of providing a critical assessment of the technology employed. Your peers may check the accuracy of your analysis, comment on the choice of materials, discuss methods to use for manufacturing and assembly and relate their experiences with similar designs on previous products. Peer reviews afford the opportunity for the synergism that makes good products better. They also provide a forum for passing on important lessons from the more experienced designers to beginning designers in a friendly, stress-free environment.

The subject of the presentation is self-evident in a design review. You are developing a product, and the content of the design briefing will deal with one or more issues regarding this development. There is a choice of topics and considerable latitude regarding the content. Although advice has been given in previous paragraphs about matching content with the interests of your audience, you must understand the importance of knowing who will be in your audience. Executive reviews serve a different purpose than peer reviews, and the content and the time allotted for the presentation is adjusted accordingly.

For every type of review, there are two **absolute** rules that you must follow. First, **you must know the material absolutely cold**. It takes an audience about two seconds to understand that you are faking it. Also, some presenters fail to rehearse adequately, and the audience often misinterprets this lack of preparation for insufficient knowledge. In either event, when they realize that you are not the expert you allege to be. Your presentation fails. You have **lost** the opportunity to effectively communicate your message. Second, **be enthusiastic**. It is all right to be calm and cool, but do not be dead on arrival. You must command the attention of your audience or they will quickly turn off. You control the attention of the audience only if you are enthusiastic and knowledgeable in presenting your material (story).

12.4   PRESENTATION STRUCTURE

There is a well-accepted structure for professional presentations. If you are familiar with PowerPoint®, a graphics presentation program (GPP) marketed by Microsoft, you are already aware of the templates included in this program for various types of presentations. An excellent structure for a design briefing is shown in the outline for a professional presentation presented below:

- Title
- Overview
- Status
- Introduction
- Technical Topics
  - First Topic
    - Background
    - Status
  - Second Topic

- - Background
  - Status
- Final Topic
  - - Background
    - Status
- Summary
- Action Items

Let's examine each of these topics individually and discuss the content that should be included on the visual aids that accompany your presentation. Recognize that the visual aids (35 mm slides, overhead transparencies or electronic projection) control the flow and the information that you will present. For this reason, the topics listed above in the context of information included on the presentation slides will be addressed.

Title Slide

The title slide[1], illustrated in Fig. 12.1, obviously carries the title of the presentation. However, it also states your affiliation and the names of the team members that contributed to the work. Because you may be reporting on the work of the entire team, it is necessary for you to acknowledge their contributions on the title slide and in your opening remarks. A brief title—ten words or less—is a good rule to follow in drafting the title for your presentation. It is also a good idea to use descriptive words in the title. For example, you could use **PRELIMINARY DESIGN BRIEFING: HOVERCRAFT SYSTEM: LEVITATION SUBSYSTEM**. Seven words and you have informed the audience of the topic (a design briefing or review), the type of briefing (preliminary), and the product being developed (a levitation subsystem for a hovercraft). A significant amount of information is successfully conveyed in seven words.

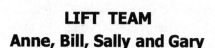

PRELIMINARY DESIGN BRIEFING
HOVERCRAFT SYSTEM
LEVITATION SUBSYSTEM

LIFT TEAM
Anne, Bill, Sally and Gary

Fig. 12.1 Illustration of an informative and descriptive title slide.

Below the title, the team name together with the name of each team member involved in the development is listed. It is essential that you share the ownership of the material covered in your presentation. Often a single member of the team covers the work of several people in a design briefing. It is not ethical to make the presentation without acknowledging the contributions of others. Students typically use first names in identifying team members, but in industry the complete names of the team members are used. If necessary, division and department affiliations are also stated.

---

[1] Reference will be made to slides in this discussion; however, the information provided is also applicable for overhead transparencies.

It is also a good idea to date the title slide and to cite the occasion for the presentation. If you make many presentations, this information is useful when referring to your files, allowing you to easily reuse some slides after selective editing and/or updating.

Overview Slide

The second slide presents an overview of the design briefing. The overview for a presentation is like a table of contents for a report. You list the main topics that you intend to cover in the presentation. In other words, you tell the audience what you plan to tell them in the next 10 to 20 minutes. The number of topics should be limited. In a focused presentation, you can cover about a half dozen topics. If you press and try to cover ten or more topics, you will encounter difficulty maintaining the attention of the audience throughout the presentation. Constrain the tendency to tell all—focus on the important issues. In a design review, the topics usually are organized to correspond with the subsystems involved in the product under development. For a component redesign, you concentrate on the issues guiding the component modification being considered. An example, of an overview slide for a preliminary design review of a lift subsystem system is illustrated in Fig. 12.2.

OVERVIEW
Preliminary Design Concepts

- Centrifugal fan selected to provide air pressure
- Hovercraft structure for air cushion is circular
  - Provides maximum lift with minimum leakage
- Out riggers used to support the two propulsion fans
  - Improved positioning for enhanced steering
- Hooks fastened to deck to provide anchors for holding deck mounted equipment
  - Auxiliary power supply, relays and RCX
- Proximity sensor mount located at the bow
- Six mill plastic sheet used for containment skirt
- Construction techniques insure robust body

February 8, 2006   LIFT TEAM   6

Fig. 12.2 The slide with the overview indicates the content in your presentation.

Status Slide

The third slide presents the product development status. You should report on the progress made by the team to date. It is a good idea to incorporate a Gantt chart showing the development schedule either on the status slide or a separate slide. The progress of the team on each task is defined on the Gantt chart and is immediately conveyed to the audience. If there are problems or uncertainties, this is the time to air the difficulties that the team is encountering. An example slide, shown in Fig. 12.3, illustrates uncertainties that the team has regarding the design of the anchoring system for holding deck mounted equipment while maintaining balance for a floating hovercraft.

Overall Status

- Team progress is excellent
  - Team is on schedule and operating with high efficiency
- Several design concepts generated for lift subsystem
- Concept selection complete on except for method to achieve perfect balance of floating structure
- Uncertainties due to positioning of deck mounted equipment to achieve required balance
  - Meeting with other teams to determine weighs and dimensions of equipment

Fig. 12.3 Status slide indicates overall progress and accomplishments while signaling potential problems.

February 8, 2006   LIFT TEAM   11

Introductory Slide

The fourth slide covers the introductory material such as background, history or previous issues. It is important that the audience understand the product, the main objectives of the development, the role the team has in that development and the most pressing of the current issues. This introductory slide permits you to set the stage for the body of the presentation that follows. The audience responds better to your presentation when they know in advance the background and the topics that you will cover. They are better prepared mentally to receive your message.

Technical Topics

The body of the presentation usually deals with five or six technical topics. In most instances, the topics selected correspond to the more critical subsystems involved in designing the product. In the redesign of a component, the topics deal with the issues arising when considering new design concepts. Avoid trying to cover a large number of topics; it is better to report on a limited number of topics thoroughly than to rush through a dozen topics with incomplete coverage. It is suggested that you use two slides for each topic. The first slide describes the progress made by the team in generating design concepts and indicates the criteria employed in selecting the best concept. The second slide covers the status of developing the selected concept. The type of information reported on this slide includes accomplishments, outstanding issues, lessons learned, etc. An example of the status of the design of a lift subsystem including a powerful centrifugal fan and a robust auxiliary power supply is presented in Fig. 12.4.

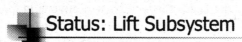

Status: Lift Subsystem

- **Design Concepts**
  - Powerful centrifugal fan delivers high pressure at a relatively high flow rate
  - Significant current requirement 10 A at 12 V
  - Auxiliary power supply provides 12 V at 2.0 A-h
  - Relay to switch from RCX to auxiliary supply
- **Design Goals**
  - Reliability of lift system
  - Sufficient power for 12 minute mission

February 8, 2006   LIFT TEAM   12

Fig. 12.4 Information presented in describing the early design of a reliable lift subsystem for a hovercraft.

Concluding Slides

After the technical topics have been covered, you must conclude the presentation with a decisive pair of slides. Suggestions for two concluding slides are presented in Figs. 12.5 and 12.6. The next to last slide in the presentation, shown in Fig. 12.5, is titled "Key Issues." This is your opportunity to identify very important issues (questions) that your team has discovered. Do not hesitate. This is the time to introduce the uncertainties and to come forward and seek help. Design reviews are not competitive. In a design review, you seek help regardless of your status or role as a member of the audience. Managers will arrange support for your team if it is required. Peers will make suggestions and introduce fresh approaches to unresolved problems that the team may find useful. The design review is a formal process in a product development in which everyone participates. Take advantage of this review to identify uncertainties and to seek help whenever your team needs assistance.

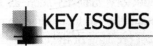
KEY ISSUES

- UNCERTAINTIES
  - Method to achieve balance of floating deck
  - Ability of auxiliary power supply to provide high current for a long mission
  - Attachment of plastic film to deck to achieve air tight skirt
- Action Plan
  - Firm up design interfaces
    - Propose placement of relays, RCX, sensor, propulsion fans and batteries to other teams
  - Test fan and power supply under load to establish mission duration limit
  - Test attachment method to insure against leakage

February 8, 2006   LIFT TEAM   13

Fig. 12.5 The "key issues" focuses the team and management on your most significant problems.

The final slide presented in Fig. 12.6 is to present your plans for the future. Management is interested in how you intend to solve the issues that you have disclosed and the amount of time and money required for the solutions. Your peers will be interested in the technical approaches that you intend to follow. You should also define any requirement that you have of anyone in the audience. If you need help from your peers or more resources from the manager, be certain this assistance is clearly indicated on this final slide. You should distinctly identify those individuals who have agreed to provide the requested support. If action items are listed, clearly identify the responsible individual and estimate the time to completion. The planning incorporated in the final slide is very important. It is the prescription for your team's get-well program.

Future Plans

- Most design concepts established
- Finalize selection of superior concepts
- Procure all materials, fans, sensors, batteries, relays, RCX, etc.
  - Build structure and test to determine if it will withstand the forces
  - Test fan and auxiliary power supply
  - Test to develop lifting capacity
  - Place all deck mounted components and suspend structure to achieve balance

February 8, 2006   LIFT TEAM   14

Fig. 12.6 The "future plans" slide gives you the opportunity to describe your approach for solving problems that the team is encountering.

12.5   TYPES OF VISUAL AIDS

Because visual aids control the content and the flow of the information conveyed to the audience, they are essential for a technical presentation. Slides provide visual information that reinforces the verbal information. With visual aids you communicate using two of a person's five senses. Take time to carefully select the type of visual aid that will most effectively convey your message.

In selecting visual aids, you are usually limited—first, in your ability to produce the visual aids within the time and financial constrains imposed on the development team and second with regard to the availability of equipment and the room in which the presentation will be made. Is the room equipped with a 35 mm slide projector, an overhead projector or a computer-controlled projector? Another important consideration is the control of the light intensity in the room. Most rooms have light switches used to control the overhead lights. However, some rooms do not have blinds and bright sunshine will cause problems with the visibility of 35 mm slides when they are projected.

Computer Projected Slides

Computer projected slides are an excellent choice if the room can be darkened. Computer projection of slides has several advantages:

- The presentation is easy to prepare using PowerPoint.
- The slides are in color are more effective than black and white illustrations.
- Advanced visual effects (slide transition and slide building feature) enhance the impact of the presentation.
- It is also easy for you to copy your presentation onto a floppy disk or a memory stick.

A computer is attached to a digital projection system that stores your presentation material. You employ a mouse to click through your slide line by line. Finally, low-cost handouts with four or six slides per page can be prepared that help the audience follow your presentation. The only disadvantage of computer-projected slides is that sometimes a relatively dark room is required for projection. With a darkened room, eye contact with the audience or even reading their body language is not possible. You lose your visual contact with the audience that is so important in reading their response to your presentation style.

Overhead Transparencies

Projecting overhead transparencies is probably the most common method of presenting visual material. Overhead projectors are relatively inexpensive (a couple of hundred dollars compared to about a thousand dollars for the computer controlled projectors). Hence, overhead projectors are available in most rooms. Overhead projectors are also the best choice if the room cannot be darkened. The projectors are sufficiently bright to provide quality images without extinguishing the lights. Bright rooms have an advantage because they aid you keep the audience alert. Overhead transparencies are also easy to make on a copy machine or a laser printer. If you are willing to sacrifice the benefits of color, utilize inexpensive black and white transparencies. If you want to use color in the presentation because it is more effective, the costs increase somewhat when using a color laser printer or a color copier to prepare the overheads. However, color ink-jet printers can be used to produce good quality color transparencies at a reasonable cost.

35-mm Slides

The 35 mm slide presentation is a third option for visual aids; however, the equipment required is not always available. Projectors for 35-mm slides are common, but in an engineering college, they usually are much more difficult to find than the overhead projector. A slide presentation is the preferred option if you have many photographs to project because the quality of colored 35-mm slides is excellent. The slides are inexpensive and easy to prepare if you have a camera and the time necessary for the film processing. It is possible to prepare colored slides from your PowerPoint files, but conversion from the digital format to the 35-mm format requires special equipment. Again, a significant disadvantage is that the room must be darkened to view the image from a 35-mm projector. Also, the relatively long distance required from the typical projector to the screen precludes the use of very small conference rooms for presentations.

Photographs taken with a high resolution digital camera are presented with a computer driven projector. These images are clear, and if the resolution of the camera was sufficient, they closely match the quality of the images projected from 35 mm slides. The ability to present both slides and photographs from the same projector is a significant advantage.

Although ineffective, sometimes speakers will employ two or three different types of projectors. Such a tactic should be avoided as switching from one projector to the other is disruptive. If possible, use only one type of projector. If you must use more than one type of projector to properly present your message, try to minimize the number of times you switch from one device to the other.

Other Types of Visual Aids

There are several other visual aids sometimes used in professional presentations. Videotapes are common; unfortunately, the small size of TV monitors often makes viewing difficult for a large audience. However, if you want to show motion or group dynamics, video clips are clearly the best approach. Video cameras are readily available and after some practice you can become a reasonably good video producer.

Motion picture films, particularly of older material with historical interest, are very effective in developing a long-term prospective. However, finding an 8-mm or 16-mm motion picture projector may be difficult—allow time in your schedule for making the necessary arrangements.

Hardware, or other materials used in the product development, is sometimes passed around the audience during a presentation. The hands-on opportunity for the audience is a nice touch, but you pay for it with the loss of attention by some members of the audience during the inspection period. It is recommended that you defer passing materials to the audience during your presentation. Instead, after the presentation, invite the audience to inspect exhibits that are placed on a strategically located table. This approach maintains the attention of the audience throughout the presentation while giving those interested in the hardware time for a much more thorough examination.

12.6 DELIVERY

Excellent presentations require well-prepared slides and a smooth, well-paced delivery. There are many aspects to the delivery part of the presentation process including: dress, body language, audience control, voice control and timing.

Attire

Let's start with dress. Broadly speaking, there are four levels of dress. The highest level, black tie for men and formal gowns for women, is a rare event and is never appropriate for design briefings. The next level, business attire for men and women, is sometimes appropriate for design briefings. A conservative Eastern company may have a dress code requiring business attire; whereas, a less formal Western company would encourage more casual attire. Casual attire should not be confused with sloppy attire. Casual attire is neat and tasteful, but without suits, ties and white shirts for men and suits for women. Although becoming more common in the workplace, sloppy attire (old jeans and a sweatshirt) is strictly taboo. You should select tasteful business or casual attire when you dress for the presentation. While it is recognized that most students these days prefer a relaxed, collegiate style of dress, you should resist the impulse to dress too casually. If you look like a homeless person, who will believe your message?

Body Language

Posture is another important element in the presentation. In the words of former President Reagan, "stand tall." Your body language signals your attitude to the audience—you are the presenter and in control. Make sure you are calm, cool and collected. Nervous gestures with your hands, rocking on the balls of your feet, scratching your head, pulling on your ear, etc. should be avoided.

Audience Control

Before you begin your presentation, take control of the audience. One approach is to pause a moment before projecting the title slide and immediately make eye contact with the entire audience. How do you look at everyone in the house? You scan the audience from left to right and then back looking slightly over their heads. Occasionally drop your eyes and make eye contact for a second or two with one individual and then another. Pause long enough for the audience to become silent (10 to 20 seconds). If someone is rude and continues to talk, walk toward them and politely ask for their attention. When you have everyone's attention, project the title slide and begin your delivery.

If you have rehearsed, you will not need notes. The slides carry enough information to trigger your memory. If they do not, you have not rehearsed long enough to remember the issues. Scripting the presentation is not recommended. People who script will eventually start to read their comments, which is a deadly practice. Rehearse until you are confident. Make notes to help you rehearse, but do not use them during the presentation.

Voice Control

When you begin to speak, be certain everyone can hear you. If you are not sure of how far your voice carries, ask those in the back of the room if they can hear. If you use a microphone with an amplifier and speakers, be careful; the tendency is to speak too loudly. Try to control the loudness of your voice within the first minute of the presentation when you introduce the topic with the title slide. The audience will read that slide with anticipation, and they will bear with you as you adjust the volume of your voice.

Do not mumble. Speak slowly, clearly and carefully enunciate each word. It is better to present your material too slowly than too rapidly. If you can improve your enunciation, you will be able to increase the rate of your delivery without losing the audience.

Don't run out of breath while speaking. Learn to complete a sentence, pause for breath (without a gasp) and then continue with the next sentence. There is nothing wrong with an occasional pause in the presentation, provided the pause is short.

Timing

If you forget some detail, do not worry. Skip it, and move to the next point that you are trying to make. If the detail is critical, count on one of your team members to raise the issue in the question and answer period. Most speakers forget some of the material that they intended to present and introduce some additional items on an ad-hoc basis. Usually, the audience will never be aware of your omissions or additions.

Studies have clearly shown that people can listen to conversation faster that the presenter can speak. The trouble with listening to someone who speaks very rapidly is not their rate of speech, but their enunciation. Many people who speak rapidly tend to slur their words, and the audience has trouble understanding poorly pronounced words. The trouble you have in listening to a person speaking too slowly is to stay with his or her message; while you are waiting for the next word, your mind goes off on a mental tangent. You anticipate that your mind will return to the speaker in time for the next word, but unfortunately the mind is sometimes tardy. You miss keywords or even complete sentences before your attention returns to the speaker.

Listening is a skill that many people have not developed. As a speaker, it is your responsibility to keep the listener, even the poor ones, on the topic. You may use several techniques to maintain the attention of the audience. Employing the screen is probably the most effective tool for this purpose. As you project the slide, walk to the screen and point to the line on the slide that corresponds to the topic that you are addressing. If a few members in the audience are coming back from their mental tangents, you reset their attention on the current topic.

When referring to the image projected on the screen—and good speakers use this technique—stand to the left of the screen as shown in Fig. 12.7. When you stand to the side of the screen being careful not to block anyone's view. Face the audience and maintain eye contact. When you must look at the slide, turn your head to the side and read over your shoulder. The 90° turn of your head and the glance at the slide should be quick because you want to continue to face the audience. Under no circumstance should you stand with your back to the audience and begin reading from the slide as if it were a script.

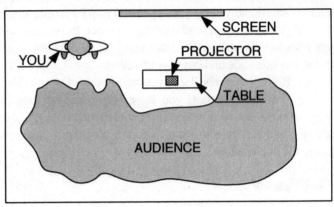

Fig. 12.7 Typical room arrangement for presentations. Position yourself to the left of the screen and face the audience.

As you develop the discussion, point to key phrases to keep the audience focused on the topic. Engage both of their senses. Deliver your message through their eyes and ears simultaneously. When you point to a line on the slide, point to the left side of the line. People read from left to right—so start them from the initial point on the left side.

As you speak, modulate your tone and volume. Avoid both a monotone and a singsong delivery. A continuous tone of voice tends to put the audience asleep. A singsong delivery is annoying to many listeners. If you have a high-pitched voice, try to lower the pitch because a high-pitch tone is bothersome to many people.

If a question is asked, answer it promptly unless you intend to cover material later in the presentation that addresses the question. The answer should be brief, and you should try to avoid an extended dialog with some member of the audience. If a member of the audience is persistent with several related questions, simply indicate that you will speak with him or her off-line immediately after the presentation. The timing of your presentation can be destroyed by too many questions. Some questions are helpful because they permit audience participation, but an excessive number of questions cause the speaker to lose control of the topic, the flow and the timing of the presentation. Too many questions cause the format of the communication to change from a presentation to a group discussion. Group discussions have a purpose; however, they are different than a design briefing.

12.7 SUMMARY

The design briefing is extremely important both to successful product development and to your career. Information must be effectively transmitted to your peers, management and to external parties involved in the project. Clear messages, which accurately define the problems that the entire development team can effectively address, are imperative. The design briefing provides the opportunity to inform management and peers of your progress and problems.

There are three types of formal communication: speeches, professional presentations and group discussions. The characteristics of these three types of communication are described. For the engineer, the professional presentation is the most important and the formal speech is the least important. The main emphasis of this chapter is on the professional presentation with a focus on the design briefing.

Preparation for a presentation is essential. Accomplished speakers often spend the better part of a day preparing for a 20-minute briefing. There are three important elements in the preparation. First, know the characteristics of your audience, and be certain that the presentation's content corresponds to their interests. Second, organize the presentation in a manner acceptable to the audience. Insure that the organization is efficient so that you use the allotted time effectively. Prepare high-quality visual aids (slides) that will help you convey your message and rehearse until you command the subject material.

A structure for the presentation is recommended that is commonly employed in design briefings. This outline includes four initial slides—title, overview, status and introduction. The body of the presentation is devoted to technical details using about eight to twelve slides. Two closure slides are employed for a summary and a listing of action items.

The type of visual aid selected is important because the visual transmission of information will markedly affect the outcome of the presentation. Computer projected slides have some significant advantages, but the availability of equipment often precludes their use. Overhead transparencies are the most common medium. However, if you have a presentation that includes many photographs and a large room in which to deliver the presentation, 35 mm slides are the most suitable medium. However, if the photographs have been taken with a high resolution digital camera or if the images have been digitized, they can be shown using a computer projection system.

The delivery makes or breaks a presentation. You have been provided with two pages of details about how to do this—and not that. However, the best way to learn to deliver a design briefing is by practicing your presentation. On your first two or three attempts, have a friend video tape the event. Then review your delivery style, technique and appearance during a trail presentation. You will identify many problems of which you were not aware. Another suggestion is to use one of your free electives to take a speech course. The experience gained in a typical communications course will not help much in crafting the content to include in a design briefing, but it will help you to develop very important delivery skills.

REFERENCES

1. Wilder, L. Talk Your Way to Success, Simon and Schuster, New York, NY, 1986.
2. Eisenberg, A., Effective Technical Communication, 2nd ed., McGraw Hill, New York, NY, 1992.
3. Goldberg, D. E., Life Skills and Leadership for Engineers, McGraw Hill, New York, NY, 1995.
4. Elbow, P. Writing with Power, 2nd Edition Oxford University Press, New York, NY, 1998.
5. Struck, W. C. Osgood and R. Angell, Elements of Style, 4th Edition, Allyn & Bacon, Needham Heights, MA, 2000.
6. Pickett, N. A. et al, Technical English: Writing, Reading and Speaking, Longman, Reading MA, 2000.

EXERCISES

12.1 Write a brief description of the characteristics describing your classmates in this course. Focus on the characteristics that will influence the language and content in your design briefings.

12.2 Prepare an outline for a preliminary design review. Include in the outline the titles of all of the slides that you intend to use.

12.3 Prepare the title slide for your preliminary design review.

12.4 Prepare an overview slide for your preliminary design review.

12.5 Prepare a status slide for your preliminary design review.

12.6 Prepare the key issues slide for your preliminary design review.

12.7 Beg or borrow a video camera. Video tape a practice presentation by a teammate. Critique his or her presentation with good taste.

12.8 Locate all of the projectors that you are permitted to borrow for the presentations to be made in this class.

12.9 Practice the delivery of your presentation employing the most suitable projector that you can borrow.

12.10 Measure and record the time required to present your design briefing. As you rehearse, continue to measure the time. When the presentation becomes more polished, the time required for the presentation should decrease. Does it?

12.11 Prepare a group of slides for a design briefing that utilizes the slide transition and slide building features of PowerPoint.

PART IV

ENGINEERING AND SUCCESS SKILLS

CHAPTER 13

THE ENGINEERING PROFESSION

13.1 WHY ENGINEERING IS A PROFESSION

You may consider engineering from two different viewpoints—as a course of study pursued in an accredited college of engineering and as an occupation. Let's discuss these two different viewpoints, because they are both important in establishing engineering as a profession.

As a student in an engineering college, you are presented with a very structured curriculum. You are required to take many credit hours of mathematics, chemistry, physics and perhaps biology in your first two years of study. Mixed with mathematics and science courses are several additional courses in engineering science. These are essentially applied science courses taught by engineering faculty members. In the last two years of the curriculum, many discipline oriented engineering courses are required. These courses provide the basic analytical methods necessary for success after graduation when you begin to practice. Also included in the final two years are engineering courses intended to provide realistic design experiences related to product development. This schedule of courses is designed to prepare you to practice engineering upon graduation. Near the conclusion of your studies, you will be presented with an opportunity to take the first of two examinations leading to a professional license to practice engineering—the Engineering Fundamental Examination. It is a good idea to take this examination before graduation because it provides you with an opportunity to test your analytical skills. It is also an attribute that you can add to your resume. Passing the examination is a certification of your fundamental analytical skills.

Your ability to practice engineering immediately upon graduation is the primary reason for such a highly structured curriculum containing so many technical courses. The on-the-job training opportunities available to you in your first position depend upon the policies of the company that is paying your salary. Some large firms provide a transition period where you serve as an engineer trainee with close supervision and rotate from one division to another to provide an overview of the company's operations. However, if you take a position with a smaller corporation, usually there is little formal training, and you are expected to earn your salary beginning with the first day on the job.

In addition to at least three years of mathematics, science, engineering science and engineering courses, you will be required to take several courses in the other colleges on campus. The university has general education requirements, which dictate that all students become familiar, if not proficient, with basic premises in the social sciences, fine arts and the humanities. The engineering profession endorses this out-of college exposure because it broadens one's perspective and encourages important assessments of contemporary issues.

Let's now consider the second viewpoint—engineering assignments undertaken after graduation. Whether you will be working as an engineering professional depends to a large degree upon the position you accept. Some engineering graduates are offered positions with investment banking and brokerage houses at very attractive salaries. Managers at brokerage houses like the analytical skills developed in engineering programs and find that engineering graduates perform well in assessing investment opportunities. These graduates pursue an interesting and lucrative career, but they are not practicing engineering.

At the other extreme, a graduate of a civil engineering program takes a position with a construction firm that designs and builds highways, bridges and large public buildings. In this position, public safety is an important consideration and licensing is essential because State laws require it. The graduate is expected to pass the Engineering Fundamental examination, and then—after several years to gain practical experience—he or she will take the Principles and Practice of Engineering examination to become licensed. During the period between the two examinations, the individual works with licensed engineers who supervise his or her work and certifies it to be accurate and correct.

For those graduates working in industry, licensing is usually less of an issue. Most corporations have a small staff of licensed engineers who certify work when certification is required. However, most of the engineers are working on products in which public safety is not an issue. In these cases, professionalism is not a matter of licensing. Professionalism is more a matter of attitude. Professionals are usually salaried and their remuneration is independent of the hours worked. They do not punch a time clock and they are expected to devote the time necessary to finish a task on schedule. Engineers assume responsibility and are committed to the project and to the development team. They cooperate with management, and freely sharing ideas and concepts to advance the welfare of the corporation and their division. They may decide to move from one corporation to another, but loyalty and respect are important considerations even after they leave.

13.2 CHARACTERISTICS OF ENGINEERING STUDENTS

Students entering colleges of engineering today are bright, with average Scholastic Achievement Test (SAT) scores ranging from 1200 to 1300 in most colleges in the U. S. In fact, the higher ranked engineering colleges attract students with average scores of 1300 or above. In addition, engineering students are usually in the top 5 or 10% of their high school graduating class with grade point averages ranging from 3.8 to 4.0. Student applications to the admissions office of the university often include strong letters of recommendation from their teachers, counselors and principals certifying the student's stellar abilities.

It is well recognized that the abilities of entering engineering students in mathematics is exceptional. SAT scores for the mathematics portion of the examination approach or exceed 700 for most students. What is less well recognized is that engineering students also have excellent verbal skills. With average SAT scores for verbal skills of about 600, engineering students usually rank above their peers in the colleges of arts, humanities and social sciences[1].

Most engineering students are men; however, 19.5% of the bachelor degrees in engineering were awarded to women in 2005[2]. The interest of women in engineering depends on the discipline as shown in Table 13.1. Statistics show that women are attracted to chemical, biomedical and industrial engineering disciplines, but not to mining and petroleum programs.

The geographic representation in the class depends strongly on whether the university is state supported or privately funded. State universities often restrict enrollment of out-of-state students to a small proportion of the class. A tuition differential is also imposed making it more expensive for out-of-state students to attend. Consequently, undergraduate classes in state universities tend to be much more homogeneous with students representing the backgrounds and cultures of the region. Privately funded universities usually do not have such restrictions and often vigorously seek students with diverse backgrounds from different states and countries.

---

[1] Data from the new three-part SAT examination is not available at this time for analysis.
[2] See Reference [6].

Table 13.1
Percentage of BS Degrees awarded to women by discipline

Discipline	Percent	Discipline	Percent
Environmental	42.1	Nuclear	20.5
Biomedical	40.4	Computer Science Outside Engineering	18.8
Chemical	37.8	Aerospace	18.8
Industrial or Manufacturing	34.9	General Engineering	18.1
Agriculture	33.7	Computer Science Within Engineering	17.9
Metallurgical and Materials	29.6	Mining	17.7
Architectural	29.5	Petroleum	17.6
Engineering Management	28.3	Electrical/Computer	16.1
Engineering Science Engineering Physics	24.5	Mechanical	14.8
Other	24.4	Computer Engineering	12.8
Civil	23.4		

Clearly, engineering students are very well prepared academically to begin their studies. They have proven themselves in high school as scholars and class leaders. They have gained the respect of their teachers, peers and family. However, the retention rate for engineering students is appalling. Probably one half or more of the entering class of well-qualified students either fail-out or withdraw from the program with most leaving in the first two years. While the exact retention rate varies from college to college across the U. S., the problem is widespread. Why then do so many students leave the engineering program without completing the requirements for a B. S. degree in one of the engineering disciplines? Answers to this question will be explored in Chapter 14.

13.3 SKILL DEVELOPMENT

Pursuing a successful career in engineering is strongly dependent on your ability to develop a number of important skills. Of course, you enter the program with many skills, but it is necessary to improve them and to add new ones to your repertoire. The engineering curriculum is designed to develop additional skills. As you proceed through the program, you will find that your ability to perform challenging engineering tasks expands each year.

The Accrediting Board for Engineering and Technology[3] (ABET) has developed a list of requirements for all of the disciplines offered in the accredited colleges of engineering in the U. S. Each department in the college must demonstrate to a team of ABET visitors that their graduates have the skills and knowledge listed in Table 13.2 to maintain the accreditation of each engineering program for which a B. S. degree is offered.

---

[3] The Accreditation Board for Engineering and Technology, Inc., 111 Market Place, Suite 1050, Baltimore, MD 21202, certifies all programs (not departments) leading to accredited engineering degrees in the U. S.

Table 13.2
Listing of knowledge and skills that engineering graduates must demonstrate.
ABET Criteria 2000

(a) Ability to apply knowledge of math, engineering, and science
(b1) Ability to design and conduct experiments
(b2) Ability to analyze and interpret data
(c) Ability to design a system, component or process to meet needs
(d) Ability to function on multi-disciplinary teams
(e) Ability to identify, formulate and solve engineering problems
(f) Understanding of professional and ethical responsibility
(g) Ability to communicate effectively
(h) Broad education necessary to understand the impact of engineering solutions in a global and societal context
(i) Recognition of the need for, and the ability to engage in life-long learning
(j) Knowledge of contemporary issues
(k) Ability to use techniques, skills, and tools necessary for engineering practice

Before examining this list, consider a somewhat different set of skills and abilities for engineering graduates prepared by a study committee for the American Society of Mechanical Engineers (ASME). The ASME listing [1] in priority order is presented in Table 13.3.

Table 13.3
Skills considered important for new Mechanical Engineers
With B. S. degrees, Priority ranking

1. Teams and Teamwork	11. Sketching and Drawing
2. Communication	12. Design for Cost
3. Design for Manufacture	13. Application of Statistics
4. CAD Systems	14. Reliability
5. Professional Ethics	15. Geometric Tolerancing
6. Creative Thinking	16. Value engineering
7. Design for Performance	17. Design Reviews
8. Design for Reliability	18. Manufacturing Processes
9. Design for Safety	19. Systems Perspective
10. Concurrent engineering	20. Design for Assembly

As you explore both of these lists, the emphasis on skill development in colleges of engineering should become evident. The importance of these skills, as determined by two major engineering societies, should guide your efforts in achieving the proficiencies necessary to insure a successful career.

Design

Most engineering graduates join corporations and participate in some way related to either product or process development. If it is a manufacturing company, the product may be a vacuum cleaner, a lawn mower, an automobile or some other item. The point is you will design, manufacture, assemble, ship and service an array of products. Design is essential to the success of product development. If the design is flawed, the product or service offered will fail, and the company and your career may suffer. If the design is optimal, the product will be profitable, the company will prosper and your career will advance. This is a simple viewpoint, but one that has merit.

Design is perhaps the most difficult of the engineering skills to acquire because it is more an art than a science. While there are well-defined design procedures that you will learn (some of them introduced later in this book), one often intuitively senses the good and bad elements of a design. Also, experience is important because design is a complex activity that must incorporate many different aspects that cross discipline lines. Examine Table 13.3 and note the various descriptors ASME used with the word design. For example, engineers design for reliability, for performance, for safety, for cost, for manufacture, and for assembly and one could add "ease of maintenance" and others to the list.

As you proceed through the engineering program, the opportunities you have to practice design will be severely limited. The emphasis in engineering education is on developing analysis methods used to predict reliability and performance of products and systems. Some programs are limited to a single capstone design experience in your final year of study. Other programs have two or three design courses where you will have the opportunity to design a new product or to modify the design of an existing product or process. Engineering programs with more than three design courses (a total of nine credit hours[4] in a program of about 128 credit hours) are rare in the U. S.

Many colleges of engineering participate in design competitions that are sponsored by the professional societies. These competitions involve the design of robots, large model airplanes, automobiles, steel bridges, and human and engine powered vehicles, etc. You are encouraged to become involved in one or more of these projects. The projects give you an opportunity to participate on an interdisciplinary design team and to gain valuable experience in the design process. Usually you can obtain credit towards your gradation requirements for this design competition by registering for a technical elective that has been established by the faculty member in charge of these design projects.

Another method for gaining design experience is by working off campus. Many students work to partially fund their expenses while pursuing their education. If you are able to find a position that involves design, manufacturing or assembly, the experience gained is probably worth more than the salary you earn. Try to obtain a position in a company that produces products rather than with a retail organization. While experience in either type of organization is valuable, the production experience is more applicable to engineering.

Teamwork

In this course, you probably will be required to participate on a development team. There are three reasons for this requirement. First, the project is often too ambitious for an individual to complete in the time available. You need the collective efforts of the entire team to develop a product during the semester. The development teams will be pressed time-wise to complete the project on schedule.

Second, you must begin to learn teamwork skills. Experience has shown educators that most students entering the engineering program are deficient in team skills. From elementary through high school, the educational process has focused on teaching you to work as an individual, often in a setting where you competed against others in your class. However, you should begin functioning as a team member where cooperation, following, and listening are as important as individual effort. Leadership is important in a team setting, but cooperation and following the lead of others are also critical elements for successful team performance.

The final reason is to better prepare you for the real world you will enter upon graduation with a B. S. degree. You will probably be assigned to a development team very early in your career if you take a position in industry. A recent study by ASME, the results of which are shown in Table 13.3, ranked teamwork as the most important skill to develop in an engineering program. Teamwork was also the first skill, in a list of 20, considered important by managers from industry. Hopefully, this course will

---

[4] The number of credit hours required varies from program to program and college to college. The 128 credit hours cited here is the typical requirement for many programs.

be instrumental in exposing you to team working skills so necessary for a successful career. A detailed discussion of teamwork skills was presented previously in Chapter 1.

Communication

Communication skills are vitally important in every aspect of your life. On the personal side you must be able to accurately convey your thoughts to your family, friends and peers. Professionally, it is just as important to communicate effectively. A great idea is of little value if you cannot express it with sufficient clarity for it to be accepted by management or your associates. With family and friends, most of your communication is by conversation (informal). Occasionally, you write letters or more frequently e-mail messages. You can be casual in communicating with friends and family because they are accommodating and overlook your shortcomings. You must be much more careful in your professional communications. Messages must be clear and unambiguous. The information conveyed must be accurate and timely. The presentation must be a concise and to the point; long rambling prose is not appreciated.

Three different modes of communication are used in engineering—writing, speaking and graphics. All three are important, and engineering versions of all three modes different to some degree from commonly accepted methods.

It is important that you enjoy writing, because engineers often have to prepare several hundred pages of reports, theoretical analyses, memos, technical briefs and letters during a typical year on the job. Communication, particularly good writing, is extremely important. Advancement in your career will depend on your ability to write well. You will be taking several courses offered by the English Department and by departments in Social Sciences, Arts and the Humanities that require writing assignments. These courses will help you with the structure of your composition and the development of good writing skills. Most of the assignments will be to write essays, term papers, or to study selected works of literature. However, there are several differences between writing for an engineering company and writing to satisfy the requirements of courses such as English 101 or History 102. These differences will be described in a Chapter 11 titled **Technical Reports**.

The design briefing is important to both the product development process and to your career. Information about the product must be effectively transmitted to your peers, management and others involved with the project. Clear messages that accurately define problems, which the development team will address, are imperative. On the other hand, ambiguous messages are often misunderstood, hinder the definition of the problem and lead to delays in implementing solutions. The design briefing provides an opportunity to review the status of a specific product development process. It also permits peers to share their ideas with you, and affords management an open forum for assessing the quality of your work and the progress made by your development team. Because the professional presentation is critically important, Chapter 12 titled **Design Briefings** has been included in this textbook. This chapter describes some valuable techniques for properly delivering your message to different audiences— strangers, peers, team members and management representatives.

Engineering graphics is the most important method of communication for presenting design concepts and details. When attempting to communicate design ideas, you will find writing and speaking insufficient to express your thoughts. A more visual technique for communicate is required. It is much more effective to present your ideas by means of drawings, sketches, pictures, and graphs of many different types. Visuals aids, such as drawings and photographs, convey your ideas quickly and with remarkable accuracy.

There are two general approaches used in preparing drawings and graphs. The first is a manual approach where drawings and graphs are prepared by hand using a few simple drawing instruments. The second utilizes a computer and suitable software programs that greatly facilitate the preparation of drawings or graphs. In this book, three chapters covering manual and computer methods for preparing drawings was presented in Part II on Engineering Graphics. An extended discussion of computer-aided design (CAD) software programs, used in preparing both two and three-dimensional engineering drawings, was included in Chapter 9. Manual and computer methods for preparing tables and graphs were presented in Chapter 10.

Design Analysis

In engineering, you will attempt to predict the performance of a product prior to its final design and construction. Obviously, you would not want to design and construct a bridge to have it fail after a few months or years in service. You must be able to predict with confidence that the bridge will not fail over its entire design life (perhaps a hundred years or more). Constructing analytical models and performing analysis is essential in making accurate predictions regarding performance. In modeling, you reduce your structure into a number of different components or subsystems that are amenable to analysis. You then apply analytical methods that enable you to accurately determine performance parameters for each model. The analytical results permit verification of the adequacy of the design. As such, analysis and design are coupled. You usually begin with a design concept; then you subject the design to an analysis. The results of the analysis are then used to improve this design enhancing its performance.

A significant portion of the engineering curriculum is devoted to developing your analytical skills. About one year, of the four-year curriculum, is devoted to courses in mathematics and science that provide the foundation for engineering courses. In most colleges of engineering, at least another year is devoted to engineering and engineering science courses where analytical and computational methods are taught. In a typical engineering program, your analytical and computational skills will be honed more than any other skill that is listed in either Table 13.1 and 13.2.

Your performance (read this as grade point average, GPA) will depend on your ability to solve problems posed on hourly examinations or on final examinations. These problems are crafted to test your understanding of a few engineering principles[5]. The mathematics involved is usually not overly complex and the physics entailed is even more straightforward. You will be provided with several useful tips on problem solving methods in Chapter 14.

Experimental Analysis

In predicting the performance of a structure or machine component, you will often conduct experiments and measure the performance parameters. While it is usually more expensive to perform experimental studies than analytical ones, the costs are warranted when you are not certain that the analytical modeling accurately reflects reality. Sometimes converting the physical reality of a product to an analytical model requires many assumptions. You will often perform carefully designed experiments to verify these assumptions and to insure the successful performance of the product. Extensive tests on almost all products are performed before they are released to the market. The purpose of these tests is to insure safety of the product, and to provide management assurance that the product will meet the guarantees specified in the warranty pertaining to performance, life and safety. In this class, you will be required to test a prototype of a product that your team has developed. The test is to determine if the product's performance meets its specification.

There are four different phases of an experimental program. The first phase is to design the experiment so that it provides the answers required to verify the modeling or to insure the adequacy of a design. The second phase is to conduct the experiment using appropriate instrumentation that will provide accurate and appropriately timed measurements of the controlling performance parameters. The

---

[5] In practicing engineering, you will find that analyses are based on only a dozen or so engineering principles. The methods developed in engineering courses may appear complex, but they are based on a very limited number of fundamental concepts.

third phase is to analyze the data obtained using suitable statistical methods that predict and bound experimental errors. The final phase is to correctly interpret the data to verify the adequacy of a design or to provide information that may be used to enhance a design.

The laboratory classes offered in conjunction with chemistry and physics courses are not designed to provide you with skills for designing experiments or for making measurements. These are highly structured laboratory experiences intended to reinforce theories and principles introduced during lecture. The curriculum of many of the engineering disciplines provides a course on circuits and instrumentation. This course introduces linear circuit theory and gives some information about sensors, transducers and both analog and digital instrumentation. A course of this type is strongly recommended because it provides some of the basic information needed to plan and design experiments.

You are encouraged to enroll in other courses that provide additional exposure to experimental methods. This may be possible with a careful selection of technical electives that are available in the third and fourth year of your program. You might also consider an assignment working in a laboratory directed by one of the professors in your department. Many professors lead significant research programs, and in conducting these studies, they often perform new and novel experiments. The experience gained in such a setting is often more valuable than that obtained in a traditional engineering measurements course.

Understanding Professional and Ethical Responsibilities

When you graduate and begin practicing engineering, you will be expected to follow a professional code of ethics. An example of the ABET code of ethics is presented in Chapter 17 titled Ethics, Character and Engineering. While there are many other codes, each endorsed by a professional society, they all are similar. Basically the codes ask you to uphold and advance the integrity, honor and dignity of the engineering profession. The codes then list fundamental principles and canons that should govern professional behavior. The fundamental principles advanced by ABET are:

1. **To be selective in the use of our knowledge and skills so as to ensure that our work is of benefit to society.**
2. **To be honest, impartial and serve our constituents with fidelity.**
3. **To work hard to improve the profession.**
4. **To support the professional organizations in our engineering discipline.**

Personal behavior out-of-class or away from the work place is also important. The results of several polls indicate that Americans are less ethical today than in previous generations. Sixty-four percent of individuals in a sample of 5000 admit to lying if it does not cause real damage. An even larger percentage of those sampled (74%) will steal providing the person or business that is being ripped off does not miss the item pilfered [3]. Many students (75% high school and 50% college) admit to cheating on an important exam [4]. Many students have not developed a moral code to use as a guide for their behavior. Relatively minor transgressions of a generation ago have been replaced with more serious problems involving mass murder, drug and alcohol abuse, pregnancy, suicide, rape, robbery and assault. Some of the reasons for these changes are explored later in Part V of this book.

Ethical behavior is important in both your professional and personal lives. Recall the golden rule—do unto others as you would have them do unto you—to guide your behavior. It is a very simple rule, and it is effective.

Committing to the Need for Life-Long Learning

Most engineering programs require successful completion of about 128 credit hours for graduation with a B. S. degree. This number is higher by about eight credit hours than the requirements for a B. S. degree from any other college on campus. Engineering educators have always required an extra effort measured in credit hours for their students prior to graduation. This belief is based on the concept that the student should be capable of practicing engineering immediately after graduation.

During the past twenty years the number of credit hours required for graduation has decreased by about 10%, while the amount of important material necessary to practice has increased significantly. Moreover, the scope of the fields in all of the engineering disciplines continues to evolve and to expand. Engineering educators recognize that they cannot increase the number of credit hours to accommodate the evolution and expansion of the knowledge base for each discipline. Instead, it becomes essential that all engineering graduates recognize the need to continue their studies after graduation.

Upon graduation many of you will take a position in industry and be challenged with a new environment and new assignments. You will be able to measure your strengths and weaknesses against this new setting. Perhaps you will conclude that you need advanced courses in your discipline, or basic courses in some other engineering discipline. With more time in practice, you will probably be given an opportunity to lead a small engineering group and become a first level engineering manager. In this case you may decide to enroll in a few courses in the business school or to pursue a masters program in business administration. No one really understands with certainty what knowledge will be important in the future[6]. The idea is to stay flexible and be prepared to devote time each week both to study and professional development.

You are fortunate that many opportunities exist for continuing education. Some companies bring instructors into its facilities to focus on topics of immediate concern to the organization. Professional societies offer short courses on timely topics in their discipline. Well-known universities offer short courses each summer with outstanding professors teaching in their specialty. Local universities offer a wide array of course on a regular basis that lead to M. S. and Ph. D. degrees in a number of engineering disciplines. Professional societies offer many well-written books that cover advances within their discipline. Distance learning opportunities abound, and this mode of education is growing rapidly.

It is essential that you recognize graduation with a B. S. degree as the beginning of your studies and not the end.

13.4   THE ENGINEERING DISCIPLINES

13.4.1 Enrollment and Salaries

The U. S. Military Academy at West Point, established by an act of Congress in 1802, was the first college of engineering in the America. Military engineering was the first discipline, but in the 1800s several other programs were introduced, and the field of engineering began its division into many different disciplines each with its special emphasis [5]. The number of different disciplines grew slowly in the 1800s and early 1900s and included:

1. Civil engineering
2. Mechanical engineering
3. Electrical engineering
4. Chemical engineering

---

[6] If someone had told me upon graduation with a B. S. degree that I would spend most of my career in research and education, I would not have believed them. At that time, I intended to design heavy-duty steel mill equipment.

    5.   Industrial engineering

In more recent years the number of engineering disciplines has continued to grow with more and more technical specialties converted into degree programs. Recently the American Society of Engineering Education (ASEE) [6] published the undergraduate enrollment by engineering discipline during the fall semester of 2005. These results are presented in Table 13.4. The total undergraduate enrollment decreased by 2% to 366,361 in the fall of 2005.

Table 13.4
Undergraduate enrollment by engineering discipline, Fall 2005

Discipline	Enrollment	Discipline	Enrollment
Mechanical Engineering	78,202	Architectural Engineering	4,392
Electrical Engineering	46,062	Metallurgical & Materials Engineering	3,577
Civil Engineering	42,007	Engineering Science Engineering Physics	3,161
Other	35,943	Agricultural Engineering	2,978
Computer Science within Engineering	27,557	Petroleum Engineering	2,131
Computer Engineering	22,620	Environmental Engineering	2,007
Chemical Engineering	21,727	Nuclear Engineering	1,562
Aerospace Engineering	15,882	Civil/Environmental Engineering	1,264
General Engineering	14,925	Engineering Management	1,239
Biomedical Engineering	14,213	Mining Engineering	626
Industrial & Manufacturing Engineering	13,065	Computer Science outside Engineering	11,750
Electrical/Computer	11,201		

    The total number of B. S. degrees awarded in engineering in 2004-2005 was approximately 73,602. A large number of degrees were awarded to graduates from mechanical, electrical and computer engineering programs; however, in recent years many departments of electrical engineering have divided their program and they are now offering two different degrees—an electrical engineering degree and a computer engineering degree. Current trends show students tending to favor the electrical engineering program when they enroll. In addition to the B. S. degrees in engineering, 40,650 M. S. and 7,366 Ph.D. degrees were also awarded in engineering in 2004-2005.

    Mechanical and civil engineering programs also have large enrollments. The curriculum in both of these programs is relatively stable in comparison with electrical and computer engineering. However, some civil engineering departments are establishing a separate degree program in environmental engineering, and this separation necessitates significant changes in the curriculum.

    A large number of engineering programs are classified as other in Table 13.4. This category contains many programs that are offered by only a few accredited colleges of engineering in the U. S. and include:

- Ocean engineering
- Marine engineering
- Fire protection and safety engineering
- Ceramic engineering
- Systems engineering
- Geological engineering

The enrollment in each discipline reflects perceived demand for the graduates and, to a small degree, the starting salaries for graduates from each program. Average starting salaries for several of the engineering disciplines with B. S. degrees are listed in Table 13.5.

Table 13.5
Starting salaries of B. S. graduates in engineering and other disciplines
Source the National Association of Colleges and Employers Salary Survey for the graduates of 2005

Discipline	Annual Salary
Aerospace Engineering	$50,701
Chemical Engineering	$54,256
Civil Engineering	$43,462
Computer Engineering	$51,496
Computer Science	$51,292
Electrical Engineering	$52,000
Industrial Engineering	$49,541
Mechanical Engineering	$51,046
Accounting	$43,909
Information Science	$43,732
Economics/Finance	$42,802
Business Administration	$39,448
Marketing	$37,832
Liberal Arts	$30,337

The difference in starting salaries for engineers in different disciplines is not large and the data differs slightly depending on its source. These differences vary somewhat from year to year with the chemical engineers and computer engineers usually leading the other disciplines by 5 to 10% and the civil and environmental engineers usually trailing the other disciplines by 5 to 10%. Starting salaries for engineering graduates are significantly greater than comparable salaries for other majors. For example, offers to business administration majors averaged $39,448 and marketing majors averaged $37,832. Average offers to accounting majors are better at $43,909. Liberal arts majors usually do poorly in starting salaries and for the class of 2005 their average offer was $30,337. The only exception to this trend was with computer science where strong demand for graduates from this program in the late 1990s produced salaries exceeding those offered to engineering graduates from several disciplines.

College graduates with a bachelor's degree in an engineering discipline have among the highest annual earnings of all major disciplines. A recent study on employment trends by economists N. Fogg and P. Harrington, and psychologist T. Harrington [7], contains the results of a study of 150,000 college students, data from the U.S. Bureau of Labor Statistics and extensive economic research and predictions. The reference outlines employment and job market trends, as well as average salaries for graduates of 60 college majors and employment projections in those categories for the next five years. Of the top eight disciplines based on mean annual earnings for graduates who hold only a bachelor's degree, seven are engineering disciplines, which include:

- Chemical engineering—$75,579
- Aerospace engineering—$73,605
- Computer systems engineering—$70,084
- Physics and astronomy—$69,612
- Electrical/electronics engineering—$68,977
- Mechanical engineering—$68,806
- Industrial engineering—$68,411

- Civil engineering—$66,126

The demand for all engineering disciplines covered by the study is expected to grow between 2000 and 2010. The disciplines that are expected to show the most growth are computer systems, aerospace and mechanical engineering. Chemical engineering is projected to have the smallest growth rate at 3%.

Median income with discipline of engineering differs considerably from the starting salaries as shown in Table 13.6. The median income is higher than the starting salaries because the engineers' compensation increases time on the job and the experience gained. The median salaries depend on the engineering discipline with higher salaries commanded by those working in aerospace and materials engineering. Those working in civil and general engineering tend to earn lower salaries. However, after about 20 years of service, salaries tend to level out for most engineers with small incremental gains from year to year in the later stage of their careers.

Table 13.6
Median Annual Salaries for Engineers by Discipline
From Engineers' Salaries: 2003 Special Industry Report,
Engineering Workforce Commission of the
American Association of Engineering Societies

Discipline	Number of Years after B. S.				
	0	5	9-10	13-16	21-25
Aerospace Engineer	—	$65,757	$75,540	$79,923	$89,143
Materials Engineer	—	—	$72,132	$83,865	$96,921
Computer Engineer	—	—	$77,656	$79,225	$84,268
Civil Engineer	$43,800	$56,200	$64,436	$67,665	$74,547
General Engineer	—	$76,160	$82,056	$80,096	$81,176
Mechanical Engineer	$51,323	$63,098	$73,146	$80,955	$88,746

13.4.2   Description of Engineering Programs

Electrical Engineering

The **electrical engineering** program provides the basic background needed to work in broad fields of electronics, communications, automatic control, computers, materials processing, electromagnetic, signal processing, and power utilization. The program places emphasis on the fundamentals of mathematics and science leading to careers in research, development, design, or operation in such diversified areas as electronics, control systems, computers, communications systems and equipment, biomedical instrumentation, radar and navigation, power generation and distribution, consumer electronics and industrial devices.

The Institute of Electrical and Electronics Engineers (IEEE), the professional society representing electrical engineering, describes the purpose of electrical engineering as:

Helping advance global prosperity by promoting the engineering process of creating, developing, integrating, sharing, and applying knowledge about electrical and information technologies and sciences for the benefit of humanity and the profession.

About 13.4 million electrical engineers work to develop a wide array of electronic products ranging from simple household appliances to sophisticated missile intercept systems. The field is so large that the IEEE has divided its organization into 10 regions, 36 technical societies, 4 technical councils, approximately 1,200 individual and joint society chapters, and 300 sections. If this discipline is of interest to you, visit the IEEE web site at http://www.ieee.org/ for a much more complete description[7].

Mechanical Engineering

Mechanical engineers apply the fundamental principles of mechanics, thermo-sciences and design to fulfill the needs of society. These principles are covered in detail in the subjects of solid and fluid mechanics, thermodynamics, heat transfer, design, systems analysis, control theory, vibrations and machine design. Mechanical engineers work on a wide variety of different assignments. One of the primary areas is the energy field where mechanical engineers are responsible for generating and distributing electricity, designing heating and air conditioning systems and developing refrigeration systems. A second significant area of activity is in the design of a wide range of product such as vehicles of all types, appliances, office equipment and machinery. Finally, many mechanical engineers work in production facilities developing manufacturing processes and overseeing production.

The American Society for Mechanical Engineering (ASME International), the professional society representing mechanical engineering, describes its mission as:

> To promote and enhance the technical competency and professional well being of our members, and through quality programs and activities in mechanical engineering, better enable its practitioners to contribute to the well being of humankind.

Mechanical engineering is one of the oldest disciplines with a broad range of industrial activities. Mechanical engineering and electrical engineering are the largest engineering disciplines. While membership in the ASME is about 125,000, employment by mechanical engineers in industry probably approaches one million. The organization of the ASME is divided into 39 technical divisions that publish technical papers and sponsor technical conferences.

Civil Engineering

Civil engineers are concerned with many societal problems including environmental quality, building and maintaining the nation's infrastructure and developing safe and rapid transportation systems. Topics of study include structures, soil mechanics, construction materials, environmental engineering, water resources, pavements and transportation. This program emphasizes fundamentals of mathematics, sciences, engineering principles and engineering methods leading to careers in engineering practice and/or graduate education. An option in **environmental engineering** or a separate degree in this discipline is often available in civil engineering departments.

The American Society for Civil Engineering (ASCE), the professional society representing civil engineering, describes its function as:

> When new developments occur in research and practice, ASCE's technical divisions and councils and their respective technical committees bring information to the membership through conferences and workshops and by publications such as manuals of practice, pre-standards, journal articles and policy statements.

The technical activities of the ASCE include more than 14 annual conferences and workshops and the publication of 10 journals. More than 5,000 ASCE members participate by serving on technical committees. Their work addresses programmatic thrust areas such as sustainable environment,

---

[7] The Web sites for all of the professional societies are listed in Table 13.7.

sustainable transportation, extreme environment and information technology and construction and materials.

Computer Engineering

Computer engineering is a relatively new discipline that is still emerging in the many colleges of engineering in the U. S. Many electrical engineering faculties are dividing the traditional electrical engineering curriculum into two separate programs—one dealing with power, electronic systems and communications and the other concerned with computers and information technology. Computer engineers work in the computer industry designing hardware and writing software. The discipline combines technical knowledge from digital electronics, signal transmission and processing with a programming emphasis of computer science.

The IEEE Computer Society, an element within the IEEE organization, has emerged as the professional society representing computer engineers. The mission statement for the IEEE Computer Society is given below:

> The IEEE Computer Society is dedicated to advancing the theory, practice, and application of computer and information processing technology. Through its conferences and tutorials, applications and research journals, local and student branch chapters, technical committees and standards working groups, the society promotes an active exchange of information, ideas, and technological innovation among its members. In addition, it accredits collegiate programs of computer science and engineering in the United States.

With nearly 100,000 members, the IEEE Computer Society is the world's leading organization of computer professionals. Founded in 1946, it is the largest of the 36 societies organized under the umbrella of the Institute of Electrical and Electronics Engineers (IEEE).

Chemical Engineering

Chemical engineering is concerned with the processing of materials and the production or utilization of energy through molecular or sub-molecular changes. Chemical or atomic reactions and physical changes are included in this field. Chemical engineers begin their process designs with well-known chemical reactions previously discovered by chemists. They use this information to design large-scale chemical plants to produce sizeable quantities of raw materials, petroleum products, pharmaceuticals, synthetics, etc.

The American Institute of Chemical Engineers (AIChE) is a nonprofit organization providing leadership to the chemical engineering profession. Representing 57,000 members in industry, academia, and government, AIChE provides forums to advance the theory and practice of the profession, upholds high professional standards and ethics and supports excellence in education. Institute members range from undergraduate students, to entry-level engineers, to chief executive officers of major corporations. The AIChE mission statement is given in the bullet listing below:

- Promote excellence in chemical engineering education and global practice;
- Advance the development and exchange of relevant knowledge;
- Uphold and advance the profession's standards, ethics and diversity;
- Enhance the lifelong career development and financial security of chemical engineers through products, services, networking, and advocacy;
- Stimulate collaborative efforts among industry, universities, government and professional societies;

- Encourage other engineering and scientific professionals to participate in AIChE activities;
- Advocate public policy that embraces sound technical and economic information and that represents the interest of chemical engineers;
- Facilitate public understanding of technical issues;
- Achieve excellence in operations of the Institute.

Fire Protection Engineering

Fire protection engineering is a unique profession that builds upon the basic tools of several other engineering disciplines. The fire protection engineer may be responsible for analyzing the level of fire safety in modern or historic buildings, nuclear power plants and aerospace vehicles; designing fire protection systems for high-rise buildings and industrial complexes; conducting post-fire investigations; and researching new technologies for fire detection and suppression.

Fire protection engineers apply engineering principles to protect people and their environment from the unwanted consequences of fire. These principles are applied to understand the nature and characteristics of fire, fire growth and products of combustion, as well as to consider the response of structures, people, processes, materials and systems to fire.

The professional society in the field, the Society of Fire Protection Engineers (SFPE), has 51 chapters in 8 countries. According to the SFPE, fire protection engineers have always been in great demand by corporations, educational institutions consulting firms and government bodies around the world. Graduating from a fire protection engineering program usually guarantees an immediate place in the workforce, most likely at a high starting salary. The median salary for fire protection engineers is significantly above that of other engineering disciplines.

The website for the Society of Fire Protection Engineers is www.sfpe.org.

Industrial Engineering

Industrial engineers design the systems that organizations use to produce goods and services. In addition to working in manufacturing industries, industrial engineers serve to insure quality and productivity in places such as medical centers, communication companies, food service, education systems, government, transportation companies, banks, urban planning departments and an array of consulting firms. Industrial engineers educate and direct these groups in the implementation of Total Quality Management (TQM) principles. Today, especially popular functions are manufacturing, health care, occupational safety and environmental management. Industrial engineering is the most people-focused discipline in engineering. Those who pursue careers in this discipline usually have strong leadership skills and a commitment to working with teams of managers, scientists and other personnel to solve important problems. They enjoy helping organizations serve human needs and accommodate many different concerns

The Institute of Industrial Engineers (IIE) is the society that serves the professional needs of industrial engineers and others involved with improving quality and productivity. Its 24,000 members stay current with new developments in their profession through Institute's life-long-learning approach, as reflected in the educational opportunities, publications, and networking opportunities offered. Members also gain valuable leadership experience and enjoy peer recognition through numerous volunteer opportunities.

Aerospace Engineering

Aerospace engineers are concerned with the physical understanding, related analyses, and creative processes required to design aerospace vehicles operating within and beyond planetary atmospheres. Such vehicles range from helicopters and other vertical takeoff aircraft at the low-speed end of the flight spectrum to spacecraft operating at thousands of miles per hour during entry into the atmospheres of the earth and other planets. In between are general aviation and commercial transports flying at speeds below and close to the speed of sound, and supersonic transports, fighters, and missiles which cruise at speeds greater than the speed of sound. Among the subjects studied are aerodynamics, flight dynamics, flight structures, flight propulsion, and the synthesis of all these principles into one system with a specific application such as a complete transport aircraft, a missile, or a space vehicle.

The nonprofit American Institute of Aeronautics and Astronautics (AIAA) is the principal society serving the aerospace profession. Its primary purpose is to advance the arts, sciences and technology of aeronautics and astronautics and to foster and promote the professionalism of those engaged in these pursuits. Although founded and based in the United States, AIAA is a global organization with nearly 30,000 individual professional members, over 50 corporate members, thousands of customers worldwide and an active international membership.

Materials Engineering

Materials engineering involves the study of mechanical, physical, and chemical properties of engineering materials, such as metals, ceramics, polymers and composites. The objective of a materials engineer is to predict and control material properties through an understanding of atomic, molecular, crystalline, and microscopic structures of materials. A materials engineer is an essential member of a team responsible for synthesis and processing of advanced materials for manufacturing. A graduate's work may vary from automobile or aerospace applications to microelectronics manufacturing. Opportunities are available through these industries in areas of research, quality control, product development, design, synthesis and processing operations.

The American Society for Materials (ASM International) serves as one of the technical societies representing materials engineering. For many years it has provided a means for exchanging information and professional interaction. Benefits to members include:

- Access to information for extending and maintaining professional skills.
- Opportunities for professional development through continuing education.
- Network locally and worldwide.
- Achieving professional recognition.

Biomedical Engineering

Biomedical engineering is the newest engineering discipline, integrating the basic principles of biology with the fundamentals of engineering. With the rapid advances in biomedical research and the economic pressures to reduce the cost of health care, biomedical engineering will play an important role in the medical environment of the 21st century. Over the last decade, biomedical engineering has evolved into a separate discipline bringing the quantitative concepts of design and optimization to problems in biomedicine. The opportunities for biomedical engineers are wide ranging. The medical device and drug industries are increasingly investing in biomedical engineers. As gene therapies become more sophisticated, biomedical engineers will play an important role in bringing these ideas into real clinical practice. Finally, as technology plays an ever-increasing role in medicine, there will be a larger need for physicians with a solid engineering background. From biotechnology to tissue engineering, from medical imaging to microelectronic prosthesis, from biopolymers to rehabilitation engineering, biomedical engineers are in demand.

The Biomedical Engineering Society (BMES) serves as the professional society representing the interests of biomedical engineers. They are a relatively new society having incorporated in 1968. Its purpose is to promote the increase of biomedical engineering knowledge and its utilization

Engineering Science and Engineering Mechanics

Engineering mechanics is a field of study that permeates all major engineering disciplines. Solid mechanics deals with analytical formulation of direct and reaction forces on structures and the resulting response of the structures to these forces. The discipline of engineering mechanics is based on applied mathematics and large scale computing. Fail-safe design is another dimension in this discipline. Because of the general applicability of this subject to many aspects of engineering design and manufacturing, opportunities are available in a broad spectrum of industries and government laboratories.

The American Society for Mechanical Engineering (ASME International), the professional society representing mechanical engineering, also represents engineering mechanics.

Agricultural Engineering

Agricultural engineering is in transition with more emphasis currently being placed on biological aspects rather than traditional agriculture topics. As such many programs carry some reference to biology in their titles—for example, **BioResource and Agricultural Engineering**. Agricultural engineers are trained to creatively apply scientific principles in the design and development of new products, systems, and processes for the conversion of raw materials and power sources into food, feed and fiber. They are also concerned with protecting the environment and worker health and safety. The diversity of knowledge and skills that agricultural engineers possess is valuable in the agricultural and agribusiness industries. The agricultural engineer develops skills in design and problem solving that are based on fundamental principles in the engineering sciences including mathematics and physics, computer tools, communication, teamwork, instrumentation and biology. Agricultural engineers are distinguished from other engineering disciplines by their commitment to meeting human and animal needs for food, feed, fiber and ensuring a sustainable, safe living and working environment. Biological and economic constraints will continue to make this a challenging career opportunity.

The American Society of Agricultural Engineers (ASAE) is a professional and technical organization dedicated to the advancement of engineering applicable to agricultural, food, and biological systems. Its 9,000 members, representing more than 90 countries, serve in industry, academia and the public.

Petroleum Engineering

Petroleum engineering is primarily concerned with producing oil, gas, and other natural resources from the earth through the design, drilling, and use of wells and well systems. The petroleum industry develops methods for conveying fluids in to, out of, or through the earth's subsurface for scientific, industrial, and other purposes while remaining mindful of the ecological needs for safety. The curriculum in petroleum engineering provides a proper balance between fundamentals and practice. Graduate engineers are prepared for life-long learning but are capable of being productive contributors immediately. Petroleum engineers are currently in high demand in the industry, and their starting salaries are consistently among the top in the nation. The curriculum includes study of design and analysis of well systems and procedures for drilling and completing wells; characterization and evaluation of subsurface geological formations and their resources; design and analysis of systems for producing, injecting, and handling fluids; application of reservoir engineering principles and practices for optimizing resource development and management; and use of project economics and resource valuation methods for design.

The Society of Petroleum Engineers (SPE), with more than 50,000 professionals from oil and gas-producing regions around the world, represents the technical interests of this discipline. Its mission is to provide the means to collect, disseminate and exchange technical information concerning the development of oil and gas resources, subsurface fluid flow and production of other materials through well bores for the public benefit. The Society also provides opportunities through its programs for interested individuals to maintain and upgrade individual technical competence in these areas. The SPE accomplishes this mission through an international schedule of meetings and exhibitions, periodicals, short courses, books, electronic publications and section programs.

Mining Engineering

Mining engineering involves prospecting for mineral deposits; planning, designing, and operating profitable mines; processing and marketing the extracted minerals; insuring safe and healthy working conditions; and protecting and restoring the land during and after a mining project. Mining engineers use technologically advanced equipment, machines, robotics and computers every day. Those in charge of mine design, plan and test mines using computer simulators before ever breaking ground. Mining engineers in management use mine scheduling software to plan mining activity when operation is underway. Surface mining operations use larger mobile equipment than any other industry in the world. Mining methods and equipment are also applied to the removal of earth and rock outside the mining industry. Each year, an individual requires an equivalent of 40,000 pounds of new minerals and energy equal to that produced by burning 30,000 pounds of coal. With each different mineral comes a different mine site and a different location, giving the mining engineer a very diverse work environment, and limitless horizons to work anywhere in the world.

The Society for Mining, Metallurgy, and Exploration (SME) is an international society of 16,000 professionals with members in nearly 100 countries. Services available to SME members include: publications, professional registration, peer-review of technical papers, future leader and college accreditation programs, meetings and exhibits, public education and SME short courses.

Nuclear Engineering

Nuclear engineering is concerned with the science of nuclear processes and their application to the development of various technologies. Nuclear processes are fundamental in the medical diagnosis and treatment fields, and in basic and applied research concerning accelerator, laser and super conducting magnetic systems. Utilization of nuclear fission energy for the production of electricity is the current major commercial application, and radioactive thermal generators power a number of spacecraft. For the longer term, electricity production based on nuclear fusion is expected to become an increasingly important segment of the field. Nuclear engineers are concerned with maintaining expertise in the design and development of advanced fission reactors, performing basic and applied research in the development and ultimate commercialization of fusion energy, developing both institutional and technical options for radioactive waste and nuclear materials management and in fostering research in nuclear science and applications, with emphasis on bioengineering, detection and instrumentation and environmental science.

The discipline is represented by the American Nuclear Society (ANS). The ANS has a diverse membership composed of approximately 11,000 engineers, scientists, administrators and educators representing corporations, educational institutions, and government agencies.

Manufacturing Engineering

Manufacturing is a prime generator of wealth and is critical in establishing a sound basis for economic growth. Manufacturing education is important in achieving and maintaining a long-term competitive position for U. S. industry in the current global economy. The program covers concepts such as design for manufacture, flexible manufacturing, new communication and information networks and the impact of automation on human experience. Achieving and maintaining a long-term competitive economic position requires students to be able to analyze manufacturing systems for overall technical, economic and environmental performance. Manufacturing engineers should also be able to quantify trade-offs between these performance characteristics in relationship to the technology deployed in the manufacturing system.

The Society of Manufacturing Engineers (SME) is the world's leading professional society serving the manufacturing industries. Through its publications, expositions, professional development resources and member programs, SME influences more than 500,000 manufacturing executives, managers and engineers. The SME has some 60,000 members in 70 countries and supports a network of hundreds of chapters worldwide

Table 13.7
Web site addresses for the professional societies

Professional Society	Web Site Address
Electrical and Electronics Engineers (IEEE),	http://www.ieee.org/
American Society for Mechanical Engineering (ASME International)	http://www.asme.org/
American Society for Civil Engineering (ASCE)	http://www.asce.org/
IEEE Computer Society	http://www.computer.org/
American Institute of Chemical Engineers (AIChE)	http://www.aiche.org/
Institute of Industrial Engineers (IIE)	http://www.iienet.org/
American Institute of Aeronautics and Astronautics (AIAA)	http://www.aiaa.org/
American Society for Materials (ASM International)	http://www.asm-intl.org/
Biomedical Engineering Society (BMES)	http://mecca.org/BME/BMES/
American Society of Agricultural Engineers (ASAE)	http://asae.org
Society of Petroleum Engineers (SPE),	http://spe.org/
Society for Mining, Metallurgy, and Exploration (SME)	http://www.smenet.org/
American Nuclear Society (ANS).	http://www.ans.org/
Society of Manufacturing Engineers (SME)	http://www.sme.org/

13.5 ENGINEERING FUNCTIONS

Engineers from all disciplines are involved in a wide variety of functions ranging from sales to maintenance. The more common engineering functions will be briefly described in the following subsections.

Design and Development

Companies continuously work on their product lines to maintain or increase their market share. Their ability to release a stream of high-quality, high-performance, reliable products that are competitively priced dictates their success and the profits they generate. **Design and development engineers** work to create new and profitable products. These "new" products are often redesigned models of an older product. Important in these redesigns are improvements in performance and reliability at lower costs. On rare occasions design and development engineers create an entirely new product that establishes a new market. Apple's IPod is a recent example of such a new product.

Design and development engineers interact with marketing personnel to establish the customer's needs. They convert the customer's needs to a product specification and then consider a large number of different design concepts that fulfill this specification. The best design concept is selected from among these options, and then the product is developed and manufactured. During this process, the design and development engineers work with manufacturing engineers, production engineers and test engineers to insure a timely and cost effective approach for producing a safe and reliable product.

Testing

Society today is increasingly litigious. It is a common occurrence for trial lawyers to seek remedies for damages due to what may be perceived as unsafe products. Product liability is a very serious concern. When a company releases a product for the market, it is imperative that it be safe. The only questions are—how safe and under what conditions? **Test engineers**, as the name implies, test products to ascertain if their performance specifications have been met. They conduct tests that enable them to predict the life of the product and the number of failures that will occur with time in service. They provide management with the data needed to determine warranty costs. During the design and development phase, they may test subsystems to provide data helpful in designing the product. Basically test engineers verify analytical methods for predicting performance. When the analytical methods are not sufficient, the test results are essential to prove the adequacy of the product.

Design Analysis

Almost every design and development team includes one or more members with excellent skills in design analysis. **Design analysis engineers** are responsible for modeling—where components from the product are converted to simplified models that may be analyzed. The analysis conducted may be in closed form where well-known mathematical formulas are used to produce results that predict performance or insure safety. In many instances, closed form solutions are not available and numerical techniques, such as finite element models, are used to generate solutions. In recent years, a large amount of specialized software has become available that enables the design analysis engineers to solve increasingly complex problems—rapidly and accurately.

The close interaction of the design analysis engineer with the design team during the development cycle is vitally important. If a problem is identified by analysis early in the design cycle, it can be corrected quickly and the costs of an error are minimal. However, if the problem goes undetected until hardware is produced and tested, the costs to correct an error are dramatically higher and the time required correcting the problem may delay introducing the product to market.

Manufacturing

A **manufacturing engineer** serves on the design and development team to provide expertise on tooling and manufacturing methods. As the components of a product are designed, many decisions are made that affect the cost of manufacturing and assembly of the product. It has been established that a major fraction of the total life-cycle cost of a product is committed in the early stages of design [8]. Quality cannot be manufactured or tested into a product; it must be designed into the product. Manufacturing engineers provide the design and development team with the knowledge of manufacturing equipment and processes available within their facilities and available from qualified suppliers. As the design proceeds, they identify manufacturing requirements and begin the design of any specialized tooling required for production. They design the manufacturing cells for producing component parts. They certify vendors who will supply externally purchased components. They design the lines, cells and the tooling used in assembly.

Manufacturing engineers also work with the quality control department to establish procedures to insure the quality of each component manufactured. They select machine tools and instrumentation to insure timely adjustments for manufacturing processes to remain within the control limits established by quality control personnel.

Production

Production engineers also participate on the design and development team. Their role is to ensure the flow of material and components to the manufacturing facilities and to the assembly lines. Often there is a need to purchase components or materials that require long lead times for delivery. In some cases, the lead-time is comparable to the design time. The production engineer must identify these items, anticipate the number required and place the order before the design is complete. After the design is complete, they work with purchasing personnel to insure delivery of components just-in-time to be used in production. They attempt to minimize the work-in-progress inventory. They work with personnel from the sales department to match production to anticipated demand for the product. They work with personnel from shipping to ensure that the product is removed from the production facilities and delivered to customers in a timely manner. Their goal is to meet the demand of the customers with a minimum of work in progress and finished product in inventory.

Maintenance

Most students own an automobile or have one available to them. Have you ever tried to travel from point A to point B and found the car would not start? This problem improves in your understanding of the need for maintenance. Manufacturing facilities often operate on a 24/7 schedule (this schedule implies the equipment operates 24 hours a day seven days a week). **Maintenance engineers** work to keep the equipment in operation. They seek to avoid breakdowns with scheduled maintenance. They plan for periodic lubrication of bearings, replacement of parts subjected to high rates of wear, replacement of belts and scheduled inspections. They develop a recording system that provides a database for predicting the time between failures and identifies the equipment that will probably fail. They design monitoring instrumentation that enables the facilities to be shut down in a controlled manner prior to failure of any critical parts. They supervise the maintenance crews that perform the repair work. If you enjoy high stresses associated with the inevitable breakdown, consider this engineering function for a career.

Sales

Sales engineers provide an interface between the customer buying technical products and the company selling them. They work with the engineers and purchasing agents representing the customer and provide technical information necessary for them to intelligently purchase a product or a system. Many engineering products and/or systems are complex with the availability of several different options. Sale engineers must know the details of the product and/or the systems, provide an overview of the options, and answer questions from the customer's representatives. People skills are as important as analytical skills in this position because the effectiveness of a sales call is often affected by personalities.

The sales engineers also seek information regarding new products from the customer. In some cases, the product line offered does not satisfy the requirements of the customer. When this occurs, it is important to ascertain the customer's requirements and the price he or she would be willing to pay for the product if it were to become available. This information is passed on the marketing representatives and is used in planning for new product developments.

Management

After demonstrating the ability to perform engineering functions, many relatively young engineers are promoted into **management** positions. If the company you are with is organized along functional lines, the first management position is usually "section head." A section is a small group of engineers (8 to 12) that specializes in a certain area such as motors, controls, transmissions, power, safety, etc. The section manager assigns tasks to the engineers in the group and provides supervision and support to aid them in performing the work. A department manager (2nd level of management) supervises several section managers. The section manager's role is a mix of both technical and administrative skills.

In companies that are organized with product development teams, program managers are responsible for the design and production of a certain product. Often an engineer becomes the program manager. He or she oversees the team, leads team meetings, makes assignments, arbitrates decisions and is responsible for the development schedule and the budget. The program manager usually reports to a vice president for development. For larger development teams, the program manager's administrative skills are usually more important than his or her technical skills.

Education

Many engineering students continue their studies after completing the B. S. degree and pursue advanced engineering degrees. Those graduating with a M. S. degree may go into industry to serve in assignments demanding a higher level of analytical skills. Others take faculty positions in educational institutions (usually a community college or a four-year college). In these educational institutions, they usually teach three or four undergraduate classes each semester. They also perform many service functions for the college and community.

Some students complete the Ph. D. and take a faculty position with a college of engineering in what are known as research universities. They are responsible for teaching, research and service in these positions. The teaching is limited to one or two courses per semester usually divided between graduate and undergraduate courses. They engage in active research programs and seek funding from external sources (federal and state governments, industry and foundations) to finance their research. They guide the studies of graduate students working in their research topic. Publication in technical journals and writing textbooks is an important part of their scholarly activities. Participation in technical societies, by presenting papers, organizing sessions and serving as officers of the society, are ways for performing community service. Many faculty members are also active consultants to both industry and government.

Consulting

Consulting engineers work on a retainer from government agencies or industry. They are independent businessmen and women who basically offer engineering services at a price. In years past, it was unusual for an engineer to select this career path. Most companies and government agencies hired their own specialists. The downsizing of companies and government has changed this practice. Many companies find that outsourcing engineering work is more cost effective. They call on engineering consultants with a given specialty as required and pay fixed prices for well-defined engineering tasks. The advances in communications—fax, cell phones, e-mail and the Internet—have enhanced the flow of information essential to conducting a successful consulting engineering business. Being in an office at a company's location, eight hours a day five days a week, is not required. A consultant can solve a problem or create an engineering document from a distance while maintaining communication with any client in any country for as long as is necessary to complete an assignment. The availability of high-speed, low-cost computers has enabled the consultant to compete effectively with an in-house analyst if the consultant has access to the same software and the data needed to perform the assignment.

13.6 REWARDS

For many years, engineering graduates have received relatively high starting salaries and most have usually found interesting positions available upon graduation. However, salaries later in a person's career depend upon many different factors including:

- Years of experience
- Level of technical and supervisory responsibility
- Type of employer
- Discipline and/or function
- Degree level
- Strength of the economy

Salaries also depend on current salary-wage policies, individual mobility, working conditions, and local and regional salary scales. Many significant factors, such as challenge, productivity, creativity, technical and supervisory responsibility and job commitment are dependent on the individual's initiative. However, other factors affect salaries that the engineer cannot control. These include company size, company policies, fringe benefits and promotion opportunities. Industry and company growth, labor-management relations and research and development funding also affect earnings. There is considerable variation, even for engineers with similar position, responsibility and experience.

It is not possible to consider all of the parameters that affect salaries in this discussion. However, the three most important factors affecting average engineering salaries— degree level, years of experience and management responsibilities are presented in Fig. 13.1 and Fig. 13.2.

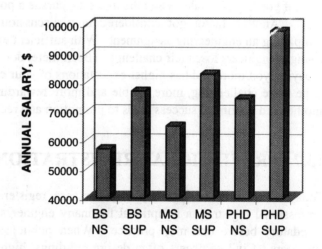

Fig. 13.1 Salaries for all engineers as a function of degree and management responsibilities.
"Engineers Salaries: Special: Industries Report, Engineering Workforce Commission of the AAES, 1997.
NS—non supervisory: SUP—supervisor

The results, presented in Fig. 13.1 and Fig. 13.2, are clear. Engineers with advanced degrees command a higher average salary than those with only a B. S. degree. Also, those with Ph. Ds earn more than engineers with a M. S. degree for the same class of work. Finally, for all degree levels, management responsibilities are rewarded with higher salaries. The differences are significant with manager's pay exceeding non-manager's pay by about 35, 28 and 25% for the B. S., M. S. and Ph. D. degrees, respectively.

The role of experience and degree level is illustrated in Fig. 13.2. All engineers gain in salary with years of experience. The gains are more rapid in the early years, with increases of about twice the increase in the cost of living regardless of the degree level. However, after about 20 to 25 years of experience, the rate of increase in annual salary decreases. In the last decade of a typical engineering career, an individual's compensation may not keep up with the increases in the cost of living. These data are averages, and you may prove to be an exception with either higher or lower rewards for your services. But the message in Fig. 13.2 is clear. Make your mark early in your career. After you reach the age of 45 to 50, significant salary gains are difficult for many engineers to obtain.

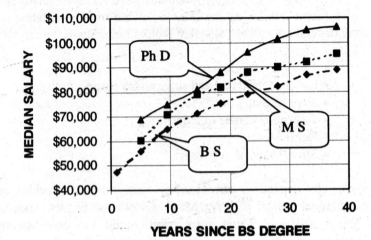

Fig. 13.2 Salaries for all engineers as a function of degree and years of experience.
Engineering Workforce Commission of the AAES, 2000

The data in Fig. 13.1 and 13.2 are from surveys that were conducted 1997 and 2000, respectively. As such, the salaries offered in 2005 and later may be significantly higher than those reported here[8]. Regardless of the exact numbers, the message is clear. If money is important to you—continue your studies and earn an advanced degree. Also, if you have the talents and the interests, pursue a position in management.

Another factor not considered in a discussion of salaries is the satisfaction derived from completing an engineering assignment. With sufficient experience, many assignments typically given to younger engineers lose their challenge. (Been there, done that.) With an advanced degree, an employer is paying you more and has higher expectations of your capabilities. Accordingly, the assignments tend to be more challenging, more visible and more rewarding. Success in one tough assignment leads to another and continued success leads to promotion and recognition.

13.7 PROFESSIONAL REGISTRATION

Today about 30% of graduate engineers are registered and licensed as "professional engineers." Professional registration is optional for many engineers unlike medicine or law where registration is mandatory before one may practice. When public safety is an issue, registration is required for engineers. Civil engineers often design buildings, highways, bridges and other structures that would endanger the public if they failed in service. Consequently, civil engineers participating in these activities must register and become licensed to practice.

You are encouraged to begin the four-step registration process for the simple reason that you do not know when in the future you will be required to demonstrate your qualifications. Sometimes evidence of your B. S. degree is sufficient for qualification, but in other instances it is not. Also as you approach graduation you are very well prepared to begin the registration process.

State boards of registration control the procedure for registration. There are minor variations from state to state, but the guide shown below is typical of the process required:

- Graduation from a four year engineering program that has been accredited by ABET.
- Achieving a passing grade on the "engineering fundamentals" examination.

---

[8] More current data for salaries in select engineering disciplines are presented in Table 13.5.

- Practicing engineering for a specified number of years to gain engineering experience. The state board of registration specifies the number of years required.
- Achieving a passing grade on the "principles and practice of engineering" examination.

The "engineering fundamentals" examination may be taken before graduation. In fact, it is recommended that you take it in either April or October of your senior year[9]. The examination questions are designed to test your skills in fundamental subjects. Because engineers of all disciplines take the same examination, it is evident that the coverage must be on subjects common to all disciplines—mathematics, physics, chemistry, and engineering science. The examination is scheduled for eight hours. After passing the examination and graduating, you receive a certificate that designates you as an Engineer-in-Training.

The "principle and practice of engineering" examination is discipline specific. The civil, mechanical, electrical, etc. engineers must pass a discipline-oriented examination. While the topics covered are more focused, the questions probe the depth of your knowledge in the discipline. After passing this examination, you receive a license as a professional engineer and a seal that may be used when certifying your work.

13.8 SUMMARY

Two reasons are cited for classifying engineering as a profession. The first is based on a highly structured curriculum that prepares you for a position as an engineer upon graduation. You are expected to be productive almost immediately after assuming a beginning position with an industrial firm. The second reason is based on the type of work performed after graduation. When practicing in industry or for the government, one is engaged in engineering activities. A professional engineer is almost always salaried and is not compensated directly for overtime. Nor are you docked for time off. Engineers assume responsibility and are committed to the project, development team and the company or agency. They devote the time necessary to complete a project on schedule and on budget.

Statistics are cited for students entering engineering colleges. The students admitted to colleges of engineering are top notch—student leaders and scholars ranking in the top 5 or 10% of their high school graduating class. Average scores for both mathematics and verbal in the SAT examinations are outstanding. About 19.5% of those graduating from engineering in 2005 were women, but the percent of women in a typical class depends strongly on the discipline (see Table 13.1).

Pursuing a successful career in engineering depends on your ability to develop a number of different skills and understandings. These include:

- The ability to design.
- The ability to perform well on a development team.
- The ability to communicate in many ways and in different settings.
    - o Writing letters, memos, reports, specifications, etc.
    - o Participating in discussions, and presenting design briefings.
    - o Preparing two and three-dimensional drawings.
    - o Preparing slide presentations.
- Performing design analysis and interpreting the effect of the results on the design.
- Designing and conducting experiments and interpreting the data acquired.
- Understanding professional and ethical responsibilities.
- Understanding the need for life-long learning and actively pursuing this goal.

---

[9] The engineering fundamentals examination is offered twice a year in April and October. It is usually offered at several of the colleges of engineering in each State.

The most popular of the engineering disciplines are described and a listing of the undergraduate enrollment by discipline is shown in Table 13.4. Starting salaries for beginning engineers by discipline shows some differences with chemical and computer engineers leading civil engineers. The technical emphasis of each of the popular disciplines is described. Professional societies representing each discipline are discussed.

Engineers of all disciplines perform many different functions in their professional activities. The tasks commonly conducted in each of these functions include:

- Design and development
- Testing
- Design analysis
- Manufacturing
- Production
- Maintenance
- Sales
- Management
- Education
- Consulting

Rewards (salaries) for engineers depend on many different factors. However, three factors—degree level, years of experience and management responsibilities are the most important. Data showing salaries as a function of these three factors are given in Fig. 13.1 and 13.2. Advanced degrees and management responsibilities are rewarded with significantly higher salaries. Experience is also important, but the salary gains due to experience are only significant for about the first 20 to 25 years. After that time, salary gains are barely equal to inflation.

Finally, the advantages of professional registration are discussed. A guide to the four-step process involved in becoming a registered professional engineer is provided. All engineering students are encouraged to take the eight-hour "engineering fundamentals" examination late in their junior year or early in their senior year.

REFERENCES

1. Valenti, M. "Teaching Tomorrow's Engineers," Special Report, Mechanical Engineering, Vol. 118, No. 7, July 1996.

2. Zhang, G. Engineering Design and Pro/ENGINEER, 4th Edition, College House Enterprises, Knoxville, TN, 2001.

3. Patterson, J. and P. Kim, The Day America Told the Truth: What People Really Believe about Everything that Really Matters, Prentice Hall, New York, NY, 1991.

4. Sommers, C. H., "Teaching the Virtues," *Public Interest,* No. 111, Spring 1993, pp. 3-13.

5. Grayson, L. P., The Making of an Engineer: An Illustrated History of Engineering Education in the United States and Canada, John Wiley & Sons, New York, NY, 1993.

6. Gibbons, M. T. "The Year in Numbers," Engineering Statistics, ASEE Web Site at http://www.asee.org.

7. N. Fogg, P. Harrington, and T. Harrington, College Majors Handbook with Real Career Paths and Payoffs: The Actual Jobs, Earnings, and Trends for Graduates of 60 College Majors, 2nd Edition, JIST Publishing, Indianapolis, IN, 2004.

8. Anon, Improving Engineering Design: Design for Competitive Advantage, National Research Council, National Academy Press, Washington, D. C., 1991.

9. Anon, "American Consulting Engineers Council Directory", American Consulting Engineers Council, 10015 15th Street, N. W., Washington, D. C. 20005.
10. Landis, R. B. Studying Engineering: A Road Map to a Rewarding Career, Discovery Press, Burbank, CA, 1995.

EXERCISES

13.1 Write a two-page paper on why engineers should be considered professionals.

13.2 Write a two-page paper comparing the engineering profession to the medical profession.

13.3 Write a two-page paper comparing the engineering profession to the legal profession.

13.4 Write a two-page paper comparing the engineering profession to the accounting profession.

13.5 Write a brief paper giving reasons that you know for why bright well-qualified students for leave engineering after only one semester.

13.6 Why do bright, well-qualified students fail courses in their first year of study?

13.7 Prepare a list of the skills that you currently have and provide a grade (from 1 to 10 with 10 being exceptional) for your achievement level in each skill.

13.8 Add to the list in Exercise 13.7 your goal for an achievement level in each skill at the conclusion of this course.

13.9 Do you believe that you will learn new skills in this course?

13.10 What is the purpose of the ethical codes endorsed by the professional societies?

13.11 Write a two-page paper describing your current ethical standards. Do you believe these standards will change during your tenure at college?

13.12 Are you committed to life-long learning? If so, why? If not, why? How do you intend to continue your studies?

13.13 Have you selected an engineering discipline? If so, state the discipline and give the reasons why you have selected it. If not, state the reasons for your delay?

13.14 If you are still concerned with selecting an engineering discipline, prepare an action plan to gain the information necessary for your decision?

13.15 If you are concerned that your first choice of an engineering discipline was not correct, prepare an action plan to evaluate your choice and to change disciplines if necessary.

13.16 For the engineering discipline of your choice, describe the mission and the activities of the professional society that represents their interests. Visit their web site. Do they have a student chapter? Do you plan to join?

13.17 Most colleges of engineering offer only a limited number of degree programs. What are the degree programs in your college? Do these programs coincide with your career objectives?

13.18 After reviewing the engineering functions, select the two that best suit your interests. Write a paper discussing why these functions interest you.

13.19 Using the data in Fig. 13.1, determine the gain in salary for each degree level that is obtained by serving in a management role.

13.20 Using the data in Fig. 13.1, determine the gain in salary for a non-supervisory position by earning an advanced engineering degree.

13.21 Using the data in Fig. 13.1, determine the gain in salary for a supervisory position by earning an advanced engineering degree.

13.22 Write a paper discussing why engineering salaries begin to plateau after about 20 to 25 years of experience. Include in your paper how you plan to avoid this problem.

13.23 Salary data is a moving target; consequently, the information provided here is at best an approximation. It is possible to obtain more up to date salary information that takes into account the city where the position is located by visiting the Wall Street Journal Website at http://www.careerjournal.com/salary. Click on the Salary Expert and perform a salary search that shows the differences in salaries for different disciplines and different management responsibilities. Also show the differences in salaries to account for the different cost of living in different locations across the United States.

13.24 Today the retirement age for most engineers is about 66. At what age do you plan to retire? Do you believe the age requirement to receive social security benefits will change prior to your retirement? Why?

13.25 Write a paper arguing either for or against professional registration for engineers. Do you plan to take the "engineering fundamentals" examination when you achieve academic status as a senior?

James W. Dally
Emi P. DiStefano
McKenzie C. Primerano

CHAPTER 14

A STUDENT SURVIVAL GUIDE

14.1 OVERVIEW FOR SURVIVAL

As a new student in the College of Engineering, you are encountering many new challenges and meeting many new friends. You have to arrange for housing, transportation and meals (the basics) in an entirely new setting. You also have to cope with a schedule of classes and find the locations of classrooms, lecture halls and laboratories in a maze of buildings scattered all over campus. Finally, you must learn to study and pass courses in a much more competitive environment than you found in high school.

This chapter will discuss some strategies for coping with the new environment and for procedures that will help ensure your successful completion of your studies for a B. S. degree in engineering. In writing this survival guide, it is recognized that you are bright and have demonstrated your scholastic and leadership abilities—otherwise you would not have been admitted to the College. Success should be insured, but it is a fact that about half of each year's incoming class does not complete the engineering program. Many students leave in the first year because of failing grades or lack of interest in the program. Some procedures to insure your success in completing the program will be discussed in this section, and then several important tools for improving your performance will be described later in this chapter.

Goals and Priorities

Many students fail to establish realistic goals that enable them to control their time and direct their energy in a manner needed to succeed in a competitive environment. Long-term, intermediate-term and short-term goals must be established to focus your activities. Otherwise your efforts are diffused, sufficient progress is not made, and one or more goals are not achieved. The goals should be written in priority order and they should be dated. Your goals should be reviewed and revised periodically with changes made to reflect both progress and realism.

Confidence

Success cannot be achieved without confidence. It is essential that you maintain a positive attitude with a can-do philosophy. If you start to doubt your ability to pass a test, fear will become a significant factor that blocks your ability to think clearly during the examination. It is a well-know fact that fear and/or anger limits a person's ability to think clearly and act promptly. One way to gain confidence is to prepare until you are certain that you understand the course material. You will then become confident in your ability to perform well on competitive examinations under pressure.

Stay Current

The clock is as important in studying engineering as it is in a football or basketball game. The course (game) begins on the first day of class and ends at the last minute allowed for the final examination. Your instructor will try to use all of that time to cover as much material as possible. There is little or no float time in an engineering course. Reading and homework will be assigned for each and every class period. Moreover, the material is cumulative—to understand the second topic it is necessary to master the first topic, etc. If you fall behind in the assignments, it is extremely difficult, if not impossible, for most students to catch up and become current.

Attend Class

As an instructor, I was amazed at the number of students cutting classes. An hour of class time costs the student, or his or her parents, more than a ticket to a rock concert. What student would throw away a ticket to a rock concert? Yet, some students throw away many opportunities to attend class over the semester. Attending class gives you the opportunity to hear the instructor's interpretation of the material and to stay up to date on assignment due dates or exam times. Attending class also gives you the opportunity to ask the instructor questions or hear other student's questions, which can often be extremely helpful. On the same note, attending class is not helpful unless you are paying attention to the lecture or questions. While attending class, you might as well make the most of your time there so that you don't have to rely solely on the textbook or other students to learn the material. You will most likely use your textbook or work with other students to study, but doing these in conjunction with attending class will maximize your opportunities for learning, and allow you to gain a deeper understanding of the course material.

Managing Time

You are accustomed to schedules because you moved from class to class following a prescribed schedule while attending high school. The schedule in college is similar except that a class period for a specific course is only scheduled for two or three hours per week instead of an hour every day. You have the **illusion** of more free time. A typical load of 16 credit hours for a semester entails only 12 or 13 hours of classroom participation and 6 to 8 hours of laboratory involvement. Unfortunately, this is misleading because reading and other assignments require significant amounts of out-of-class time. A rule followed by many instructors is to adjust the material and class assignments to require a **time factor of 3 to 4 on the credit hours for the class**. This adjustment means that if you are taking a three-credit hour class, you should expect to spend 9 to 12 hours each week to attend class, complete the assignments and study for the examinations. Your 16 credit hour schedule really involves a time commitment of **48** to **64** hours per week. Time management becomes important when considering the actual time needed to succeed in engineering courses.

Developing Good Habits

It is self evident that good study habits are essential to successfully completing an engineering degree program. Yet many successful high school students have never developed good study habits. Competition in high school was not intense and you could meet or exceed your teachers' expectations without much out-of-class work. The academic competition is much more intense in a college of engineering and your instructors' expectations are higher than you can imagine. It is essential that you establish and follow a personalized schedule for study periods for each of your classes. The time allocation must be sufficient for you to complete the class assignments. Your schedule must insure that you do not fall behind in each subject.

In addition to good study habits, you should develop good personal habits. These include eating a well balanced diet and exercising regularly. Arrange your schedule to enable you to sleep six to eight hours each night. You should wake up ready to face, if not the world, certainly your instructors. It is important to be well rested and to stick to a daily schedule. Finally, remember that drinking excessively may affect your ability to maintain good academic performance.

Managing People

While you are not yet an engineering manager, you will still interact with a number of people, and will find it necessary to manage your relations with them. A short list of these people includes your instructors, roommate, classmates, teammates and friends on and off campus. Maintaining relationships requires time, and as you will soon determine, time is a precious commodity. Your instructors should command priority; fortunately they are easy to manage. All you have to do is go to class, and perform well on examinations. Your instructor will respond with good grades. You may also meet many of your friends in classes. This social bonding is an important part of the educational process. Education should be fun and classmates help make it so.

Teammates differ to some degree from classmates, because a team is a more formal group that has been organized to perform a well-defined task such as the design of a prototype of some product, process or system. Teammates also bond socially and often remain friends long after the team has been disbanded. Friends on and off campus are important to your social life. Pursuing an engineering degree will take most of your time and energy, but it is essential that you allow some time for socializing.

Your roommate and dorm neighbors will play a central role in your life on campus. Living in close proximity and sharing space dictates that a set of rules be established regarding community behavior. There are many decisions to be made involving music, visitors, study time, quiet time, lights out, entertaining, etc. You are encouraged to discuss a schedule with your roommate and neighbors that will provide an environment conducive for study. Also important is quiet time to enable you to sleep and rest. Performance and health both depend upon having a good night's rest. Medical experts recommend eight hours of sleep to avoid illnesses and to perform at your peak capability.

Seeking Help

Compared to high school, the university is huge with tens of thousands of students and countless buildings scattered over a very large area. Orientation held in the summer preceding fall admission is recommended. These orientation sessions will provide some of the answers to your questions. Information will be provided about the engineering programs, the locations of the buildings, dormitory life, availability of food and entertainment, etc. However, after you arrive and begin classes, many unforeseen questions will arise. Where do you seek reliable answers?

If your questions are about dormitory life, find the **Resident Assistant (RA)** and he or she will provide guidance. The RA's are carefully selected; they relate well to new students and often anticipate questions. The point is for you to ask questions at your earliest opportunity. The answers will alleviate your concerns.

If your questions are about the engineering program, find the **Office for Freshman Advising**. This office, usually found near the Dean's office, is staffed with professional counselors. They know the details of all the programs in the college, and they understand the needs of the students entering the program. They can provide valuable advice about program requirements that will ease your entry into the program.

If you have questions about a class you are taking, check the instructor's schedule. He or she will have a few hours each week reserved for office hours. You may stop at his or her office during these times and find your instructor. He or she should be willing to take the time to discuss your concerns. Sometimes meeting with the instructor during regularly posted office hours is not possible. In these instances, contact the instructor by phone or e-mail and arrange an appointment. The instructors will help you understand the course material, but you should be prepared to ask clearly phrased questions that define the problems you are encountering. The instructor will be willing to respond to your questions, but he or she will not be willing to tutor you.

14.2   TIME MANAGEMENT

Weekly Schedules

The college schedule is a weekly affair. You may have classes every day of the week, for about 15 weeks, before the final examination period. It makes sense to prepare a weekly schedule showing the time and location of your classes similar to the one shown in Fig. 14.1.

WEEKLY SCHEDULE							
FALL SEMESTER 2006							
NAME		ADDRESS			PHONE NUMBER		
TIME	SUN	MON	TUES	WED	THUR	FRI	SAT
MORNING							
6:00							
7:00							
8:00		ENES 100		ENES 100		ENES 100	
9:00		ENES 100		ENES 100			
10:00		MATH 140	MATH 140	MATH 140		MATH 140	
11:00							
12:00		CHEM 135		CHEM 135		CHEM 135	
AFTERNOON							
1:00		ENGL 101		ENGL 101		ENGL 101	
2:00			CHEM 135				
3:00							
4:00							
5:00							
EVENING							
6:00							
7:00							
8:00							
9:00							
10:00							
NIGHT							
11:00-6:00							

Fig. 14.1 A possible class schedule for first semester electrical engineering students.

The recommended load of 13 credit hours for the fall semester of first year students is less than the typical load of 15 to 18 credit hours for the second and subsequent semesters. The lighter load is to provide you with the opportunity to adjust to a new academic environment. While the times for classes on the schedules will differ from one student to another, the number of credit hours and the number of contact hours will probably remain about the same. The number of contact hours[1] (16) exceeds the number of credit hours because laboratories and recitation periods are not always counted with the same weight as lecture hours.

The next step in time management is to schedule study periods. How much time should you schedule? The general rule most instructors use in designing assignments is that the **average** student

[1] Contact hours are the number of hours you attend class, recitation or laboratory periods in a given week.

spends two hours studying for each hour of lecture and another hour studying for each hour of laboratory or recitation. This rule implies that a student should schedule $3 \times 2 + 1 \times 1 = 7$ hours of out-of-class study time for Chemistry 135 each week. For English 101 an average student would schedule $3 \times 2 = 6$ hours of out-of-class time for study and writing. Adding the study time for all four courses required this semester gives a total of 27 hours that should be devoted to study for the **average** student. The word **average** is stressed. You may be brilliant and be able to perform well with less than the recommended 27 hours. Or you may be having trouble with one or more courses and find 27 hours is not sufficient. The weekly schedule shown in Fig. 14.2 provides for 27 hours of study time.

WEEKLY SCHEDULE							
FALL SEMESTER 2006							
NAME		ADDRESS			PHONE		
TIME	SUN	MON	TUES	WED	THUR	FRI	SAT
MORNING							
7:00	OPEN	BRKFAST	BRKFAST	BRKFAST	BRKFAST	BRKFAST	BRKFAST
8:00	OPEN	ENES 100	STUDY	ENES 100	OPEN	ENES 100	OPEN
9:00	BRKFAST	ENES 100	STUDY	ENES 100	OPEN	STUDY	OPEN
10:00	OPEN	MATH 140	MATH 140	MATH 140	OPEN	MATH 140	STUDY
11:00	OPEN	LUNCH	STUDY	LUNCH	OPEN	STUDY	STUDY
12:00	LUNCH	CHEM 135	LUNCH	CHEM 135	LUNCH	CHEM 135	LUNCH
AFTERNOON							
1:00	OPEN	ENGL 101	STUDY	ENGL 101	OPEN	ENGL 101	SOCIAL
2:00	OPEN	STUDY	CHEM 135	STUDY	OPEN	STUDY	SOCIAL
3:00	OPEN	STUDY	STUDY	STUDY	OPEN	STUDY	SOCIAL
4:00	OPEN	STUDY	STUDY	STUDY	OPEN	STUDY	SOCIAL
5:00	OPEN	OPEN	OPEN	OPEN	OPEN	OPEN	WORKOUT
EVENING							
6:00	DINNER	DINNER	DINNER	DINNER	DINNER	DINNER	DINNER
7:00	STUDY	OPEN	OPEN	OPEN	STUDY	SOCIAL	SOCIAL
8:00	STUDY	OPEN	OPEN	OPEN	STUDY	SOCIAL	SOCIAL
9:00	STUDY	OPEN	OPEN	OPEN	STUDY	SOCIAL	SOCIAL
10:00	STUDY	OPEN	OPEN	OPEN	OPEN	SOCIAL	SOCIAL
11:00	OPEN	OPEN	OPEN	OPEN	OPEN	SOCIAL	SOCIAL
NIGHT							
12:00-6:00	SLEEP	SLEEP	SLEEP	SLEEP	SLEEP	SLEEP	SLEEP

Fig. 14.2 A complete weekly schedule for a first semester student.

Many assumptions were made in developing the schedule presented in Fig. 14.2. The first assumes you are a morning person because all mornings are packed with classes and study except for Sunday. There are two reasons for carrying a heavy schedule in the morning. First, most people think more clearly in the morning after a good night's rest. Second, if you encounter difficulties in completing an assignment during a study period, time is available to recover later in the day.

Every hour in the weekday morning has been filled[2]. If you have a period or two between classes, you may want to schedule it for study. Usually it is easy to find a quiet desk in the library and complete an assignment in mathematics or chemistry in an hour. On the other hand, you may decide to work with a small study group in one of the student lounges.

The weekly schedule shows four of the seven afternoons devoted to classes or study periods. Two afternoons are completely open and Saturday afternoon and evening are scheduled for social

---

[2] All mornings are filled except for Thursday, which is an open day. Many students find it necessary to work during the academic year. This schedule has been arranged with one open day to enable part time employment.

events. It is important to schedule time for socializing with your friends, just remember to keep everything in moderation. The open periods (unscheduled time) provide much needed flexibility in the schedule. On some weeks you will find the assignments much more demanding than other weeks. When this situation occurs, you will be able to use the open periods to study. Open time before bed time has been scheduled for five evenings of a typical week. A total of 36 hours of open time is available. There is a significant amount of flexibility to add study periods or even to work part time if necessary.

The schedule also reflects the authors' concern for your physical well-being. An hour has been set-aside for each of three meals a day, seven days a week. Nutrition is important to your health throughout life. Eat regularly and avoid junk food. You will feel better and perform better. Sleep is also as important to your well being as eating is. Seven hours a night is reserved for sleep, but you may need more to stay healthy. You will begin to have frequent colds[3] if the amount of your sleep and rest is not sufficient. Regularity is also important in establishing beneficial sleep patterns. Go to bed at the same time each night and arise at the same time each morning. A sleep period from 12 PM to 7 AM is scheduled.

Attending class and studying is sedentary. You will lose your body tone without interrupting this pattern of class, study and sleep. Try going to the gym and working out at least three times a week. The schedule provides ample open time for working out every day.

You may find that the schedule shown in Fig. 14.2 is not suitable. That's fine. It is not necessary to impose this schedule on you. However, it is imperative that you prepare a weekly schedule that is more suitable, and one that accommodates time for:

- Attending all your classes, laboratories and recitations
- Sufficient study time at suitable times (27 hours is a minimum suggested)
- Adequate uncommitted (open) time to provide needed flexibility in your schedule
- A measure of social time to keep your spirits high
- Enough time for meals, sleep, work outs, and mentally preparing for the day

Daily Schedules

You will prepare a weekly schedule each semester. It is a guide to your daily activity, but it does not contain the detail needed to use your time efficiently. A daily schedule provides the detailed guide for allocating sufficient time to tasks essential to your success in each course. To provide information on preparing a daily schedule, consider a typical Monday as defined in the weekly schedule of Fig. 14.2. A new format for the daily schedule is constructed as illustrated in Fig. 14.3.

Let's review this example. Each hour of the day is scheduled and tasks are assigned. At 7:00 AM, you rise, shower, etc. and plan the day. Next you walk[4] to the dining hall and enjoy breakfast. At breakfast you arrange for a study group to meet after the English class in the afternoon. At 8:00 AM you arrive on time for your ENES 100 lab and engage as an active team member in planning the design of the hovercraft. At 10:00 AM you attend your mathematics class before walking to the dining hall for an early lunch.

---

[3] If you are in good health and resting your body sufficiently, you should not experience more than two or three colds each year, and they should be of short duration.

[4] Remember to allow for time to walk between buildings in preparing your daily schedule. Total walking time probably approaches an hour each day for students on large campuses with scattered buildings.

DAILY SCHEDULE					
MONDAY, OCTOBER 16, 2006					
NAME					
TIME	**MONDAY**				
MORNING		**AFTERNOON**		**EVENING**	
7:00	PREPARE	1:00	ATTEND ENGL 101 CLASS	6:00	DINNER
	BREAKFAST	2:00	STUDY	7:00	OPEN
8:00	ATTEND ENES 100 LAB		WORK TWO MATH PROBS	8:00	OPEN
9:00	ATTEND ENES 100 LAB	3:00	STUDY ENES ASSIGNMENT	9:00	OPEN
10:00	ATTEND MATH 140 CLASS		COMPLETE READING	10:00	OPEN
11:00	LUNCH	4:00	STUDY CHEM 135 ASSIGNMENT	11:00	OPEN
12:00	ATTEND CHEM 135 CLASS		COMPLETE PROBLEMS	**NIGHT**	
		5:00	WORKOUT & SHOWER	12:00-7:00	SLEEP

Fig. 14.3 An example of a daily schedule for a typical Monday.

Following lunch, you cross campus and arrive on time for your chemistry class scheduled for 12:00 PM. After this class, you proceed to the library and find a quiet desk. You have planned three hours (from 2:00 to 5:00 PM) to complete the problems assigned in the MATH 140 class and to begin an assignment for ENES 100 and CHEM 135. Unless you encounter difficulties, this amount of time should be sufficient. At about 5:00 PM, you are saturated with mathematics, engineering and chemistry and call time out. You walk over to the gym, dress in running gear and jog for two miles. After showering and dressing, you walk over to the dining hall for dinner.

Your daily schedule indicates that the evening is open. Should you relax or study for a while after dinner? You have efficiently used the three hours of study scheduled during the day to work on assignments in three of your four classes; however, you have not completed the assignment for your mathematics and chemistry classes. To make this decision, examine your class schedule for Tuesday. You have three hours set aside for study in the morning, another three hours later in the day. You can relax Monday evening or take on a part time job if you are able to consistently complete your assignments during the study periods scheduled during the day.

While the class structure is consistent from week to week, the assignments vary from day to day. Sometimes the assignments can be completed quickly with little effort. Other times term papers are due and the library research and writing requires much more time than the usual assignment. A few times each semester an hourly examination will be scheduled in each course. The time to prepare for the examinations is often significant, particularly for those students who fall behind. Because of the changing demands for time from one course to another, it is vital that several open periods are available in the daily schedule to provide you with an opportunity to put in the extra effort needed for success.

14.3 MONEY MANAGEMENT

Money or the lack thereof, is often a problem for many students. It is a difficult problem to be discussed in a textbook because of the large variation in the circumstances of the students and their personal spending habits. To simplify the discussion, let's divide a student's money requirements into two categories—mandatory and discretionary. The mandatory category is the funding necessary for tuition, fees, books and room and board. Reference to a university catalog gives an estimate of these mandatory costs as shown in Table 14.1. An example of a typical in-state-student attending a State university is given. The total costs increase by about $5,000 to $6,000 dollars for non-resident students attending a State university. The mandatory costs for attending most privately financed universities often range from $30,000 to $40,000 dollars.

Table 14.1
Annual estimated mandatory costs
In-state student living on campus[5]

Item	Amount
Tuition and Fees (in State)	$8,000
Room and Board	$8,200
Personal	$2,500
Books and Supplies	$1,000
Transportation	$1,300
Total	$21,000

If you live in the vicinity of the university, you may choose to live at home and commute. If this is the case, you will probably drive, car-pool, or use public transportation to travel from home to the university. In this case, the room and board costs in Table 14.1 may be eliminated (if it is absorbed by your parents); however, commuting costs, not listed in Table 14.1, are incurred. Commuting cost vary widely from zero if you are driven to campus to about $3,500 per year[6] if you drive an older used car and have a long distance to travel to and from the university. A commuting cost of $2,000 per year is assumed in the estimates for the commuting student in Table 14.2 to arrive at a total annual mandatory cost of $13,500.

Table 14.2
Annual estimated mandatory costs
Student commuting to campus

Item	Amount
Tuition and Fees (in State)	$8,000
Books and Supplies	$1,000
Personal	$2,500
Commuting Costs	$2,000
Total	$13,500

The next category is discretionary costs where you have a choice about expenditures. Of course you will need spending money—the question is how much? If you are on a room and board plan, it probably does not include three meals a day seven days a week. You will find it necessary to purchase a few meals at the local eateries each week. You will also find it necessary to finance your social life. Social costs vary widely depending on your choice of entertainment and tastes. A night at the movies with a soft drink and a slice or two of pizza is a low cost evening. Taking a date to dinner at a good restaurant followed by a rock concert will dent your budget big time. The amount allocated for discretionary costs is a decision you should make with your parents if they are funding a large portion of your college expenses. They will want to help, but their resources are not infinite.

If you are working your way through college, the hours you must work increase with the amount of your discretionary spending. As you commit more hours to the workplace, your study time decreases and you endanger your chances for success. Sometimes working students find it necessary to withdraw from courses during the semester and accept a reduced course load that lengthens the time to complete a B. S. degree from four to five or more years.

---

[5] These estimates, taken from the University of Maryland website, are for the 2006-2007 academic year.
[6] This estimate is based on driving 5,000 miles per year at an average cost of $0.40 per mile for gasoline, insurance and maintenance.

Budgeting

Depending on your circumstances, attending a State supported university is going to cost from about $13,500 to $21,000 per year. For this discussion, assume that you have an agreement with your parents and know the level of support that they can provide and when they will make their contribution. You will find the university is not a generous organization. They will demand their money[7] and if it is not forthcoming, you will not be in good standing with the University. Clearly, you have two problems.

1. Raising $13,500 to $21,000 per year.
2. Insuring adequate cash flow so that the funds will be available at the correct time.

Planning is essential to your money management and developing a budget is the first step in managing your funds. To provide an example of budgeting, let's assume that your parents take care of the cost of tuition and fees. You are to handle the cost of books, commuting expenses and personal funds (spending money). The first step is to prepare an estimated budget of annual expenditures. You list your financial requirements, make many decisions about books and supplies, commuting costs and estimate out-of-pocket expenditures at $83 per week or $2,500 for a 30 week academic year. Your budget of estimated expenditures is listed in Table 14.3. Clearly, the listing of estimated expenditures indicates a $5,500 problem.

Table 14.3
Budget of estimated expenditures for a commuting student

Requirements	Amount
Books and Supplies	$1,000
Commuting Costs	$2,000
Pocket Money	$2,500
Total	$5,500

Let's assume that you worked during the summer and managed to save $3,500 of your after tax earnings. That's great because your shortfall has been reduced to $2,000. How do you raise the $2,000? Your parents are tapped out; they have their hands full with their own financial concerns. You have three options:

- A scholarship
- A student loan
- A job

Without a doubt, a scholarship is the best way for helping to resolve your problem. Some scholarships are based on merit (your high school record) and others are based on financial need. Scholarship awards usually do not happen without you actively seeking financial aid. Contact the **Office of Student Financial Aid** at your university and be prepared to fill out a number of forms. There are a large number of scholarships available and you may qualify.

If a scholarship or grant is not possible, you may wish to consider a student loan. The Office of Student Financial Aid administers all types of federal, state and institutional financial aid programs. This office will not magically recognize your needs. Financial aid does not happen automatically; you

---

[7] Near the beginning of a typical undergraduate catalog, the university describes its charges for tuition and fees. They also clearly indicate that "all charges incurred during a semester are payable immediately. Returning students will not be permitted to complete registration until all financial obligations to the university—including library fines, parking violations, and other penalty fees and service charges—are paid in full."

must make it happen by applying and following up to make certain that all of the forms have been processed.

Your third option is to work and earn money while you pursue your degree. Many students are forced to take a part time job to pay part of their college expenses. If you decide to work, you may be able to arrange a part time position with a local company or on campus for 10 to 15 hours per week. Working more hours than this amount would seriously cut into your scheduled open and/or social time, and endanger your chances for successful completion of a full load during the semester.

Another option for working to help pay your college expenses is to enroll in the cooperative education program. Students enrolled in this program alternate semesters of full time work with semesters of full time study. The program enables a student to gain professional experience, integrate theory and practice, confirm career choices and help finance their education. Students are eligible to participate in the cooperative program at any time during their studies; however, most employers seek students with sophomore standing or higher.

Let's continue with the budgeting example. Suppose you have been successful in obtaining a $900 scholarship, in arranging a $1,000 student loan, and landing a part time job for 12 hours a week that yields $80 per week after taxes ($80/week × 30 weeks = $2,400 per academic year). Your parents are pleased with the scholarship award and agree to give $900 in cash as a tuition reimbursement. You have arranged for an income of $4,300 for the academic year. The next step is to budget the flow of cash into and out of your checking account[8]. Both monthly and weekly budgets are recommended to track the flow of your funds. A monthly budget for a hypothetical commuting student is shown in Table 14.4.

Table 14.4
Monthly budget—September, 2007

Monthly Income		Monthly Expenses	
Item	**Amount**	**Item**	**Amount**
Balance Checking Acct.	$3,215	Books	$400
Wages	$ 320	Supplies	$ 50
Cash gifts	$ 80	Car Pool Payments	$110
		Meals	$ 90
Total Income	**$ 400**	Entertainment	$120
		Miscellaneous	$ 50
Income less Expenses	**$400 – $820 = ($420)**		
		Total Expenses	**$820**
New Balance	$2795		

The monthly budget for September shows expenses exceeding income by $420. Is this cause for concern? If you examine the expenses, you will note that books and supplies total $450. This is much higher than usual because of the beginning of the fall semester. Next month, you do not plan on purchasing any books, and expenditures on supplies will only be budgeted at $15. Next month's income will be planned to be $320 and expenses planned to be $385. In next month's budget, expenses will exceed income by $65.

The budget provides $50 for miscellaneous because one always encounters unforeseen expenses. However, the miscellaneous items ordinarily should not exceed 10% of the total expenses budgeted for the month.

In an analysis of the budget, you should always try to identify unusual items. On the income side of your budget, you will find a cash gift of $80. The money came from an older brother who

---

[8] It is assumed that you have established a checking account to accommodate cash flow requirements prior to enrolling in college.

responded to your tale of woe about the high price of textbooks. This is an unusual event and should be recognized as such. The gift will not be part of your regular budgeting for next month.

Considering the unusual items on both the income and expense sides of your budget, you can conclude that the negative cash flow in October will be much smaller, but the cash flow will remain negative. If you seek to balance the monthly budget, it will be necessary to work more hours to increase income or to entertain more economically to decrease expenditures.

Have you ever approached the end of a week (or month) and discovered you were broke? How did you spend that much money? Did the cash evaporate? What happened? To avoid this difficulty, it is suggested that you keep a weekly budget and a daily record. The weekly budget, shown in Table 14.5, is a short-term version of the monthly budget. Use a period of one week because many payments are made on a weekly basis.

Table 14.5
Weekly budget—October 7 to 13, 2007

Weekly Income		Weekly Expenses	
Item	Amount	Item	Amount
Balance Checking Acct.	$ 3,120	Supplies	$ 15
Wages	$ 80	Car Pool Payments	$ 25
		Lunch 5@$4 ea	$ 20
Total Income	$ 80	Entertainment	$ 30
		Miscellaneous	$ 10
Income less Expenses	$ 80 – $100 = ($20)	Total Expenses	$100
New Balance	$ 3,100		

The weekly budget shows that the total expenses exceed the total income by $20. It appears that the money in your checking account will be sufficient to carry you until the summer when you can take on a full time job and begin to save money again. The question is if you are able to stay within the budget, week after week? A daily record of all of your expenditures will help.

To develop a guide for your daily record, let's define discretionary expenses as the sum of the expenditures for lunch, entertainment and miscellaneous which is $60 for the week of October 7 to 13 on our weekly budget. The expenditures budgeted for the car pool payment and supplies are required and not discretionary. For Friday, let's suppose you record your daily expenditures as shown in Table 14.6. Your expenditures for the day and a modest evening out were $24.25. This may sound reasonable when compared to the $65 for discretionary spending, but the week has seven days. If you would like more entertainment (and everyone does), it will be necessary to cut back on other expenditures. The daily record gives the data necessary for you to examine what you buy with your money. What could you give up with out great pain and sacrifice? That coffee and pastry for $1.95 tastes good, but it is junk food that does little for you other than adding empty calories. The coke and pizza is in the same category. Take in the movie and avoid the munchies. You will save money, meet your budget and avoid gaining unwanted weight.

The advantages of the daily record of expenditures are:

1. Provides accurate data on expenditures that reveals your spending patterns.
2. Gives a daily reminder of your budget constraints as you record your purchases.
3. Enables you to compare daily expenditures with your weekly budget.

It is recommended that you maintain a weekly budget and a daily record of expenditures. The budget and record of expenditures will not necessarily relieve your financial difficulties, but they will enable you to deal with them more effectively.

Table 14.6
Daily record of expenditures
Friday, October 12, 2007

Item	Amount
Coffee and Bagel	$1.95
Lunch	$4.75
Pizza and Coke	$2.80
Share in Fuel Costs	$5.00
Movie with Snacks	$9.75
Total	$24.25

Loans and Credit Cards

Many students are eligible for student loans that are guaranteed by the Federal government. The benefit of this is that you avoid the risk of not being able to pay back the lending agency. As a student you can arrange a loan with a low interest rate that is repaid after graduation when you should be earning a significant salary. If you need to use a loan to finance your college education, a low interest rate, deferred payment, government-sponsored loan is the best alternative.

The worse alternative is to accumulate credit card debt. Credit cards are used by almost everyone today because they offer a convenient way of paying for anything without carrying around a pocket full of cash. You also do not need to go the bank to cash a check. Credit cards offer free **short-term** credit. The **short-term** nature is emphasized. After receiving the monthly statement that summarizes the purchases charged to the credit card, you have two or three weeks to pay the bill. However, if you do not pay on time, the interest charge is very high—usually about 18 to 21% of the balance. In addition, there are extra fees charged for late payments. Suppose you carry an average credit balance of $5,000 on a credit card for a year at an interest rate of 21%. The cost of this credit card loan for a year is $1,050. The interest on a student loan for $5,000 at a rate of 5% is only $250.

Use a credit card freely only when you are certain that you have enough money in your checking account to pay off the entire balance, on time, each and every month. If you need to use a loan to pay for education, a student government loan is the best option.

Scholarships

Scholarships were discussed previously in this section. You must actively apply for scholarships if you wish to receive one. It does not hurt to apply for as many scholarships as possible for which you qualify. There is no scholarship advisor who automatically awards scholarships to students on the basis of merit or need. Visit the Office of Student Financial Aid and talk to their staff about scholarship information. There are many scholarships based on merit. If you have an excellent high school record, you may qualify for one of these merit scholarships. If your personal income is limited, you may qualify for an **Institutional Scholarship, a State scholarship or a Federal grant.** Ask for financial aid[9], fill out the forms and submit them on time. Usually these scholarships or grants are for one year, and it is necessary for you to reapply each year.

If you cannot manage to obtain a scholarship or grant, it may be possible to obtain a low interest federally-subsidized loan. These loans must be repaid. However, there is a very low interest rate and

---

[9] A personal note—out of ignorance, I failed to ask for financial aid until the second semester of my junior year. I worked long hours in a factory to finance my education at a private university. When I did visit the Office for Student Financial Aid and explained my situation, a tuition scholarship for my senior year was made available almost immediately and life became much easier.

you do not need to pay the loan back until after you graduate when you should have a significant income. To obtain one of these loans complete a free application for **Federal Student Aid** and submit it. The deadline for submission is February 1. If you are eligible for financial aid and find it necessary to use a loan to pursue your education, a subsidized loan will be the best option and will reduce the cost of borrowing money.

14.4 PEER MANAGEMENT

As you move through the program, you will meet hundreds of fellow students. Some of these students will be no more than faces in a common classroom, while others will become close friends. These close friends may mold your life style and behavior. Peer pressure is a recognized factor in affecting behavior. You will have an established set of behavioral and moral patterns before you arrive on campus. As you develop friendships in the classroom, dormitories, fraternities, sororities, gymnasium, etc., your behavioral patterns will change. In fact, you will become associated with several groups of friends each having different lifestyles, interests and personalities. For example, you may have a group of friends from a design team in an engineering class. You may have another group of friends from the dorms and still another from a recreational basketball team.

These groups of friends make up your social life which is important for you to have while pursuing an education. It is healthy to be involved with friends, but there are two dangers of which you should be aware. First, the time involved in pursuing social activities with your friends should not interfere with any of your scheduled study time. You must remember that while pursuing an engineering degree, schoolwork comes first and then social commitments.

Second, the behavior of some social groups may cause concern. Before you become deeply involved with a social group you should compare the group members' moral standards with yours. If members of the group break the law or do illegal drugs it would be in your best interest to find a different social group. It is easy to find trouble if you associate with the wrong people.

14.5 STUDY HABITS

Good study habits are important to successfully completing an engineering degree. A study habit is a routine that you establish to prepare for class. There are several important aspects of study habits which affect your performance. First is timing. Many students prefer to study in the morning when they are rested and their minds are fresh. Others prefer to study very late at night when it is quiet. You should choose the time of day that is the most effective for you. Establishing several blocks of study time each week is an important scheduling habit to develop.

Organization is another important feature for effective study. You will be carrying loads of four to six courses each semester with different sets of requirements and different types of assignments. In addition to your time management schedule, it is important to list all assignments. Organizing the assignments, assigning priorities and making time estimates and commitments for each course are critical tasks. Time management, as discussed previously, is an important element to organization; however, priorities must be set so that assignments are completed and handed in on time. If you list all assignments, tasks, and due dates at the conclusion of each class, it is easy to review your notes at the end of each day and establish priorities for your time.

Where you study is just as important as when you study. Single desks in the library or computer labs are a good place to study. It is quiet with little or no chance of interruptions. If you form a small study group, tables in the library will serve your purpose. You may talk quietly and discuss problems and approaches to be used in the solution. You may find your dorm room suitable if your roommate is cooperative. If not—head for the library. Student lounges may be suitable in some instances, but they are often noisy with students talking, eating and moving about.

Developing smart study methods is also important. Most engineering students spend much of their study time working homework problem assignments. A typical assignment for a class may include three or four problems. The instructor probably intends the students to spend 15 to 20 minutes per problem. Students understanding the material will be able to complete the assignment in about an hour. However, if you have difficulty with one or more problems you will need to devote additional time to complete the work. How much time should you spend before asking for help? You probably should not devote more than about 30 minutes trying different approaches for the solution to a difficult problem. If you have not solved the problem by then, seek help from your TA, instructor, or a study group. Study groups will be discussed later in this chapter.

Prepare neat homework solutions. If your homework is illegible or has many pencil/pen scratch marks on it, it may not be accepted. If you know what you are doing, most solutions can be completed with about a dozen lines and a diagram or two. Plan your approach and execute the answer showing all work. If you do not show all of your work, you may not get credit either. If you make a mistake, either erase completely or rewrite the solution. Always use a pencil because it is easy to erase after you have make a mistake. Ballpoint pens are not appropriate writing instruments for most students.

14.6 PROBLEM SOLVING METHODS

From your first semester until graduation, you will be required to solve problems that require accurate numerical answers. A large percentage of your grade in science, mathematics, and engineering courses will be your performance in solving these problems. It is important for you to master a proven technique to solve problems. It is recommended that you follow an eight step plan as indicate below:

1. Write the problem statement as you begin the solution.
2. Carefully read and understand the problem.
3. Identify the unknown quantity that you are being asked to determine.
4. Write the rule or equation you plan on using in your solution.
5. List the information provided and prepare sketches as required.
6. Execute the solution.
7. Add units as required to the final result.
8. Check the solution for numerical errors.

Let's consider an application problem in a Calculus I course to illustrate this approach.

Step 1: Write the problem statement.⇒⇒Through how many revolutions does a bicycle wheel with a radius of one-foot turn when the bicycle travels one mile[10].

Step 2: Carefully read the problem and understand the physics involved.

Step 3: It is clear that the unknown quantity is the number of times the wheel turns (without slipping) as the bicycle travels a distance of one mile.

Step 4: The equation to obtain the solution is $S = N\,(2\pi R)$.   (14.1)

Step 5: List the known information pertaining to the problem as:
  $S = 1$ mile $= 5280$ ft is the distance traveled.
  N is the number of turns made by the wheel, which is the unknown.

---

[10] Problem 40, page 57 of <u>Calculus with Analytic Geometry</u>, by Robert Ellis and Denny Gulick, Sanders College Publishing, New York, 1994.

R = 1 ft. is the radius of the wheel.

Step 6: Execute by solving Eq. (14.1) for N to obtain:

$$N = \frac{S}{2\pi R} = \frac{5280 \frac{ft}{mile}}{2\pi \frac{1 ft}{rev}} = 840.3 \frac{rev}{mile} \qquad (a)$$

Step 7: Add the required units to the numerical answer. Checked the unit conversions and added rev/mile with the numerical result.

Step 8: Check the numerical result. You should mentally perform an arithmetic estimate of the answer and check if it is reasonable. Also, plug the numbers into the calculator a second time to verify the result. Many careless errors are found with this simple procedure. With repeated practice the eight steps will become a routine that you will follow with little thought. You will also find that checking your results for accuracy and for units is important because errors of both types are common.

14.7 COLLABORATIVE STUDY OPPORTUNITIES

Collaborative study is encouraged in college. However, you must recognize the difference between collaborative study and academic dishonesty. Academic dishonesty would be if you blindly copy the work of a fellow student. Collaborative study is when two or more people work together sharing ideas and discussing approaches for solving problems. Collaboration allows you to find errors in solutions and then to discuss these errors. Everyone involved benefits from the discussion. The individual with the correct solution increases the depth of his or her understanding by explaining the proper approach. The individual with the wrong answer discovers the correct approach from a fellow student. The student works the problem again and has another chance to learn and to succeed. By students teaching each other in this manner everyone's depth of knowledge is increased.

Regular meetings of small study groups are beneficial and reinforce study habits. The group bonds with time, and learning becomes a social experience. Respect, confidence and esteem grow as group members become comfortable with one another.

Forming small (two to four members) study groups for each of your classes is suggested. Meet at a scheduled time after all members of the group have had an opportunity to work on an assignment independently. Compare results, share ideas and discuss difficulties. Seek knowledge from one another to determine the correct approach for solving each and every type of problem. If all members of the study group are encountering difficulty, seek help from your teaching assistant (TA) or your instructor.

14.8 WORKING THE SYSTEM

Believe it or not, there are resources and strategies in the College of Engineering which are designed to help you succeed. Professional counselors and Engineering advisors provide advising for first year students. They will listen to your problems—academic or personal and will provide valuable advice. However, these counselors are only vaguely aware of your existence. You are one among hundreds of students entering the college each fall, and they may not be aware of your concerns. You must seek them out, arrange an appointment, figure out what your questions are, and ask for help. They often can find solutions for your academic worry or problems.

Your instructors also want you to succeed in their courses. They have organized the course material in order to maximize your learning opportunities over the duration of the semester. They have scheduled preliminary and final examinations for the course that provide you with several opportunities to demonstrate your level of understanding. They assign, collect and correct many homework assignments that allow you to develop problem-solving skills important in performing well on examinations. To perform well in a course you must be fully engaged which includes:

- Prepare in advance for every class
- Attend class
- Arrive on time
- Stay awake and alert
- Ask questions when you become confused
- Complete homework assignments correctly and on time
- Stay current with all assignments

If you participate fully in the educational process and are having difficulties, arrange an appointment to discuss your problems with the instructor. He or she will have office hours posted. Seek the help offered, but do not go to the instructor's office and say, "I don't understand." Be prepared with questions about specific principles that are causing you difficulty. Show your work and indicate where you went astray. Your instructor will help you with specific questions, but he or she will not tutor you during office hours.

14.9 PREPARING FOR EXAMINATIONS

If you attend class, understand the material presented in class, and successfully complete all of the homework assignments, preparing for an examination is not difficult or time consuming. Exam preparation becomes time consuming if you have fallen behind in your daily preparation for the class and find it necessary to cram. Stress caused by cramming can be avoided if you stay current. Let's assume that you have studied regularly and are reasonably current with the material. What should you do to prepare for the examination?

A confident student with a good set of notes should try to write the examination by anticipating the instructor's questions. Anticipating the type of questions on the examination is not only possible; it is usually easy to do. In the instructor's class, he or she will emphasize a few topics indicating their importance. The instructor will make statements like:

- "This is really very important."
- "If you remember anything remember this."
- "I really like this derivation."
- "This is the most important principle in the course."

The instructor will usually dwell on the important topics and move quickly through the ones considered of less importance. Review you notes and place a star beside the passages where the instructor signaled either importance or his or her interest. In the typical five or six week period for a preliminary examination, you may find about a half a dozen stars. Write a question about the topics you have identified as important and ask one of the members from your study team to solve the problems on your exam. Discuss this process with your study team and arrive at a dozen questions that the team solves. After a few examination experiences, you will gain skill at anticipating the instructor. You will be able to predict the form, if not the substance, of at least half of the examination questions in advance. Sample exams and past examinations for the course should also be reviewed.

14.10 STRATEGIES FOR TAKING EXAMINATIONS

There are strategies that you may exercise in taking an examination that will enhance your performance. Let's discuss some of them:

Confidence: Study and learn until you become confident that you understand the material and that you will perform well on the examination. If you believe you will screw up the examination, you may not be comfortable enough with the material. Confident does not mean cocky, but rather comfortable with the subject and the instructor.

Calm: To be calm, you are at peace with yourself. You know that you have prepared for the examination. You are not nervous because you have demonstrated the ability to perform well on the homework assignments. You look forward with anticipation to the opportunity to demonstrate your problem solving skills.

Plan and Select: Before beginning the examination, read each problem in the entire examination. Be certain that you understand each problem. If not, ask the instructor for clarification. Then arrange an order to complete the problems from the easiest to the most difficult. Many instructors will arrange their examinations in this fashion with the easy problems first to give you confidence and the more difficult problems later to give range to the grades. Other instructors will arrange problem chronologically with the first problem covering material studied early in the examination period and the last problem covering much more recent material.

Most hourly examinations are limited to a small number of problems. The instructor has timed the examination so that prepared students will have enough time to complete it. In many cases, students with experience and a deep knowledge in the subject can complete the examination in 15 to 20 minutes. You have 50 minutes, which is usually more than enough time to complete your solutions, **but not enough time to be studying or learning during the examination**. Divide one-half of the time for the exam (25 minutes) by the number of problems on the examination to determine the amount of time to spend on each problem. Suppose there are five problems on the examination. Then you should plan on spending five minutes to write the solution to each problem, and hold 25 minutes in reserve.

Execute: Begin with the easiest problem because you know the approach for its solution. If your initial assessment of the difficulty of the problem was correct, you successfully complete the solution in a few minutes and can begin the solution of the next easiest problem. You are ahead of schedule and full of confidence. You continue in this manner until, say problem No. 4, when you encounter a problem and become stuck. When you have spent five minutes on problem No. 4, stop your efforts and move to the next problem. You should return to this problem only if time remains at the end of the examination. Continue solving problems and completing solutions until your initial time amount of 25 minutes is up.

At the end of an examination, use 15 of the remaining 25 minutes and return to the problem causing difficulty. Review the problem statement and your solution to the point of where you got stuck. Perhaps you will find an error, a missing formula, or an omission and be able to complete the solution. If not write the instructor a brief note explaining your difficulty. He or she may reward your efforts with generous application of partial credit.

Checks: Use the final ten minutes of the examination for checking your results and making sure that all of the numerical answers have appropriate units. There are three different mental checks you should go through to detect errors. The first is a simple judgment check. Does the number look reasonable or is it obviously too large or too small? Common sense often is a good guide to finding obvious errors.

The second is a numerical check. Examine the equation and make a mental estimate of the result. With practice you will soon be able to estimate the answer to within ± 20%. Next, run the numbers in the equation through the calculator again. Rerun the numbers until the answer is confirmed. In many instances, a button may have been pressed by mistake and errors result. Recalculating eliminates these careless errors.

The third check is for dimensional consistency. Sometimes errors occur because you fail to recognize the need for unit conversions or you do not make the conversions correctly. By checking an equation for consistent units, it is possible to locate and correct these errors. Equations used in the solution of engineering problems are independent of the units of measurement. Consider an example from physics where the distance S that a baseball falls after is dropped from a very tall building is to be determined. You seek the distance traveled by the ball after it has fallen for 6 seconds. If you assume[11] that the drag due to air resistance is negligible, the equation governing the distance the ball falls is given by:

$$S = gt^2/2 \qquad (14.2)$$

where g is the gravitational constant and t is time of the fall

When applying this equation, you must insure that it is dimensionally homogenous by using units that are consistent on both sides of the equation. To illustrate the various options you have in dealing with Eq. (14.2), consider the three following cases:

Case 1: Use U. S. Customary units of feet for length and seconds for time. The gravitational constant in this system of units is 32.17 ft/s^2. Accordingly, you write Eq. (14.2) as:

$$S = gt^2/2 = (32.17 \text{ ft/s}^2)(6 \text{ s})^2/2 = (32.17 \text{ ft/s}^2)(36 \text{ s}^2)/2 \qquad (a)$$

The time unit "s" cancels out on the right hand side of Eq. (a) and the remaining unit represents a length expressed in feet. Therefore:

$$S = 579.1 \text{ ft} \qquad (b)$$

Case 2: Use U. S. Customary units of inch for length and seconds for time.

The gravitational constant in this system of units is g = (32.17 ft/s^2)(12 in./ft) = 386.0 in./s^2. Accordingly, you write Eq. (14.2) as:

$$S = gt^2/2 = (386.0 \text{ in./s}^2)(6 \text{ s})^2/2 = (386.0 \text{ in./s}^2)(36 \text{ s}^2)/2 \qquad (c)$$

The time unit "s" cancels out on the right hand side of the equation and the remaining unit represents a length expressed in inches. Therefore:

$$S = 6948 \text{ in.} \qquad (d)$$

Case 3: Use SI units of meter for length and seconds for time.

The gravitational constant in the SI system of units is 9.807 m/s^2. Accordingly, you write Eq. (14.2) as:

$$S = gt^2/2 = (9.807 \text{ m/s}^2)(6 \text{ s})^2/2 = (9.807 \text{ m/s}^2)(36 \text{ s}^2)/2 \qquad (e)$$

---

[11] The drag is not negligible. It is assumed to be negligible to simplify the mathematical solution leading to Eq. (14.2). As your skills develop, drag forces will be considered and you will have the opportunity to solve the more complex mathematical expressions that result when the drag force is introduced.

The time unit "s" cancels out on the right hand side of Eq. (a) and the remaining unit represents a length expressed in meters. Hence:

$$S = 176.5 \text{ m} \tag{f}$$

A careful check of units on each side of the equation will allow you to locate errors caused by using inconsistent units when evaluating equations.

14.11 ASSESSMENT FOLLOWING THE EXAMINATION

Errors on an examination can be grouped into three classes.

1. Careless mistakes
2. Execution errors
3. Concept errors

Careless mistakes are when you clearly understood the material but you lacked focus. You may have had the correct numbers substituted into the correct equation but the wrong answer for the final result. You may have accidentally reversed digits in a number or forgot to specify the units. It is recommended that you slow down, and increase your concentration on the problem. The more time you spend checking the results, the more careless errors you will find and correct before the end of the examination.

Execution errors occur when you understand the principles covered on the examination, but lack skill in applying these principles to one or more specific problems. Examples include leaving out a force in a physics problem or using inappropriate compound combinations in an equation representing a chemical reaction. Execution errors indicate a need to solve more and different homework problems. You already have the talent necessary to complete your degree since you were admitted to the College of Engineering. However, it is crucial for every student to practice solving homework problems. The more problems you work correctly before the examination, the better prepared you will become for problem solving during the examination.

Concept errors are much more serious. They occur when you do not use the correct principle in your attempt to solve a problem. Concept errors indicate that you do not understand the material. These errors signal that you may have difficulties in earning a passing grade in the course. Hopefully, the instructor will identify this type of error with a comment on the graded examination. If an explicit comment is not made, the instructor may signal his or her concern by giving you zero credit for your unsuccessful efforts toward a solution. If you have concept errors it is important that you seek help from the instructor as soon as possible. It will be necessary to devote more time to the course until you have mastered the material.

It is important to carefully review your performance following an examination. The instructor will usually devote class time to show the correct procedure for solving every problem on the examination. You will have the information necessary to study your mistakes. Classify them and adapt your study habits to improve your performance. If you find an execution error, ask questions to your study group or your instructor until you clearly understand the correct solution to the problem.

Unfortunately, many students do not attempt to learn from their errors. It is essential that you understand why you made the error because it would be senseless to lose points again for the same error. Engineering is a discipline that builds from a basic foundation to more advanced subjects. If you do not have a strong understanding in the foundation, your performance in the more advanced subjects will be severely impaired. You will get more and more behind. Make it your business to identify your errors on an examination, correct your procedures and **never repeat the same error**.

14.12 SUMMARY

An overview for your success as a student in a College of Engineering has been presented. Success as a student of engineering depends on setting goals and priorities, developing confidence in your ability to perform and being dedicated to staying current in each subject. Also important is attending class, managing time, developing good study habits, managing your relationships with many new people and seeking help when the need arises.

Time management is based on schedules with time carefully designated to all activities including attending class, studying, open or flexible time, social meetings, and maintaining your physical well being. Weekly schedules have been described in considerable detail and a sample schedule is given. A daily schedule is also presented that allocates time for specific tasks for the entire 24-hour day. A weekly and daily schedule is very important in college because most students have so many time commitments.

Many students find themselves short of funds, and they resort to working to cover the shortfall. Working is great if you can manage the time without hurting your performance in any of your courses. However, some students cannot manage time well and often drop a course or two before the end of the semester. By working, the time it takes to complete the B. S. degree[12] may be lengthened. Clearly time management and money management are related. Procedures for estimating costs for commuting and resident students have been described in monthly, weekly, and daily budget recordings. The use of credit cards has been discouraged by illustrating the cost of a credit card loan.

A brief discussion has been given on peer management. You will be interacting with several new groups of friends in the next few years. Your relations with some of these groups will be very helpful. Study groups and design teams provide opportunities for collaborative learning and support. Groups formed to workout in the gym or ball fields help maintain your physical well-being. Most social groups provide you with enjoyment which is very important to experience during your education. Try to avoid excessive time committed to social activities and avoid groups that cause behavioral concerns. It is important to compare the ethical standards of a group with your own set of standards. Participate in a group in which you are comfortable.

Study habits important to your success have been discussed. Suggestions have been made for when and where you study. Advice has also been given on how to study smart and how to prepare homework assignments that are acceptable and respected when reviewed by your instructor. Study groups have been encouraged because of the many benefits of collaborative study.

An eight-step approach to problem solving was reviewed. While eight steps may seem tedious when you are trying to hurry through an examination, experience shows that the process actually saves time by reducing errors. The eight steps include:

1. Write the problem statement as you begin the solution.
2. Carefully read and understand the problem
3. Identify the unknown quantity that you are being asked to determine.
4. Write the rule or equation you plan on using in your solution.
5. List the information provided and prepare sketches as required.
6. Execute the solution.
7. Add units as required to the final result.
8. Check the solution for numerical errors.

An example was provided demonstrating the problem solving approach.

---

[12] The average time to complete a B. S. degree in engineering is about 4.8 years in State supported colleges.

Collaborative study opportunities were described. Study teams or design teams are encouraged for all of your classes. Meet at scheduled times after each team member has had an opportunity to work independently on an assignment, and then compare and discuss approaches and results. Student to student learning is effective. There will be occasions when you cannot determine how to solve a problem on your own but your study partner does know how (or vice versa). Respect, confidence and esteem grow as the group members bond and become comfortable with one another.

There are resources and guidelines in the College of Engineering to help you succeed in your studies. The counselors, instructors and teaching assistants are ready to help you. However, they usually will not initiate the process. It is your responsibility to seek help. Procedures for helping yourself and for seeking help from others in the College have been described.

Next, techniques for preparing for an examination have been discussed. The most important factor in preparing for an examination is to stay current daily and avoid the need to cram. Finally, strategies to follow in taking an examination were described. They include:

- Be confident
- Be calm
- Select problems and plan your approach
- Manage your time on each problem
- Execute the solution
- Check the results

Follow these strategies when you take your next examination and determine if it is helpful.

REFERENCES

1. Combs, P. and J. Canfield, <u>Major in Success: Make College Easier, Fire Up Your Dreams, and Get a Very Cool Job</u>, 3rd Ed., Ten Speed Press, Berkeley, CA, 2000.
2. O'Brien, P. S., <u>Making College Count: A Real World Look at How to Succeed In and After College</u>, Graphic Management Corporation, Green Bay, WI, 1996.
3. Hanson, J., <u>The Real Freshman Handbook: An Irreverent and Totally Honest Guide to Living on Campus</u>, Houghton Mifflin, Boston, MA, 1996.
4. Carter, C., <u>Majoring in the Rest of Your Life: Career Secrets for College Students</u>, 3rd Ed., Farrar Straus and Giroux, New York, NY, 1999.
5. Davis, L., <u>Study Strategies Made Easy</u>, Specialty Press, New York, NY, 1996.
6. Dodge, J., <u>The Study Skills Handbook: More Than 75 Strategies for Better Learning</u>, Scholastic Trade, New York, NY, 1995.
7. Luckie, W. R., et al, <u>Study Power Workbook: Exercises in Study Skills to Improve Your Learning and Grades</u>, Brookline Books, Cambridge, MA, 1999.
8. Tufariello, A., <u>Up Your Grades: Proven Strategies for Academic Success</u>, vgm Career Horizons, Lincolnwood, IL, 1996.
9. Hjorth, L., <u>Claiming Your Victories: A Concise Guide to College Success</u>, Houghton Mifflin, Boston, MA, 2000.
10. Jensen, E. and T. Kerr, <u>Student Success Secrets</u>, 4th Ed., Barron's Educational Series, Hanppauge, New York, NY, 1996.

EXERCISES

14.1   Prepare three different lists of goals—short term, intermediate and long term. Define the time for accomplishment of each set of goals and provide your motivation for setting each goal.

14.2   Do you believe that confidence is important to your success in achieving the goals stated in Exercise 14.1? Explain your response to this question.

14.3   Is it possible to raise your confidence level? How?

14.4   Describe your plan for staying current in each and every subject for which you have registered this semester.

14.5   Did you cut class in high school? Do you plan to attend all of your classes this semester? If you cut a class, how do you plan on mastering the material covered in that class?

14.6   How many total clock hours per week hours do you intend to devote to your studies this semester? How many credit hours are you carrying? How many study hours are you planning to spend for each credit hour earned?

14.7   Write a critical assessment of your study habits—good and bad.

14.8   Identify the friend who you find most helpful as you adapt to college life. Identify the friend who causes you the most difficulty in adapting. Write a short paper describing the reasons for your determination. Also indicate your plans for the future of these relationships.

14.9   Locate the Office for Freshman Advising and the offices of each of your instructors. The instructors should have their office hours posted. Did you find these postings and did you make a note of them?

14.10   Prepare a weekly schedule for your activities this semester similar to one shown in Fig. 14.2. Also list time reserved for class, study and social activities. How many open hours were you able schedule?

14.11   Describe the provisions in your schedule for taking care of your physical and mental health.

14.12   Describe the consequences of becoming ill for three weeks in the middle of the semester.

14.13   How many hours do you sleep each night? Do you maintain a regular schedule for retiring and rising?

14.14   Prepare a daily schedule for each day of the week. Identify your open periods and discuss a strategy for using them when an examination is scheduled.

14.15   Discuss your balance between working, studying and playing over the weekend.

14.16   Suppose you are working 16 hours per week, prepare a weekly schedule. Also list what time has been reserved for class, study and social activities. How many open hours were you able schedule?

14.17   What is a strategy for preparing for examinations for students working 16 hours per week?

14.18   How many hours do you believe you could work per week without significantly affecting your performance in class?

14.19   Prepare a listing of your mandatory educational expenses.

14.20   Prepare a listing of your discretionary educational expenses.

14.21   Describe the rationale for each item listed under discretionary educational expenses.

14.22   Prepare a budget for your costs for this academic year. Are you living on campus in a dorm? Are you commuting to school?

14.23   Does your budget of annual expenditures indicate a shortfall or a surplus? If it is a shortfall, how do you plan to raise the funds necessary to balance your budget?

14.24   Prepare a monthly budget showing your income and expenses.

14.25   For the next week, prepare a daily record of expenditures.

14.26   Analyze your expenditures for the past week using the data revealed in your daily records. Is it possible to reduce your spending without severe sacrifices?

14.27   Do you have one or more credit cards? Do you diligently pay-off the outstanding balance each and every month?

14.28   What is the annual interest rate on each of your credit cards?

14.29   Suppose you have an average outstanding balance of $1252 with a credit card that carries an annual interest rate of 21%. Determine the monthly interest charges and the total interest charges accumulated over the year.

14.30   Do you know the procedure to apply for scholarships, grants or financial aid?

14.31   Have you discussed your financial situation with the counselors in the Office for Student Financial Aid?

14.32   Do you have a friend who sometimes behaves in ways that cause you concern? What actions cause you concern? What do you do when you observe this behavior? Do you plan on continuing this relationship?

14.33   How many blocks (two to three hour periods) of study time have you scheduled each week?

14.34   Where do you study? Which location do you find the most conducive? Why?

14.35   Have you established a procedure for controlling the amount of time devoted to a typical assignment? Why not?

14.36   Using the problem solving approach outlined in Section 14.5, solve the following problem: A 36 inch diameter pipeline services two cities 87 miles apart with fuel oil. Determine the number of gallons of fuel oil required to initially fill the pipeline.

14.37   Using the problem solving approach outline in Section 14.5, solve the following problem. A V-notch weir, illustrated in Fig. Ex 14.37, is used to measure the rate of flow of water in an open irrigation channel. The relation used to convert the measured height "h" to flow rate Q is given by:

$$Q = (8/15)(2g)^{1/2}\tan(\theta/2)\, C_D\, h^{5/2}$$

Determine the flow rate and express the result in terms of gallons per minute (GPM) if the head h of the flow through the weir is measured at 14 inches and the angle $\theta = 70°$. Note that the drag coefficient for the weir is C_D = 0.60, and g is the symbol for the gravitational constant.

Fig. Ex 14.37   V NOTCH WEIR

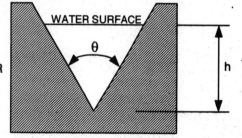

14.38   Have you joined a study group (team) for this class? If so, when and where do you study?

14.39   Which course do you consider the most difficult this semester? Have you joined a study group associated with this course?

14.40   Write a short paper describing the benefits of collaborative learning for your study efforts. What are some of the disadvantages?

14.41   What is your role in your favorite study group? What is your role in the study group you least enjoy?

14.42   Write a short paper describing how you "work the system?"

14.43   Have you ever visited your instructor's office during office hours? Is he or she ready to receive you? Did you find the help provided useful? Do you plan on visiting again when you encounter difficulties with the course material?

14.44   Do you stay current in the mathematics course you are taking? If not, why not?

14.45   Which is the most difficult course you are taking this semester? How much time do you allocate for study in this course? Is that sufficient? What other strategies do you employ to improve your performance other than study time?

14.46   What actions do you take in the week before the scheduled date of an examination to prepare?

14.47   What actions do you take in the day before the examination to improve your performance?

14.48   Write a short paper describing your strategies for taking an examination?

14.49   Write a short paper describing your actions after an examination is returned to you with a grade of 70.

14.50   Write a short paper describing your actions after receiving a grade of 78 on an examination.

14.51   Write a short paper describing your actions after receiving a grade of 53 on an examination.

14.52   Write a short paper describing your actions after receiving a grade of 92 on an examination.

14.53   Explain the methods you use to check answers during an examination?

14.54   Do you find yourself rushed when taking an examination? Which subjects cause you the most difficulty in completing the examinations on time? Explain why you are so rushed on these examinations.

14.55   What actions do you plan for managing your time on examinations more effectively?

PART V

ENGINEERING AND SOCIETY

CHAPTER 15

ENGINEERING AND SOCIETY

15.1 INTRODUCTION

It should be made clear from the beginning of this chapter that engineering and society are closely joined. Engineering is an integral part of society, and consequently, engineering must be treated as a social enterprise as we continue to develop sociotechnical systems. When engineers develop a new product or improve an existing system, we change society. The magnitude of the change differs from one product to another, and the effect on society may be trivial or enormous, but be assured that there is an impact. For example, when a few companies introduced a random orbit sander a few years ago, the time, effort, and cost of producing flat, smooth, scratch-free surfaces was reduced. These sanders helped several hundred thousand workers in the finishing business to produce high-quality coatings on furniture, cabinets, counters and floors. The results also benefited many millions of customers who purchase furniture each year or remodel kitchens. This new product had a modest, but beneficial, impact on society. There was, however, a small negative effect in that some emissions to the environment occurred during manufacturing of the product and virgin resources were consumed. Overall though, society has clearly gained by the introduction of this product.

When transistors were first developed from new ultra-pure semiconductor crystals, the impact was revolutionary. While the first transistor was developed in the mid 1950s, the integrated circuit was introduced in 1959, and the microprocessor was released in 1971, the technological revolution resulting from the introduction of these three products in the microelectronics industry is still underway. Micro-electronic based products are markedly affecting the way in which society lives today. The media refers to the current period as the "information age". New products, which are introduced almost daily, are driven by technology. Products based on new advances in microelectronics will continue to change the way everyone works and plays for the foreseeable future [1].

You often hear about new products and changing life styles. Television, radio, Websites and even newspaper advertising keep everyone well informed of the availability and prices of both old and new products. However, you rarely hear much about the actual technology involved even when the product is new and revolutionary. The media and the public usually are not interested in learning about the technology employed in a new product; this is unfortunate, because it would be easier to introduce new or modified systems if the public were more technologically literate.

Engineers are, in part, responsible for this apparent wall between the public and the understanding of technology. Engineers often speak in such technical terms that those few individuals interested in our work become confused. As engineers, we must clearly recognize the need to effectively communicate to society. We can begin by learning how to explain why and how products work in terms more easily understood by the general public. We must also be more enterprising in understanding society and their changing views relative to technology. Lastly, we need to think on a more global scale. Engineers develop sociotechnical systems that are often large and complex. These systems rely on hundreds of products and many different types of services to function efficiently. Even though, as an engineer, you may be dedicated to developing a single component, subsystem or product that is only a small part in a much larger system, you should understand the entire system. Those development teams that have a thorough understanding of all aspects of the complete sociotechnical system develop the superior designs.

Let's consider an example of a sociotechnical system—the commercial air transport system. Do we have technical products involved in this system? **Yes**, literally hundreds of advanced state-of-the-art products. Do we provide technical services for the airlines? **Yes**. Does an engineer working for General Electric in the development of a new jet engine need to understand what Boeing is doing in the design of a new aircraft or modifications of an existing aircraft? Does that same engineer need to know about the sound levels that the public will accept as the plane lands and takes off? Is the engineer working for Boeing required to understand the causes of each and every airline accident? Again, **yes**.

From these simple questions, it should be evident that sociotechnical systems are extremely complex. It will require significant effort for you to understand the detail of the interaction of the many products incorporated in these systems with the public in general. Nevertheless, engineers that understand and appreciate the improvements, which both the customer and society seek, will develop the most successful products.

Finally, as an engineer, you must understand that the systems developed are adaptive. They are dynamic and change with time in response to several societal and economic forces. First, society changes. The requirements of society and the customer in the 1950s are much different than either will accept today. The customers, a segment of society (more on this fact later), typically expect that the new model of the product will be much improved over the older model and be lower in price. Others in society, not necessarily the customers, will expect (perhaps demand) that you manufacture the product without degrading any aspect of the environment.

An example is useful to distinguish between the customer and society. Let's consider the automobile as the product. The customer of an automobile demands style, comfort, mobility, convenience, affordability, reliability, performance and value. Society requires (through federal regulation) safety, fuel efficiency and reduced emissions and less pollution. The fact that the requirements from the customer and society often conflict should be recognized. Conflicting requirements challenge engineers to create improved designs that advance the state of the art in all areas of technology.

The governments (Federal, State, County and City) are also involved in the relationship between engineers and society. Various government agencies frequently issue regulations that affect both the sociotechnical systems and the product. Consider the commercial air transport system again. Obviously, safety is a very serious issue. The Federal Aviation Administration (FAA) is responsible for issuing regulations to insure safe operations of commercial aircraft. However, 1996 was not a good year for the FAA or the commercial airlines because several very serious accidents occurred with significant loss of life. As a consequence of a fire in the cargo bay, which caused the crash of a Boeing 737 aircraft in the Florida Everglades and the loss of all aboard, the FAA is now requiring the airlines to retrofit all planes with a fire suppressant system. Regulations from government agencies drive a portion of the changes to any sociotechnical system.

This understanding of the relationship between engineering and society is relatively new. If you examine the five-volume tome, *A History of Technology*, [2] published by Oxford University Press over the period from 1954-58, you will find a chronological, deterministic catalog of technological development. However, the editors almost completely ignore the relation between engineering and society. In his review of this tome, Hughes [3] states that the editors belatedly conclude their five-volume treatise with an afterthought essay on technology and the social consequences. Today, the relationship is more clearly recognized although educators have not yet managed to include adequate treatment of this very complex relationship in a typical engineering curriculum. A few examples of the relationships between engineering and society will be described in this chapter. For a more complete treatment, read reference [4].

15.2   ENGINEERING IN EARLY WESTERN HISTORY

Technology has impacted the way that we live since the beginning of recorded history. Long before the first engineering college was founded, intelligent, hard-working men[1] worked as engineers developing new products and processes. You may study history and cite many developments that changed, and for the most part, benefited society.

Prior to about 6,000 BC, there was little evidence of technology. Men and women were hunters and gatherers. They scrounged for food and lived in caves if they were lucky and shacks or huts if they were not. The significant developments in the period from 6,000 to 3,000 BC were in agriculture with the domestication of animals and the cultivation of grains. With time, it was possible to insure a relatively stable, if limited, food supply and people began to congregate together in villages and small cities.

The first technology involved the building of roads, bridges and boats for local transportation. People clustered in villages and small cities wanted to transport food, water and other materials over short distances. The geographic regions where technology was evident were extremely local. Early Western history involved only Mesopotamia, Egypt, Greece and then Rome. In a few cities, there were aqueducts to transport water and some sewerage. In most regions, only the ruling class lived well. The great monuments of Egypt and Greece were built with slave labor. In fact, slaves (people who were on the losing side in a war or those born into slavery) provided most of the power required for building, mining and agriculture for many thousand more years. The rise of religion is often credited with the reduction in the prevalence of slavery. However, the steam engine developed by Thomas Savery, Thomas Newcomen and James Watt in the early 18th century signaled the beginning of the end of the need for power provided by humans (slaves and other unfortunate souls in bondage). As society enters the 21st century, the development of the low-cost, high-performance computers, a wide variety of specialized software programs and effective communication 24×7 by the Internet, portends another beginning of the end for many employees performing repetitive tasks or clerical work.

The Egyptians

The Egyptians built many huge monuments and were masters in moving large blocks of stone from distant quarries to a construction site and then elevating these blocks to their respective positions in gigantic structures. Some estimates indicate that 20,000 to 50,000 men were required to drag 20 to 50 ton blocks of stone used in the construction of these monuments. In spite of the majestic grandeur and durability of the monuments, the new innovations introduced by the Egyptian engineers were not outstanding. They constructed temples in a simple fashion using only uprights (columns) and cross pieces (short, deep beams). While the arch was known during this period, it was constructed with mud bricks and never adapted to stone construction. Classical treatments of the history of technology tell us very little about the impact of these developments on society. However, it is apparent that the construction of a pyramid or a temple required a significant segment of the working population for many decades. This effort probably did little to improve the welfare of even the **free** segment of the population. It is also evident that the effort required of the slaves to move the countless large, heavy stones must have been brutal.

---

[1] Historical accounts of engineering achievements indicate that the profession was male dominated. The author is sensitive to the gender issue in engineering education, but it must be recognized that the presence of women in the profession is very recent when considered relative to a historical span of about 8000 years.

The Greeks

Although they had very capable engineers, the Greeks are better known for their contributions to science. Thales, Euclid and Aristotle are recognized for their contributions to the development of the basics of geometry and physics. This new knowledge of geometric relations was applied almost immediately in architectural engineering. However, scientific understanding during this period was so limited that nearly 2,500 years elapsed before this knowledge was expanded sufficiently to be useful to the engineering community.

Greek engineers built cities, roads, water (hydraulic) systems and some machinery. Archimedes and Hero were innovators who used the pulley, screw, lever, and hydraulic pressure to develop cranes, catapults, pumps, and several hydraulic devices. The Hellenistic period brought better buildings and local roads that improved living conditions for the upper class. However, the construction of the city infrastructure was still performed largely by slaves because the use of other forms of power was nonexistent. The Hellenistic world was divided into extremely independent city-states. Politics of the period did not foster cooperation between these governmental entities; as a consequence, very few long roads were built. Transportation was largely by sea in boats equipped with oars powered by a galley of slaves. Some of the boats were fitted with a square sail, but these were so poorly designed that they were effective only on down-wind tacks.

The Romans

The Roman engineers followed the Greeks, and while they were considered by some historians to be less inventive, they were masters of detail. As the Roman Empire spread from the Middle East to Scotland, the Roman engineers built long roads connecting the distant cities; roads so well constructed that they were in service for many centuries. The large cities constructed in the conquered lands had paved streets and heated homes with water supplies and sewer systems. The Roman engineers had little or no theoretical knowledge, but they had developed practical methods of construction that served their purposes well. During this period, the subjects of statics and mechanics of materials were not well understood. The theory used to design beams would not be developed for another 1,700 years. The Roman engineers based their construction on experience, and they used very large safety factors. These empirical methods, while crude by today's standards, sufficed to produce many bridges, roads, aqueducts and cities. A great bridge over the Tagus River in Spain supported a road nearly 200 m long with six spans 60 m above the river. Considering it was completed at about 100 AD, this bridge was a remarkable engineering feat. The durability of the Roman construction is well documented by their structures some of which are still in use today—2,000 years after they were placed in service.

The Roman engineers deviated from the Egyptian and the Greeks in their construction techniques by using much smaller building stones and bricks. They were able to produce massive structures with the smaller blocks by utilizing mortar and cement. The use of smaller stones, bricks and cement clearly impacted society. Buildings could be constructed with much less effort because thousands of slaves were not necessary to move a single block of material. Life for the slaves clearly improved although the construction work was still very demanding. Also, the upper classes benefited from the more rapid construction of buildings and houses for their cities.

While the Roman engineers were outstanding in the construction of civil infrastructure, they did very little to advance the use of power in driving the few machines in use during that period. Humans were still used in most instances to power pumps, mills and cranes. In rare cases, water wheels were used at power mills to grind grain, but these were of the undershoot design and much less effective than the overshoot design that was developed later. While references on the history of technology are quiet on the issue, the life of a worker powering a pump or a mill must have been very difficult.

Eventually, the vast Roman Empire crumbled, and the world entered a period that the historians classify as medieval civilization from 325 to 1300 AD. The fact that barbarians stripped and destroyed the cities does not imply that progress stopped, but it slowed progress to a snails pace. When Rome collapsed, the Byzantine Empire was the next civilization to follow. Significant engineering achievements were made in this period in developing dome structures and curved dams. The author has omitted discussion of these developments to keep this historical treatment brief.

Early Agriculture and the Use of Animals

In spite of the slow development of engineering and the lack of the construction of monuments during the Middle Ages, an early agricultural revolution occurred. White [5] has described the changes that occurred as agriculture moved from the dry sandy soil of Italy into the heavy alluvial and wet soil of northern Europe. The plows that were effective in the light sandy soils would not turn the sod in the heavy, moisture-laden soil of the north. A new, heavy-wheel-plow was invented that would cut through the grass sod and turn over the soil. However, this plow required eight yoked oxen to provide the large force necessary to pull it through the heavy sodded soil. Very few farmers of the day owned eight oxen. They had to combine their land, share their oxen and cooperate in plowing and planting. The combination of farmers and their land holdings eventually led to a manorial society.

It may be difficult to imagine that the invention of a new plow made such a major change on the way people lived. Today farmers are able to produce all of the food needed in the U.S., while exporting excesses in significant quantities, with less than 2% of the population. However, before the advent of engines and motors (stream, gasoline and electrical), men and women struggled to grow enough food to keep from starving. In the Middle Ages, 90% of the population worked in agriculture, and they had much less to eat than we do today.

Oxen were used as a source of power for plowing and transport on the farms. However, oxen are very slow and they consume large quantities of food. At that time, horses were of limited use because of inadequate harnesses and saddles and the frequent problem of splitting hoofs. (Early harnesses fitted about the horse's neck tended to strangle the animal.) These problems were alleviated when a newly designed horse collar, which transferred the load to the shoulders of the horse, was introduced and when iron horse shoes were fitted to their feet. Horses gradually replaced the oxen because they were 50% faster, worked longer and required less food.

In early warfare, the horse was not very effective in frontal attacks because stirrups for the rider did not exist. Consequently, horse mounted warriors were of limited value because they could not adequately thrust their lance while staying mounted on the horse. With the advent of the stirrup, the horsemen could stand-up, lean forward and thrust the lance through the defenders' shields without losing their seat. The simple addition of stirrups to saddles had a profound influence on society. Feudalism evolved with an aristocracy based on warriors conquering and defending landholdings. A new battle strategy with armored knights mounted on large strong horses was dominant for several centuries [5].

In early history, small technological improvements produced major changes in the form of government, the type of society and the way people of all classes lived. It is remarkable that for about 5,000 years, humans were the prime source of power for most activity. Progress in agriculture and engineering proceeded together to provide more food, better housing, roads and bridges and eventually horse- and oxen-powered plows and wagons.

15.3 ENGINEERING AND THE INDUSTRIAL REVOLUTION

Civilization has always been power limited; even today, space travel is severely constrained by the inadequacies of rocket power. Until the last three centuries, animals, windmills and waterwheels provided the only power available for agriculture, construction, milling, mining and transportation. Progress was often slow because power was not available when and where it was needed. Life, in the power starved 17th century, was rather bleak for the general population. The people worked from dawn to dark as farmers or craftsmen. Only those endowed with landed estates and/or money lived what you would consider a modestly comfortable life.

The Development of Power

The first of several breakthroughs in developing new sources of power occurred in the 18th century. In Great Britain, Savery, Newcomen and Watt developed stationary steam engines. While these early steam engines were woefully inefficient (only 0.5%), they were produced in relatively large numbers (500 existed in 1800) and were employed to pump water and power textile and flour mills.

Because the steam engines were replacing horses as power sources, James Watt conducted experiments to measure the power a "brewery horse" could provide. His measurements indicated the brewery horse generated 32,400 (foot-pounds/minute) of power. The horse's output was rounded to 33,000 (foot-pounds/minute) and this value is still used today as the conversion factor for one horsepower (HP).

Most of the 18th century was devoted to innovations directed toward improving several different models of steam engines. Applications were largely limited to stationary power sources because engines, of that day, were large, heavy and inefficient. However, with improved metallurgy and machining methods, it was possible in the 19th century to reduce the size and the weight of the engines while maintaining their output. These improvements permitted the steam engine to be adapted to power riverboats, ocean-crossing ships, rail locomotives and later even a few steam-powered automobiles.

The development of the steam engine and its adaptation to several modes of transportation formed the foundations of the industrial revolution. The life style of almost everyone was markedly changed with the emergence of steam-powered factories and much more effective transportation. Fewer people were employed in agriculture, but many more were working under appalling conditions in factories and mines. Twelve-hour days, seven days a week, with only Christmas as a holiday during an entire year was the norm. Manufactured products, such as clothing and housewares, were available to a larger share of the population at reduced prices, but the margin between earnings and the money necessary for a factory worker, miner or farmer to stay alive remained very narrow. Steam power had relieved many thousands of men and horses from brutally hard labor, but they were released from one dull job for another. On a more positive note, the progress in agriculture and industry enabled the population to triple during this period (1660 to 1820).

15.4 ENGINEERING IN THE 19TH AND 20TH CENTURIES

For the eight thousand years of recorded history prior to about 1800, the advance of technology was extremely slow. Agriculture dominated society with most of the population growing food that was consumed locally. Factories were being developed and simple products needed on the farm, in the home or by the military were produced. Long distance transportation of goods was usually limited to sailing ships on the seas or horse drawn barges on rivers or canals.

During the 19th and 20th centuries, technology literally exploded. The many scientific discoveries of the preceding millennium expanded knowledge sufficiently to allow countless engineering applications. To illustrate this explosion, let's define modern technology to consist of the following five elements:

1. Abundant available power.

2. Available food in sufficient quality and quantity to insure a nutritious diet for everyone.
3. Transportation by air, land and water with associated infrastructure to provide safe, rapid, convenient and inexpensive access.
4. Communication between everyone, anywhere and anytime at a reasonable cost.
5. Information (knowledge) storage, retrieval, and processing in seconds at any location, anytime and anywhere at negligible cost.

While the advances in technology have been astonishing, particularly in the last 50 years, there are still significant deficiencies. Rocket power for launching space vehicles currently constrains space travel. Traffic jams in large cities extend the working day for many millions of people commuting to their jobs. Today more than 42,000 people die in accidents on the highways every year. Travel on airlines is becoming unreliable, difficult and often unpleasant. Communication in many parts of the world is still not available to the public. Fifteen years ago the author vacationed near a small fishing village in Mexico and learned that the radiophone at the tourist hotel was the only phone in the entire area. However, the recent introduction of cell phones has greatly improved communication in underdeveloped countries. The cost of cell phone towers is much less than the expense of installing land lines required for conventional phone service. Many people are just beginning to explore the different ways to use the massive amount of information that has recently become available to those fortunate enough to own a personal computer, a modem and a broadband connection to a server.

You may debate the exact year of the invention. However, the exact date is rarely important because an invention usually evolves into a commercially successful product after several iterations with improvements or modifications by several designers. You should not worry about the exact dates. Instead, recognize the continuous progress and observe the shift in emphasis from one period to another. In power, technology evolved from animals, to steam, to internal combustion (gasoline) and to nuclear energy. Power, from gasoline fueled internal combustion engines became mobile, and electricity was widely distributed. In transportation, technology evolved from the horse and buggy and sailing ships to automobiles, airliners, nuclear powered submarines and space shuttles. In communications, technology progressed from the pony express to the telegraph, telephone, facsimile machines, cellular telephones, pagers and the Internet. In the information business, technology is currently evolving from a hard copy library system to a digital multimedia information source available at any time, to everyone, at nearly zero cost.

Have these advances in technology changed society? It is believed that they have had a profound beneficial effect. Farmers produce an overabundance of food with less than 2% of the population working in agriculture. Factory workers manufacture most of our consumable products less than 20% of the work force. The workweek is shorter, and we have many more affordable products. Most families own at least two cars and home ownership is at an all time high. Concerns are expressed regarding the improvements in the standard of living over the past 20 years, particularly for those people at the lower end of the income spectrum. Technology has probably not helped this group as much as others. Global trade has permitted many of the semi-skilled tasks to be moved to third world countries, where the cost of labor is a small fraction of the labor costs in developed countries. Consequently, many of the previously well-paid factory jobs in manufacturing in the U.S. have been moved offshore and this process has negatively affected those with limited marketable skills.

15.5 GREATEST ENGINEERING ACHIEVEMENTS—20TH CENTURY

Recently the National Academy of Engineering (NAE) working with 27 professional engineering societies complied a listing of the 20 greatest achievements in engineering during the 20th century. What made these achievements great? They are notable because of the very significant benefits of each accomplishment to all members of society. An excellent publication outlining the engineering achievements that transformed our lives is presented in Reference [7].

1. Electrification—in the 20th century, widespread electrification gave us power for cities, factories, farms, and homes and forever changed our lives. Thousands of engineers made it happen, with innovative work in fuel sources, power generating techniques and transmission grids. From streetlights to supercomputers, electric power makes our lives safer, healthier and more convenient.

2. Automobile—perhaps the ultimate symbol of personal freedom, the automobile is a showcase of engineering ingenuity and the world's major transporter of people and goods.

3. Airplane—with its speed and efficiency, air travel makes personal and cultural exchange possible on a global scale.

4. Water Supply and Distribution—water is our most precious resource, and improvements in its treatment, supply and distribution have changed our lives profoundly, virtually eliminating waterborne diseases in developed countries and providing clean and abundant water for communities, farms and industries.

5. Electronics—electronics provide the basis for countless innovations including digital cameras, iPods, TVs, and computers to name only a few. Electronic products improve the quality and convenience of modern life.

6. Radio and Television—radio and television were major agents of social change in the twentieth century, entertaining millions and opening windows to other lives to remote areas of the world and to history in the making.

7. Agricultural Mechanization—tractors, combines and hundreds of other agricultural machines increase the productivity of farms, reduce the need for manual labor and improve our ability to feed our citizens and others in the world.

8. Computers—perhaps the defining symbol of 20th century technology, the computer has transformed businesses and lives around the world, increased our productivity and opened up access to vast amounts of knowledge.

9. Telephone—a cornerstone of modern life, the telephone brings the human family together and enables the communications that enhance our lives, industries and economies.

10. Air Conditioning and Refrigeration—air conditioning and refrigeration make it possible to transport and store fresh foods and adapt the environment to human needs. Once luxuries, they are now common necessities which greatly enhance our quality of life.

11. Highways—vast networks of roads, bridges and tunnels connect our communities, enable goods and services to reach remote areas and encourage economic growth.

12. Spacecraft—the human expansion into space is perhaps the most awe-inspiring engineering achievement of the 20th century. The development of spacecraft has thrilled the world, broadened our knowledge of the universe and led to thousands of new inventions and products.

13. Internet—in a few short years, the Internet has become a vital instrument of social change, transforming business practices, educational pursuits and personal communications. By providing global access to news, commerce and vast stores of information, the Internet brings people together and adds convenience and efficiency to our lives.

14. Imaging—From tiny atoms to distance galaxies, modern imaging technologies have expanded the reach of human vision with tools such as X-rays, telescopes, radar and ultrasound.

15. Household Appliances—household appliances have greatly reduced the labor involved in everyday tasks, giving us more free time and enabling more people to work outside the home.

16. Health Technologies—medical imaging devices, artificial organs, replacement joints and biomaterials are a few of the advanced health technologies that improve the quality of life for millions.

17. Petroleum and Petrochemical Technologies—petroleum and petrochemicals provide fuel for cars, homes and industries and the basic ingredients for products ranging from aspirin to zippers.

18. Laser and Fiber Optics—lasers are used in supermarket scanners, surgical devices, satellites and other products. Today fiber optics provide the infrastructure to carry huge amounts of information via laser light, spurring a growing revolution in telecommunications.

19. Nuclear Technologies—the harnessing of the atom led to dramatic changes in the 20th century by providing a new source of electrical power, improving techniques for medical diagnosis and treatment and transforming the nature of war forever.

20. High-performance materials—from the building blocks of iron and steel to the latest advances in polymers, ceramics and composites, engineering materials are used in thousands of important applications.

15.6 NEW UNDERSTANDINGS

In this discussion of the historical development of technology, we have been optimistic about the very positive effects of technology on society. In the past the public accepted, without serious questions, all forms of technology and the risks that were involved. However, the situation has changed to a remarkable extent in the past 40 years. Our leadership (engineers included) has, to a large degree, lost the public trust. An excellent example of this loss pertains to the technology employed in the generation of electricity using nuclear power. Since the Three Mile Island incident that occurred in Pennsylvania in 1979, the general public in the U.S. has opposed nuclear power[2]. There is a valid public perception that the technological risks associated with nuclear power were significantly understated. John O'Leary, a deputy director for licensing at the Atomic Energy Commission (AEC), stated that "the frequency of serious and potentially catastrophic nuclear incidents supports the conclusion that sooner or later a major disaster will occur at a major generating facility" [8]. This statement was made several years prior to the accident at Three Mile Island. In spite of this warning of the likelihood of a serious accident, federal, state or local governments have not planned for orderly emergency evacuation of nearby residents or for containment measures to limit the possible contamination of many of power reactors.

The Chernobyl accident, which occurred in the former Soviet Union in 1986, proved that O'Leary was correct. The No. 4 reactor at the Chernobyl facility exploded sending a radioactive plume in the air so high that an alert nuclear plant operator in Sweden detected it. Many died in this radioactive explosion and many more have died since then from radiation poisoning. High radiation levels affected countries as far away from Chernobyl as Italy.

The nuclear industry and the government agencies regulating it, world wide, have not been candid with the public. Nor has the public been very interested in an intelligent debate—resorting instead to large, ugly demonstrations to express their displeasure. The public needs an accurate and honest assessment of the risks of any nuclear power project by the industry, the regulators and experts. The risks must be weighed against the benefits to society. Sometime in the future fossil fuels will be come unavailable, scarce or costly. Indeed, we have recently witnessed a sharp increase in the price of crude oil following the hurricane Katrina which devastated the Gulf Coast and New Orleans. When a serious fuel crisis occurs, we may find that nuclear power is the only viable alternative for base load power generation to replace carbon burning plants. Regulations pertaining to carbon emissions may also limit burning of coal or oil to generate electrical power. Natural gas, favored by both industry and the public to generate energy while minimizing environmental impact, is in short supply and its cost has been increasing.

The purpose here is not here to argue the case for or against nuclear power. Nuclear energy is used to illustrate the fact that understating or poorly assessing risk brings a severe penalty—loss of public trust followed by a ban on technical development in the subject area.

As engineers, we must significantly broaden our perception regarding technological risk. Today, many engineers dismiss failure due to human, operator or pilot error because they are not machine errors. This attitude is **absolutely wrong**. The **human operator and the machine** constitute a system and affect

---

[2] In the past few years many environmentalist who opposed nuclear energy have recognized that generating base load electricity with nuclear energy is preferable to its generation by fossil fuels such as coal or natural gas. Nuclear power plants do not emit CO_2 that contributes to global warming.

public safety accordingly. The machine and its operator must be included in system design since they both constitute an interacting set of weaknesses and capabilities [9].

Another new understanding pertains to geography. Prior to the Second World War (WW–II), international trade was very limited. As a country we produced what we consumed, mined our own minerals and pumped our own oil. Geography was very important because it constrained trade by law and by distance. After the war, laws were changed and trade agreements (GATT, NAFTA, etc.) that lowered or eliminate tariffs were arranged. International trade was encouraged so much so that the U.S. now consistently imports more goods and materials than it exports. The annual deficit in the trade balance for the U. S. is expected to exceed 800 billion dollars in 2006. This deficit represents the equivalent to more than 20,000,000 high-income jobs paying $40,000 a year.

Previously, the cost of shipping limited trade among countries separated by large distances. However, following WW-II many technological improvements were implemented in the shipping business: the super tanker was developed—greatly improving the efficiency of shipping huge quantities of oil. Also very large containerized ocean vessels were developed reducing the cost and time of loading and unloading cargo. Large efficient diesel engines were installed on the ships permitting increased speed and significant reductions in shipping time and costs.

With the time and costs of shipping greatly reduced and the legal barriers to trade removed, the country's commerce is conducted in a global marketplace. Engineers compete worldwide to develop world-class products or services. Products designed in the U.S. may be produced anywhere in the world, or products designed in Europe or Asia may be produced in the U.S. Today, design and production activities are located to minimize costs and to maximize benefits to the customers and/or to the corporate entities.

The third new understanding is in the role of the governments (federal, state, county and city). Prior to WW-II, governments were small and taxes were relatively low. The primary role of the federal government was to insure national security. The state governments worried mostly about transportation infrastructure, and the local governments concerned themselves with education. Today, the situation is markedly different. Governmental agencies, at all levels and in response to their interpretation of the law (either old or new), issue regulations which pertain to topics of concern to society including the protection of the environment, safety, health and energy conservation.

You can debate the wisdom of many of the regulations, but that is not the point. The regulations currently exist and new ones will continue to be issued with an alarmingly frequency. As an engineer, you must be prepared to serve society-at-large as reflected by these governmental regulations. If you question the wisdom of some of the regulations, the best approach is to become involved in the political process so that you may have the opportunity to influence them. Engineers tend to resist constraints because they limit the freedom of their designs and add to the cost of the product. You may think of regulations as government imposed constraints, but they are really barriers imposed by society-at-large. Society controls the regulatory process because it has the voting power to change the decisions of the politicians drafting the laws and of the bureaucrats formulating the regulations.

15.7 BUSINESS, CONSUMERS AND SOCIETY

Engineers serve three constituencies—business, the customer and society-at-large. The business leaders (management) look to engineers to develop products, services and processes that meet global competitive challenges. Consumers seek more convenient, reliable, enhanced and value-laden products at reduced prices. Society-at-large, through elected politicians, trial lawyers and public interest groups demand action leading to solutions of problems involving safety, health, energy conservation and preserving or improving the environment. It is an overlapping set of constituencies, because society encompasses business, governments and consumers. To discuss the issues raised by serving three constituencies, consider the U.S. automotive industry as an example. Marina Whitman has described many aspects regarding the three-way demands. Because this discussion is based on her excellent contribution, you may want to study reference [10] in depth.

Business demands are simple although they are often difficult to meet. Any business must be profitable to continue to exist. Losses can occur over the short term, but if the red ink is prolonged the company is forced by its creditors into bankruptcy proceedings. Making a well-defined profit every quarter is the usual goal of management. For many years following WWII, it was relatively easy for U.S. businesses to make money. Bombing and shelling during the war had destroyed most of the manufacturing capability and a significant part of the infrastructure in every developed country in the world except in the U.S. It was easy to be a leader when the competition was without factories and adequate infrastructure. For the automobile industry, the golden years began to erode in the late 1960s, and survival became a life or death battle in the 1980s. Problems for the auto industry came from global competition (mostly Japan). The Japanese firms provided much more reliable automobiles at lower costs than those produced in the U.S.

To show the extent of the market penetration by imported automobiles, recall that the imports accounted for only 4% of the market in 1962, but they controlled 44% of the market by 2005[3]. This devastating loss of market share by US car makers was not limited to the auto industry; several other industries including optics, consumer electronics, ship building, steel, etc. suffered even more serious losses.

Engineers and management in the U.S. firms had lost the competitive edge after decades of modest competition from the Europeans. The Japanese automobile companies (Toyota, Honda, Nissan, etc.) provided more affordable, reliable, appealing, fuel efficient and value-laden cars than either the American or European companies. In recent years, the loss market share has been slowed by improved management and technical methods (concurrent design and systems engineering[4]), but regaining the market share of the 1960s will be a very long and difficult process.

The second constituency, the consumers, has been well recognized in the past decade. The importance of keeping the customer pleased is foremost in the minds of both management and engineering. Systematic methods for establishing the consumers' needs have been developed and placed into practice [11]. Many companies in the U. S. currently follow a product development process that begins with the customer interviews and proceeds through all of the phases of product development. You do not want to underestimate the importance of this second constituency; however, engineers have usually kept the customer in mind during the design of new products. The newer methods [11, 12] tend to insure that the customers' requirements are more accurately defined by the product specification before beginning a development.

The third constituency, society-at-large, generates the most challenging requirements. Society-at-large is represented by government agencies and a variety of public action groups. Officials of the government agencies (bureaucrats) produce regulations that define requirements for automobiles that may affect safety, fuel conservation or the environment. The public action groups create pressure and influence politicians in setting public policy that eventually generates new requirements.

Engineers often have some difficulties with regulations and/or proposals from the public action groups. Lawyers write the laws and public servants write the regulations. Because lawyers and public servants are often technically illiterate, they sometimes seek technical support prior to drafting binding regulations. However, government bureaucrats, prior to a thorough study by knowledgeable representatives of the public and engineering communities, sometimes place regulations into effect. The difficulty with this procedure is that the demands of the customer, including style, comfort, affordability, reliability, performance, convenience, etc., may be in serious conflict with societal imposed regulations on safety, fuel efficiency, emissions, etc.

A very recent example illustrates what many believe to be an ill-advised regulation [13]. The National Highway Transportation Safety Administration (NHTSA) mandated requirement for dual airbags in every vehicle licensed in the U.S. Most people consider air bags a valuable safety feature because tens of

---

[3] The market share of U S automakers is expected to drop to 50% by 2009.

[4] Systems engineering means many different things to different groups. It is defined here as a cost and time efficient method to specify, design, develop and integrate all of the individual subsystems so that the total system for the product meets the requirements of customers, business and society-at-large.

thousand lives have been saved in serious frontal accidents since they have been mandated. However, many children and small adults (mostly women) have been killed by rapidly inflating airbags. NHTSA has reported that from 1990 to 2002 280 people have died by rapidly deploying air bags in low severity crashes since the use of this safety feature was mandated [14]. More than 90 percent of the passenger airbag fatalities have been children and infants. They account for about 60 percent of all airbag related deaths. Small women are also at risk particularly if they cannot position themselves far enough away from the airbags. The statistics cited above are confirmed deaths due to rapidly deploying airbags. In addition, there were many unconfirmed air bag related fatalities where NHTSA did not send an investigative team to the accident scene to confirm the cause of death. The number of deaths reported by NHTSA did not include those that might have been killed by the air bag in high speed crashes.

Newer airbags trigger slightly less violently; nonetheless, passengers must remain at least 25 centimetres (10 in) from the bag to avoid injury from the bag in a crash. Injuries such as abrasion of the skin, hearing damage (from the sound during deployment), head injuries, and breaking the nose, fingers, hands or arms can occur as the airbag deploys. In 1990, the first automotive fatality attributed to an airbag was reported, with deaths peaking in 1997 at 53 in the United States. The first gas-inflated airbag was produced in 1994, with sensors and low-inflation-force bags becoming common soon afterwards. Dual-depth airbags appeared on passenger cars in 2005. By that time, deaths related to airbags had declined, with no adults deaths and 2 child deaths attributed to airbags that year. However, injuries remain fairly common in accidents with an airbag deployment.

In a recent ruling, that modifies the existing standard, NHTSA reduced the impact speed of the crash test from 30 to 25 MPH [14]. NHTSA will also require vehicles be tested using a family of dummies—an average size man, a small woman and 1, 3 and 6 year-old children. This new standard will be phased in over a three-year period beginning in 2004. Hopefully these changes will enable the protection of large adults while eliminating the dangers of air-bags to small women and children.

The killing of women and children by a safety device is a **very serious concern**. Moreover, the regulation prohibits anyone other than the car's owner to disable even the passenger side air bag[1]. If you are a concerned parent, you can disable it yourself, but few owners have the required skills. Also, if you ruin the mechanism and/or the bag, the cost of replacement may approach $1,000. Because of their high costs, air bags have replaced stereo systems as the theft of choice if your car is robbed.

The issue is not whether you should have air bags or not. They have proven to be a very effective safety device for most adults either driving or riding as passengers. The issue is with the rigidity of the federal mandates and the failure of the regulators to:

1. Update the regulations to accommodate different behavior of drivers.
2. Provide more design flexibility in the requirements.
3. Provide a means of permitting the driver to activate or deactivate the passenger side bag.

The air bag regulation was written in 1977 when only 14% of the population even bothered to use their seat belts. By 1994, when the seat belt mandate was fully effective, 49 of 50 states were enforcing seat belt laws and usage had risen to 68%. While seat belts do not eliminate the need for air bags, the belts increase the time allowed for the deployment of the air bag. The NHTSA did not change the air bag

---

[1] A recent ruling from the NHTSA allows auto dealers to install an on-off switch to control the passenger side air bag; however, NHTSA requires an extremely complex process for drivers to obtain a waiver, which is necessary before the dealer will perform the necessary modifications to the deployment system. First, you have to obtain a NHTSA request form. Then you must swear—under threat of criminal penalty—that you are too short to be safe with an air bag deployment, or cannot always put the children in the back seat, or have one of five NHTSA approved medical conditions. If you option for the medical conditions you must have a letter from a medical doctor confirming the condition. To date, relatively few people have fought the bureaucratic battle with NHTSA and obtained waivers. As of October 30, 1999, 57,183 requests have been authorized, but only 11,195 switches have been installed.

regulation to accommodate the increased usage of seat belts until early in 1997. Responding to the controversy about the risk due to air bags, NHTSA issued a new regulation permitting automakers to install airbags 25% to 30% less powerful than the current bags. The new regulation gives the automakers two years to implement the change. In the meantime on many cars manufactured before 1999, air bags continued to be deployed at a speed of 200 MPH to protect an unbelted adult male. This is great for the big guys; however, for kids or smaller women, a 200 MPH kick in the face is extremely dangerous. When seat belts are used, more time is available to deploy the air bag, and the velocity of deployment may be reduced from the dangerous speed of 200 MPH.

Industry representatives recognized these problems and warned the NHTSA of the dangers to children. Rather than change the regulation to provide more flexibility in the design of air bag systems, NHTSA drafted a warning label and industry added instructions in the owner's manuals explicitly cautioning against placing a front facing child safety seat in the car's front seat. The manuals indicate that it is safe to place a child in the passenger seat if the child is buckled-in with a seat belt, **but** the seat must be in its full rear position. The manual also states that children should never lean over with their faces near the air bag cover while the car is in motion. The warning labels and instructions in the owner's manual are a result of poor design that was forced by an inflexible regulation. Currently safety devices (air bags) are mandated for automobiles that are dangerous to small women and children. Warning labels and instructions in manuals reduce the consequences of litigation (there is another law indicating that the user must be warned of possible dangers), but they are totally inadequate to protect children who are often too young to read. The solution is simple—young children should not ride in the passenger seat.

It would be relatively easy to modify the design of the control system to include an on/off switch with a warning light on the control panel of the automobile. With this feature the driver could deactivate the passenger side air bag whenever he or she wished to do so. Another design approach entails a modification of the pneumatic system to reduce the explosive force with which air bags deploy. This modification would probably have taken place several years ago when seat belt usage finally became widespread. However, the federal regulation did not permit modification then, and only recently has it allowed modification after the death of 256 men, women and children[6].

Another problem with air bags that is not understood by the public is related to sodium azide that is often used as the propellant for inflating the bag. Sodium azide is very toxic. When the air bag is deployed, you may be contaminated; the usual practice following an air bag deployment is to take the person or persons involved to the hospital for decontamination.

The public is concerned, if not angry. The NHTSA has received many letters from owners requesting the waivers required to have trained mechanics at the dealers to disconnect the dangerous air bags. Many mechanics refuse to disconnect the airbags for fear of future litigation. A recent survey of 700 repair shops indicated that two-thirds refused to install the on-off switches. There appears to be a significant amount of confusion about the remedy. Recently the NHTSA has provided a listing of auto dealers willing and certified to install switches on their Website.

The current disarray is alarming. The interaction of the government agencies representing the public-at-large, business and the consumer is woefully inadequate. The public and business needs solutions often in a relatively short time. Yet, the regulatory system is too cumbersome to provide rapid solutions. The awkwardness comes from inflexibility on the part of the government and inadequate cooperation between the automakers. Competition on the part of business to increase market share is understandable, but safety features should be excluded from the normal competitive secrecy. The government agencies must become much more flexible and recognize that they cannot decree away all danger.

---

[6] Since 1990, 256 deaths have been caused by airbags inflating in low severity crashes, most of them in older model vehicles. These deaths include 86 drivers, 12 adult passengers, 135 children and 23 infants. Of the 86 drivers killed by airbags 64 were women and 22 were male. These statistics were reported by Insurance Institute for Highway Safety as of April 2005.

The difficulties between the government and business are not new. An interesting summary of the regulatory relationship between government and business was recently published by Jasanoff [15]. She is quoted below:

> Studies of public health, safety and environmental regulation published in the 1980s reveal striking differences between American and European practices for managing technological risks. These studies show that U.S. regulators on the whole were quicker to respond to new risks, more aggressive in pursuing old ones and more concerned with producing technical justifications for their actions than their European counterparts. Regulatory styles, too, diverged sharply somewhere over the Atlantic Ocean. The U.S. processes for making risk decisions impressed all observers as costly, confrontational, litigious, formal and unusually open to participation. European decision making, despite important differences within and among countries, seemed by comparison almost uniformly cooperative and consensual, informal, cost conscious and for the most part closed to the public.

The assessment by Sheila Jasanoff gives us clear direction. It is important to reduce the adversarial relationship between business and government. As engineers working for both government and business, we should be an important element in the future to produce a regulatory system with more flexibility, cooperation, cost consciousness and consensus.

15.8 CONCLUSIONS

The standard of living in the U.S. will be determined by the interplay of three powerful influences:

- New and rapid technological advances.
- Business (management) response to global consumer demands.
- Social demands as evidenced by legislative and regulatory requirements.

Engineers have always provided the leadership in technological advances, and they are expected to continue to provide business and the public-at-large with cutting-edge, world-class technology.

Business requires capital, technology and astute management to remain competitive. In addition to providing the technological base for a company, engineers frequently serve in management. If a management career path appeals to you, plan on extending your education in a business school. The combination of a Bachelor of Science degree in engineering and a Master of Science in business provides a very solid foundation for a career path in a technically oriented business.

In the past, engineers usually have not been actively engaged in societal issues. They lack patience in dealing with the public and become irritated by the public's lack of technical literacy. They fume at the inefficiencies of government agencies and their lack of flexibility. They become outraged at the arrogance of bureaucrats and at their lack of concern for time and costs. They withdraw from public forums and focus on new developments and products. In the meantime, the public-at-large grows wary of many important sociotechnical systems and the government resorts to litigious processes that impede real progress.

It is unfortunate that engineers have not been a significant force in dealing with societal issues. When society, business and the customer create conflicting demands, engineering expertise is essential in crafting well-balanced solutions. The conflicting demands provide new opportunities and complex challenges for engineers. To take advantage of these opportunities, the engineers of tomorrow will require enhanced communication skill, greater disciplinary flexibility, a better understanding of the mechanisms of regulatory agencies and a much broader perspective relative to societal demands.

REFERENCES

1. Friedman, T., <u>The World is Flat</u>, Farrar, Straus and Giroux, New York, NY, 2005.
2. Singer, C. E. J. Holmyard, A. Hall, and T. Williams, Eds. <u>A History of Technology</u>, Oxford University Press, New York, NY, five volumes, 1954-1958.
3. Hughes, T. P., "From Deterministic Dynamos to Seamless-Web Systems," <u>Engineering as a Social Enterprise</u>, ed. Sladovich, H. E., National Academy Press, Washington, D. C. 1991, pp. 7-25.
4. Sladovich, H. E., ed. <u>Engineering as a Social Enterprise</u>, National Academy Press, Washington, D. C. 1991, pp. 7-25.
5. White, L., Jr., <u>Medieval Technology and Social Change</u>, Oxford: Clarendon Press, NY, 1962
6. Kirby, R. S., S. Withington, A. B. Darling, and F. G. Kilgour, <u>Engineering in History</u>, Dover Publications, New York, NY, 1990.
7. Constable, G. and B. Somerville, <u>A Century of Innovation: Twenty Engineering Achievements that Transformed Our Lives</u>, Joseph Henry Press, Washington, D. C., 2003.
8. Ford, D. F., <u>Three Mile Island: Thirty Minutes to Meltdown</u>, Viking, New York, NY, 1982
9. Adams, R. McC. "Cultural and Sociotechnical Values," <u>Engineering as a Social Enterprise</u>, ed. Sladovich, H. E., National Academy Press, Washington, D. C. 1991, pp. 26-38.
10. Whitman, M. v. N., "Business, Consumers, and Society-at-Large: New Demands and Expectations," <u>Engineering as a Social Enterprise</u> ed. Sladovich, H. E., National Academy Press, Washington, D. C. 1991, pp. 41-57.
11. Clausing, D. <u>Total Quality Development: World-Class Concurrent Engineering</u>, ASME Press, New York, NY, 1994.
12. Schmidt, L., et al, <u>Product Engineering and Manufacturing, 2nd Edition</u>, College House Enterprises, Knoxville, TN, 2002
13. Payne, H. "Misguided Mandate," Scripps Howard News Service, Knoxville News-Sentinel, January 12, 1997, p. F-1.
14. Anon, Q & A: Airbags, Insurance Institute for Highway Safety, http://www.hwysafety.org/safety_facts.qanda/airbags, 2004.
15. Jasanoff, S. "American Exceptionalism and the Political Acknowledgment of Risk," Daedalus, Vol. 119, No. 4, 1990, pp. 61-81.

EXERCISES

15.1 Consider a product that you or your family has purchased in the past month or so. Write a brief paper describing both the positive and negative impacts of that product on society.

15.2 The Federal Aviation Administration enacted a regulation affecting the mailing of packages weighing more than 16 ounces. The regulation requires you to present the package to a postal clerk for mailing and prohibits the mailing of the package from a postbox. Write a paper covering the following issues:

- Describe the regulation in more detail.
- Why is the FAA writing a regulation affecting the U.S. Post Office?
- Do you believe that the regulation will be effective for its intended purpose? Please give arguments supporting your viewpoint.
- What actions will the post offices (40,000 of them) have to take to make the regulation effective? What actions do the post offices actually take with regard to the regulation?
- Will these actions be costly? Estimate the costs to both the public and the individual. Assume that a postal clerk is paid $14/h and that an individual considers his or her free time worth $14/h.
- Give your assessment of the cost to benefit ratio for this regulation?

15.3   Suppose that you were a slave in the time of Rameses II and were assigned to the task of constructing his statue. For those not up to date on Egyptian statues, this one weighs 1,000 tons and is 56 feet tall. Write a paper describing your daily tasks.

15.4   Horses were not used extensively to relieve man from brutal work or to provide a significant advantage in military actions until nearly 1,000 AD. Write a brief paper describing both the social and technical reasons for the very long time required to effectively employ horses in either military or commercial enterprise.

15.5   Suppose that you were living on a manor in England in about 1,300 AD. Describe your lifestyle and indicate how technology affects your work and play. Select in which of the two classes of society that you existed.

15.6   Why was California subjected to rolling blackouts in the summer of 2002?

15.7   Write a paper comparing the lifestyles, as you imagine them, for a man or woman living today and living in the year 1,000. Did technology make a difference in the quality of life?

15.8   Find the Website for the National Academy of Engineering and click on Greatest Achievements for the 20th Century. Select one of the 20 achievements and prepare a paper describing the history of that particular achievement.

15.9   Find the Website for the National Academy of Engineering and click on Greatest Achievements for the 20th Century. Select one of the 20 achievements and prepare a paper describing the time line for that particular achievement.

15.10 Why is the public-at-large opposed to power generation with nuclear energy? Is the public-at-large correct in their collective assessment? Explain why you formed this opinion.

15.11 In the past few years the public's perception of nuclear power is slowly changing because of the publicity associated with global warming. Why is nuclear power slowing gaining acceptance because of the fear of global warming?

15.12 Why does a global marketplace exist today? Does global trade improve our standard of living? What does our current trade deficit have to do with wages for factory workers? Does technology help or hinder the balance of trade deficit?

15.13 If you were the director of the National Highway Transportation Safety Administration (NHTSA), what action would you take regarding the public concern about air bags?
   • Explain the reasons for your actions.
   • Explain how you would convince the major automakers to agree with you.
   • Prepare an outline you would follow in the press conference announcing this action.
   • Describe how you would handle a public interest group disagreeing with your ideas.
   • Describe how you would handle the public interest groups that agree with you.

15.14 Prepare a paper describing the information provided by NHTSA on air bags. A significant amount of information is given on its Website at http://www.nhtsa.dot.gov/airbags.

15.15 From the NHTSA Website, find the Website for the Insurance Institute for Highway Safety and examine the page on Airbag Statistics. Determine the ratio of deaths caused by air bags to the estimated number of lives saved by air bags. Write a position paper about actions that should be taken in the future regarding air bags to insure public safety.

15.16 Recently side airbags have been installed in some automobiles. Are these side air bags a threat to small children seated near a door? What has the NHTSA done to mitigate this threat?

15.17 What social science courses should you take during your undergraduate program to broaden your prospective and aid you in dealing with sociotechnical issues?

15.18 Is there a technical elective in your program of study that deals with sociotechnical issues?

15.19 Prepare a paper discussing the issue of air quality regulations and the design of automobiles.

CHAPTER 16

SAFETY, RISK AND PERFORMANCE

16.1 LEVELS OF RISK

When engineers design a new product or modify an existing one, an element of risk is often introduced in society-at-large. Sometimes the risk is minimal, and the resulting damage from a malfunction or failure is small. However, in some instances the risk may be large and the damage may be catastrophic.

Two simple examples are in order to illustrate the concept of risk with attendant benefits to society. Suppose you design a new model of an electric, pistol-grip drill that is powered from the standard 120 volt, 60 cycle single-phase power supply from a local utility company. The drill operator is at risk from an electric shock, although the probability for injury is very small. It is your responsibility as an engineer to minimize this risk while maintaining the advantage in performance of a new improved model of an electric-powered drill.

In the late 1950s a new aircraft—the Lockheed L-188 Electra powered by turbine driven propellers—was introduced and placed in service by several airlines. Three of these aircraft crashed, due to structural failure in flight, before the Federal Aviation Administration (FAA) grounded the entire fleet of Electras. An investigation revealed that a torsional resonance occurred under certain operating conditions causing the engines to vibrate violently and break away from the wing. This was a very serious engineering error that caused the death of over a hundred innocent passengers and significant losses of property. Clearly, this example illustrates a case where the risk far outweighs the benefits of performance of a new aircraft design. It also illustrates the seriousness of design errors and inadequate testing of prototypes of products that pose danger to the public.

Acceptable Risk

How do you protect an operator from electric shock? The operator is holding an electric motor with 120 volts across the armature coils and the field coils in his or her hands. The answer in this instance is to build the case for the drill from a tough, durable plastic that is structurally strong and an excellent electrical insulator. The case keeps the operator from touching any part of the electrical circuit powering the motor. In fact, Black & Decker® a major manufacturer of small power tools, uses double insulation to provide two independent insulation barriers to prevent the operator from making contact with any part of the electrical circuit. Operators may still manage to receive an occasional shock, but it will not be easy. Accidental contact of the operator's hands with a live electrical circuit is a rare event. In fact, the risk of electrical shock is so small that a worker routinely picks up a power tool and employs it to perform some task without even a passing thought about the possibility of receiving an electrical shock. Operating a power tool does involve risk, but it is acceptable because the operator is not consciously concerned about his or her safety while using the tool.

Voluntary Risk

Let's move up the risk ladder and consider flying in a commercial airliner from point A to point B, some 1,200 miles distant. Is it safe to make this trip? Most people appreciate that it is reasonably safe to fly commercial airliners, but they realize that there is a small probability of a crash. The statistics show the probability of a fatality p_k in an airline accident is about one in a billion passenger miles flown ($p_k = 1/10^9$). Hence, if you make the round trip from point A to B, your probability P_k of being killed due to an airline accident is:

$$P_k = s\, p_k = (2)(1200)/10^9 = 2.4 \times 10^{-6} = 0.00024\%$$

where s is the number of miles in a round trip.

This (2.4 chances in a million) is a very low probability for a fatal accident; consequently, most people do not worry much about the possibility of dying in a crash when they board an airliner. However, if you are a sales engineer flying 100,000 miles a year, every year for 20 years, your total mileage accumulates to 2×10^6 miles. Your probability of being killed in a crash increases to:

$$P_k = 2 \times 10^6/10^9 = 2 \times 10^{-3} = 0.2\%$$

The probability has increased significantly if you accumulate the miles flown by a traveling professional over a 20-year period. Knowing there is one chance in 500 that you will die in an airline crash, would you still want to be a sales engineer flying weekly to meet with customers? Could you tolerate this level of risk? Would you be apprehensive?

Everyone has some tolerance for risk. People weigh the speed, convenience, cost and risk of flying against that of traveling by train or by car. In almost all instances, travelers select the plane for long distances and the car for short distances. Travelers select the train only in those rare instances when it combines relatively good service (high speed with frequent trains) to convenient (downtown) locations. Most people do not contemplate the risks involved in travel because, except for driving a car, fatal accidents are rare events.

Unfortunately, fatal accidents in automobiles are not rare events. Preliminary estimates by the National Highway Traffic Safety Administration (NHTSA) showed that 43,443 occur on U.S. highways in 2005. In addition about 2,669,000 serious injuries occur in these accidents. Death on the highway is the leading killer of Americans 1 to 35 years old. It is clearly more dangerous to drive than to fly when you compare the probability of fatalities and/or injuries on a per mile basis. How do you handle the higher risk associated with driving? Rationalization—I am an excellent driver, and it will not happen to me. Not necessarily a true statement, but the rationalization of one's superior driving skills alleviates the worry and concern about a fatality or injury producing accident. The real risk remains.

When driving, you are an active factor in determining the risk. Driving habits (high-speed, reckless steering, tailgating, etc.) and driving skills affect, to a large degree, the level of risk involved. When flying, you are a passive participant. You have voluntarily decided to fly and sometimes even have a choice of airlines. You may pick the airline with the best safety record. (Southwest Airlines is the best with only one accident in their entire 30 plus year history). However, even knowing the facts, you are placed at risk and have little choice in the matter.

Involuntary Risk

Suppose you live 40 miles northeast of a nuclear power plant. Is 40 miles sufficiently far away to avoid serious fallout from an explosion involving the nuclear reactor at the power plant? Remember the winds are usually from the southwest, so that the fall-out plume will be pointed in your direction in the event of an accidental release of gasses containing radioactive particles.

The risk associated with radioactive fall-out, air pollution, toxic chemicals, polluted ground water, etc., is a serious concern to almost everyone; people dislike being exposed involuntarily. They expect governmental regulators to control the environment and to reduce the level of the risks of accidental exposure. Unfortunately, governmental agencies are not always effective in protecting the population from these exposures. Accidents have occurred in the control of reactors, and citations for safety violations by inspectors of the Nuclear Regulatory Commission are commonplace. From a public perception, the risk of a serious malfunction of a nuclear reactor became unacceptably high in the U.S. and many other European countries. After the accident at the Three Mile Island facility in Pennsylvania in 1979, the commercial nuclear industry died in this country. Some plants that were under construction were completed and others were converted to fossil fuels, but no new nuclear plants have been started since that time. With the more recent and much more serious accident at Chernobyl in 1986, new plans for power generation using nuclear energy in the U.S. were placed on hold until very recently. Two years ago TVA began to work on a nuclear power plant that was nearly complete prior to a stop work order issued after the accident at the Three Mile Island plant. The new nuclear plant is scheduled to go on line in 2007. In addition about a dozen utilities have indicated that they are currently planning to submit applications for construction and operating licenses for new nuclear power plants in the next two years.

There is a significant difference in the acceptable level of risk depending on whether the risk-producing activity is voluntary or involuntary. The degree to which one personally controls the risk is also extremely important. Activities like skiing, scuba diving, horseback riding, hang-gliding, mountain climbing, dirt biking, etc. carry extremely high levels of risk. Yet, intelligent people swarm to the ski resorts and pay large amounts of money to potentially break their bones. Why? They have voluntarily decided that the thrill or other pleasures derived from the activity is worth the risk. Also, they control the level of risk. They can choose the "bunny" slope and minimize the risk or the "black diamond" slope to maximize the thrill.

Engineers must recognize that some level of risk is involved in almost all the products produced. It is imperative that this risk be minimized while maintaining an acceptable level of performance. Also, it is essential that this risk be acceptable to customers and to society-at-large. An appropriate balance between risk and performance must be achieved[1]. It is important that engineers, business and governmental regulatory agencies cooperate to provide a realistic assessment of the risk level. Finally, the public should be made aware of the risks involved and should not be astonished by news releases describing the gruesome details of victims involved in these relatively rare accidents.

16.2 MINIMIZING THE RISK

An engineer minimizes risk by preventing failure of each and every component in the system. This task is not easy as there are several different ways in which components can fail. Parts fail by breaking and aging, by corrosion or fatigue, by overload and burning, etc. In some instances, you may anticipate these failures and replace the parts before they malfunction. This controlled replacement of finite life parts is known as preventative maintenance. In most cases, engineers try to design each component so that failure will not occur during the anticipated life of the product.

Safety in Design

A complete treatment of methods to avoid failure would require more coverage than a single chapter or even an entire textbook. However, an analytical procedure for designing tension members will be introduced to provide an illustration of a conservative design that insures safety. These tension members will be sized

---

[1] Cost is an issue when balancing safety, risk and performance. As a guide to cost, the Environmental Protection Agency (EPA) uses a worth of $6.1 million for what economist call a "value of statistical life" in their cost benefit analyses for new regulations.

(made sufficiently large) so that they exhibit an acceptable safety factor. In other words, they will accommodate a service load higher than anticipated in service over the product's life cycle.

Let's begin by introducing a tension member as shown in Fig. 16.1.

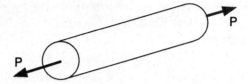

Fig. 16.1 A tension member is a long thin rod subjected to axial load P.

When the tension rod is subjected to an axial load P, an axial stress σ develops that is uniformly distributed over the cross section of the rod. This uniformly distributed stress acting on the rod is illustrated in Fig. 16.2.

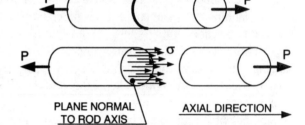

Fig. 16.2 The stress σ is distributed uniformly over the cross-sectional area of the tension rod.

The magnitude of this stress is determined from the formula:

$$\sigma = P/A \qquad (16.1)$$

where σ is the stress given in units of pound/square inch (psi) or newton/square meter (Pa).
P is the load expressed in units of pound (lb) or newton (N).
A is the cross sectional area of the rod specified in square inch (in^2) or square meter (m^2).

Two examples demonstrating the technique for computing stresses in axially loaded tension members are presented below:

EXAMPLE 16.1

If the axial load P applied to a circular tension rod is 10,000 lb, determine the axial stress σ that is developed if the rod is ½ in. in diameter.

Solution: The cross sectional area is given by:

$$A = \pi r^2 = \pi(0.25)^2 = 0.1963 \text{ in}^2$$

From Eq. (16.1), the axial stress σ is:

$$\sigma = P/A = 10,000/0.1963 = 50,930 \text{ psi}$$

where psi is the abbreviation for lb/in^2.

At this stage of the analysis, it is not possible to interpret the significance of the result for the stress σ = 50,930 psi. More information is needed pertaining to the strength of the material from which the tension member is fabricated. When the strength of the rod's material is known, it is possible to make a comparison of the magnitude of the stress relative to this strength and remark on its significance.

In this first example, U.S. Customary Units were used to express the axial load in pounds, the cross sectional area in square inches and the stress in psi or pounds per square inch. In the next example, the International System of Units (SI) will be employed where the load is expressed in newton, the cross sectional area in meters, and the stress in pascal.

EXAMPLE 16.2

If the axial load P imposed on a circular rod is 50,000 newton (N), and the radius of this rod is 5 mm, determine the axial stress σ.

Solution: First, determine the cross sectional area A as:

$$A = \pi r^2 = \pi(5 \times 10^{-3})^2 = 78.54 \times 10^{-6} \, m^2$$

Then, from Eq. (16.1), the axial stress is given by:

$$\sigma = P/A = (50 \times 10^3 \, N)/(78.54 \times 10^{-6} \, m^2) = 636.6 \times 10^6 \, Pa = 636.6 \, MPa$$

where the abbreviation Pa stands for pascal—the unit used for stress in the SI system.

A pascal is equal to a newton per square meter (N/m^2). When calculating stresses in the SI system, it is common to obtain very large numbers for the stress when σ is expressed in pascal. For this reason, MPa (mega pascal) is often employed as the unit for σ where $MPa = 10^6 \, Pa = N/(mm^2)$.

The stress on a tension member can be accurately determined if you can ascertain its size and the load that it must support. The calculated stress is a number with associated units—either psi or MPa. As a number, it is not very informative until the stress is compared to the strength of the material from which the tension member is fabricated.

Let's suppose that the tension rod used in Example 16.1 is machined from very strong alloy steel exhibiting a yield strength of 100,000 psi. Do you believe the tension rod is safe? Will it retain its shape under load? Will it fail by yielding (stretching under load and not recovering completely when the load is removed)? These questions may be answered by making a simple comparison of the applied stress σ to the strength S of the material. In Example 16.1, the axial tensile stress $\sigma = 50,930$ psi and the strength of the material from which the rod was machined is S = 100,000 psi. The comparison shows that the yield **strength** of the rod is **greater** than the applied **stress**; therefore, the rod **will not fail** by yielding or breaking. But how safe is the rod? You know it will not fail; however, some measure of the safety of the rod should be established. Is the safety of the rod marginal, or is the rod too safe? Why would any structural element be too safe?

Safety Factor

To respond to the safety issue for the rod, determine the ratio of the strength to the stress and define this ratio as the safety factor SF:

$$SF = S/\sigma \tag{16.2}$$

For the stress imposed on the rod described in Example 16.1, the safety factor determined from Eq. (16.2) is:

$$SF = 100{,}000/50{,}930 = 1.96$$

The safety factor is unitless. The value of $SF = 1.96$ indicates that the tension rod is almost twice as strong as it must be to resist yielding under the applied load. You have a reasonable safety factor and can be confident that the rod will perform safely in service. Your only concerns are the accuracy with which the load has been predicted and the quality control for the material used in manufacturing the tension rods. If the factors listed below are satisfied:

* You have extensive experience with the application.
* You are certain of the magnitude of the load.
* You have confidence in the manufacturing division in tracking their materials.
* You have verified the strength of the materials received from suppliers.

Then you usually can be satisfied that a safety factor of about two is adequate. However, if you are not certain about the load and/or the materials, a safety factor of two is not sufficient. You should increase the safety factor to accommodate your ignorance of either the applied load or your lack of control over the material employed. Safety factors ranging from two to four are commonly employed in design. Safety factors of less than two require considerable care, expense and expertise. The use of relatively low safety factors is justified only for very high-performance applications or for components that are produced in very high volume. In these special situations, the engineering analysis, prototype testing, quality control inspections, maintenance inspections and documentation must be extensive.

Margin of Safety

The margin of safety MS is sometimes used to describe the degree of safety incorporated into the design of a structural component. The margin of safety should not be confused with the safety factor as they are different quantities. The margin of safety is defined as:

$$MS = (S - \sigma)/\sigma = SF - 1 \qquad\qquad (16.3)$$

Using the results from Example 16.1 and Eq. (16.3), the margin of safety is given by:

$$MS = 1.96 - 1 = 0.96 \ \text{ or } \ 96\%$$

The tension rod has a margin of safety of 96%, which implies that the strength of the material exceeds the applied tensile stress by 96%. If the load on a tension member can be determined, it is easy to size (adjust the cross sectional area) of the rod to provide any margin of safety that is deemed necessary to satisfy management's concerns and to meet professional obligations to society-at-large.

16.3 FAILURE RATE

Sometimes components fail by burning out, aging or wearing out—not necessarily by breaking or yielding (excessive permanent deformation). When conducting a failure analysis for wear or aging, the safety factor is not a relevant parameter. You must cope with wear, aging or burn out by using other methods. To begin, let's recognize that all components do not wear out at the same time. Some people drive their car 25,000 miles before the brakes wear out and others may drive 60,000 miles or more. Some people can use their personal computers for several years before a component fails; yet others have problems after only a few months in service. To analyze these differences in service before failure, a mortality curve for both mechanical and electronic components is introduced in Fig. 16.3a and Fig. 16.3b, respectively.

To determine the failure rate FR, records of the performance with time-in-service of a large number of components are maintained. You begin your record of performance with N_0 components that are placed in service at time t = 0. After some arbitrary time t, some number N_f will have failed and the remainder N_s will have survived. At any arbitrary time, you may write:

$$N_f + N_s = N_0 \qquad\qquad (16.4)$$

With time, N_f increases and N_s decreases, but the sum remains constant and equal to N_0. The failure rate FR is defined as:

$$FR = N_f / (N_0 \times t) \qquad \text{when } t > 0 \qquad\qquad (16.5)$$

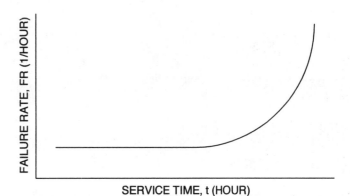

Fig. 16.3a Mortality curve for mechanical components showing failure rates with time in service.

When the results of Eq. (16.5) are plotted with respect to time, the component mortality curve presented in Fig. 16.3a is obtained. For mechanical components, where testing is performed on the assembly line, the failure rate is small when the components are relatively new. However, the failure rate increases non-linearly after some time in service because parts begin to wear, corrode or they are abused.

The failure rate for electronic components is usually characterized by the well-known bathtub mortality curve presented in Fig. 16.3b. This curve exhibits three different regions of interest—each due to a different cause. The high failure rates in the first region, $t < t_1$, are due to components manufactured with minute flaws that were not discovered during inspection and testing at the factory. In the center region of the curve, for $t_1 < t < t_2$, the failure rate FR_0 is small and nearly constant with time. Clearly, this is the best region to operate if you are to maximize the reliability of the product. Near the end of the service life, when $t > t_2$, the failure rate increases sharply with time as the components begin to fail due to the effects of aging.

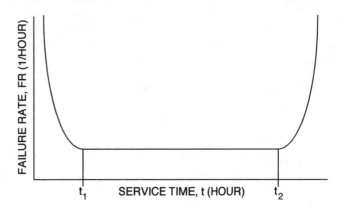

Fig. 16.3b Mortality curve showing the failure rate for electronic components with time in service.

For high-reliability systems, it is important to eliminate the inherently defective components associated with early failures. For mechanical systems, inspecting and testing the system's performance to eliminate flawed components produced in manufacturing accomplishes this objective. This task is not easy for the electronic component manufacturers because the feature sizes on the components are so small that inspection requires very high magnification and detailed examination of relatively large areas. Even with elaborate inspection schemes, not all of the flaws can be detected. A better procedure to eliminate the few inherently flawed components is to conduct what is known as burn-in testing prior to incorporating the components into an electronic product. With burn-in testing, all of the components that are produced are operated for a time t_1 (as defined in Fig. 16.3b). The flawed components burn out (fail) and are eliminated. The surviving components that will function with a much lower failure rate FR_0 are used in fabricating high reliability products.

Clearly, it is important to design components with low failure rates. If the failure rates are low, then the time between failures will be high. In fact, manufacturers sometimes cite what is known as the mean time between failures (MTBF) in their product specifications. The MTBF and the failure rate are related by:

$$MTBF = 1/FR \tag{16.6}$$

You should understand the concept of component failure rates and how they vary over the life of the component. How is this information about the failure rate used to establish the reliability of a specific component to perform without failing over the anticipated life of a product? This topic is discussed in the next section.

16.4 RELIABILITY

Component Reliability

To answer the question of component and system reliability, you must introduce the concept of probability or chance of occurrence. In considering probability, let's return to the test conducted to measure the failure rate and use the data collected for N_f, N_s, and N_0. The probability of survival P_s, which varies with time t, may be determined from:

$$P_s(t) = N_s(t)/N_0 \tag{16.7}$$

And the probability of failure P_f is given by:

$$P_f(t) = N_f(t)/N_0 \tag{16.8}$$

Both the probability of survival and the probability of failure are functions of time. With increased time in service, the probability of survival decreases and the probability of failure increases.

From Eqs. (16.7) and (16.8), it is evident that:

$$P_s(t) + P_f(t) = 1 \tag{16.9}$$

Substituting Eq. (16.8) into Eq. (16.9) yields:

$$P_s(t) = 1 - \frac{N_f(t)}{N_0} \tag{16.10}$$

If Eq. (16.10) is differentiated with respect to time t and the result is rearranged, it is possible to write:

$$\left[\frac{N_0}{N_s(t)} \right]\left[\frac{dP_s(t)}{dt} \right] = \left[\frac{-1}{N_s(t)} \right]\left[\frac{dN_f(t)}{dt} \right] = -FR(t)dt \tag{16.11}$$

Consider the term $[1/N_s(t)][dN_f(t)/dt]$ and note it is the instantaneous failure rate FR(t) associated with a sample size N_s at time t. Next, integrate Eq. (16.11) and assume that FR(t) = FR_0, is a constant. The relation obtained gives the probability of survival as a function of the failure rate.

$$P_s(t) = e^{-(FR_0 t)} \qquad (16.12)$$

where e = 2.71828 is the exponential number. Let's consider an example to demonstrate the method for determining the reliability of a mechanical component.

EXAMPLE 16.3

Suppose a company maintenance records show the number of failures of a certain component over an extended period of time. Analysis of this data indicates that the failure rate is essentially constant with 1 failure per 10,000 hours of service. Does that sound good to you? Will the reliability be adequate? Let's determine the probability of survival (the reliability) of this mechanical component with time.

> **Solution:** Substituting the failure rate FR_0 = 1/10,000 h into Eq. (16.12), yields the results for $P_s(t)$ shown in Table 16.1.

Table 16.1
Reliability $P_s(t)$ of a mechanical component
with increasing service life (FR$_0$ =1/10,000 h)

Time (1000 h)	Time* (years)	$FR_0 \times t$ (unitless)	Reliability $P_s(t)$
1	0.5	0.1	0.905
2	1	0.2	0.819
5	2.5	0.5	0.606
10	5	1.0	0.367
20	10	2.0	0.135
50	25	5.0	6.738×10^{-3}
100	50	10.0	4.540×10^{-5}
200	100	20.0	2.061×10^{-9}

*The conversion from hours to years of service life is based on 2,000 hours/year.

Reliability varies markedly with service life. For a short service life, say a year, the reliability is reasonable with 81.9% chance of surviving. However, for long life, say 10 years, the reliability drops to only 13.5%. For very long life 50 to 100 years, failure is almost certain. While the failure rate of 1 in 10,000 hours appears satisfactory in an initial assessment, the reliability resulting from this failure rate is disappointing. You cannot be certain of a service life of 10,000 hours prior to failure. In fact, there is only a probability of 36.7% to survive for 10,000 hours. If a reliability of 90% is required for a service life of 10,000 hours, the failure rate FR_0 must decrease to about 1 failure in 100,000 hours.

System Reliability

The failure of a component may or may not cause failure of a system. When a system is comprised of several components, its reliability will depend on the probability of failure of the individual components and their arrangement. There are two possible arrangements—series or parallel. Consider your automobile to illustrate both the series and parallel arrangement of components. For lighting the highway during the night, an auto is equipped with two headlights. This is a parallel arrangement because the design incorporates two identical components (the headlights) that perform nearly the same function. If one headlight burns-out, you can still drive. Your visibility is impaired to some degree and the system has been compromised; however, it has not failed.

On the other hand, suppose you intend to start your engine. Consider the components involved in this action:

* Ignition switch
* Battery
* Wiring
* Solenoid relay
* Starter motor
* Starter motor clutch and gear
* Engine ignition components (spark plugs and points)
* Fuel availability
* Fuel pump
* Fuel lines
* Fuel injectors

If any of these components should fail, the engine will not start when you turn the key. A complete series of successful components is required for the system to function. To start your engine, every component must properly function because the failure of a single component will result in a failure of the entire system.

Reliability of Series Connected Systems

Let's consider the reliability of a system involving three components that are in a series arrangement, as shown in Fig. 16.4.

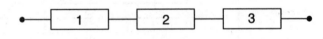

Fig. 16.4 A system comprised of a series arrangement of three components.

Because all three of the components in this series arrangement must operate successfully for the system to perform satisfactorily, the probability of survival of the system (P_s^s) is a product function given by:

$$P_s^s = P_{s1}\,P_{s2}\,P_{s3} \qquad (16.13)$$

where the superscript s refers to the entire system of components.

If you combine Eqs. (16.12) and (16.13), it is possible to express the system reliability in terms of the failure rate of the individual components as:

$$P_s^s = e^{-(FR_1 + FR_2 + FR_3)t} \qquad (16.14)$$

When the number of components in a system is increased with a series arrangement, the system reliability is decreased markedly. Also, the system reliability continues to decrease with time in service.

EXAMPLE 16.4

Consider the case where n components are arranged in a series where n is a variable that increases from 1 to 1000. Let's also assume that the reliability of each of the n components is the same ($P_{s1} = P_{s2} = \ldots\ldots = P_{sn}$) at some time during the operating life of the system. Evaluate the system reliability for three different values of the component reliability—$P_s = 0.999, 0.99$, and 0.90.

Solution: The results obtained from Eq. (16.13) are shown in Table 16.2.

Table 16.2
Reliability of a system with n series arranged components
as a function of component reliability P_s

Number of Components, n	$P_s = 0.999$	$P_s = 0.990$	$P_s = 0.900$
1	0.999	0.990	0.900
2	0.998	0.980	0.810
5	0.995	0.950	0.590
10	0.990	0.904	0.349
20	0.980	0.818	0.122
50	0.950	0.605	*
100	0.905	0.366	*
200	0.819	0.134	*
500	0.606	*	*
1000	0.368	*	*

* System reliability of less than 1%.

Examination of the results presented in Table 16.2 clearly shows the detrimental effect of placing many components in a series arrangement. For components with a high reliability ($P_s = 99.9\%$), the system reliability P_s^s drops to about 90% with 100 series arranged components. However, with components with a lower reliability ($P_s = 90\%$), the system degrades to a reliability of less than 1% when the number of components approaches 50. The lesson here is clear—if you connect a large number of components together in a series arrangement to develop a system, then the individual component reliability must remain extremely high over the entire service life of the system.

Reliability of Systems with Parallel Connected Components

Recall the description of a highway lighting system on an automobile with a pair of headlights. This system is an example of a parallel arrangement because both headlights must fail before the system fails. It is possible to drive with one light, although the State Troopers might warn the driver of the dangers of doing so. The human body has several parallel systems; you have two hands, two eyes, two legs, two feet and two ears, etc. This redundancy increases a person's ability to function in the event of a mishap.

A system with a parallel arrangement of three identical components is represented with the model shown in Fig. 16.5.

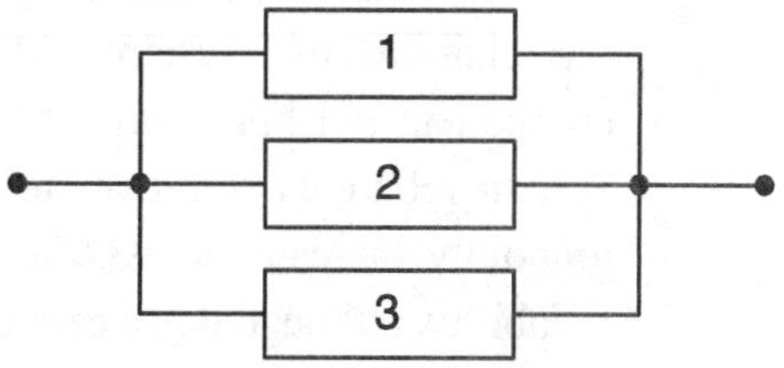

Fig. 16.5 System with a parallel arrangement of three identical components.

These parallel systems are considered to be redundant because all of the components must fail before the system fails. Accordingly, the relation for the probability of system failure is written as:

$$P_f^s = P_{f1} \, P_{f2} \, P_{f3}$$

(16.15)

Substituting Eq. (16.9) into Eq. (16.15) gives:

$$1 - P_s^s = (1 - P_{s1})(1 - P_{s2})(1 - P_{s3}) \qquad (16.16)$$

Expanding Eq. (16.16) and reducing the resulting expression gives:

$$P_s^s = P_{s1} + P_{s2} + P_{s3} - P_{s1} P_{s2} - P_{s2} P_{s3} - P_{s1} P_{s3} + P_{s1} P_{s2} P_{s3} \qquad (16.17)$$

Let's examine Eq. (16.17) to determine if redundancy (the number of parallel components) improves reliability.

EXAMPLE 16.5

Suppose the probability of success P_s for all the components in a parallel system is the same. However, P_s is considered as a variable increasing from 0.1 to 1.0. Show the effect of component redundancy by considering a system comprised of one, two and three components in parallel.

Solution: The results obtained from Eq. (16.17) are shown in Table 16.3.

Table 16.3
Effect of the degree of redundancy on reliability
$$P_{s1} = P_{s2} = P_{s3}$$

Component Reliability	Single Component	Two Components	Three Components
0.10	0.10	0.19	0.271
0.20	0.20	0.36	0.488
0.30	0.30	0.51	0.657
0.40	0.40	0.64	0.784
0.50	0.50	0.75	0.875
0.60	0.60	0.84	0.936
0.70	0.70	0.90	0.973
0.80	0.80	0.96	0.992
0.90	0.90	0.99	0.999
1.00	1.00	1.00	1.000

An examination of the results presented in Table 16.3 shows that the degree of redundancy improves the reliability of a system with a parallel arrangement of components. If you have relatively poor component reliability, say 60%, and employ two of these components in a parallel arrangement, the system reliability improves to 84%. Add a third component to the parallel arrangement and the reliability increases to 93.6%. Adding more components in parallel will continue to improve the reliability, although it is a case of diminishing returns.

The improvement in reliability by employing parallel (redundant) components in designing a system is a distinct advantage. However, one rarely if ever, enjoys a free lunch. The corresponding disadvantages are the increased costs and the added power, weight and size of the system. These are significant disadvantages, and redundant design is used only when component reliability is too low for satisfactory system performance or when a failure produces very serious consequences. For a more complete discussion of component and system reliability, see reference [1].

16.5 EVALUATING THE RISK

When the space shuttle Challenger exploded during launch on January 28, 1986, a Presidential Commission [2] was established to:

1. Review the circumstances surrounding the accident and determine the probable cause or causes for the explosion.
2. Develop recommendations for corrective or other actions, based on the Commission's findings and determinations.

Richard Feynman, a Nobel Prize winning physicist from the California Institute of Technology, was appointed to the Commission. Dr. Feynman, near the end of his career and his life, took the appointment very seriously and devoted his entire time and energies to the investigation. One of his many contributions during the review was to determine the probability of failure of the space shuttle system.

During the investigation, Dr. Feynman asked three NASA engineers and their manager to assess the probability of failure P_f of a mission due to the failure of one component—the shuttle's main rocket engine. Secret ballots by the engineers indicated two estimates with $P_f = 1/200$ and one with $P_f = 1/300$. Under pressure, the program manager finally estimated the risk at $P_f = 1/100,000$. There was such a large difference between the manager's and the engineer's assessment that Dr. Feynman concluded that "NASA exaggerates the reliability of its products to the point of fantasy" [3].

The safety officer for the firing range at Kennedy Space Center, who had been under considerable pressure to remove the remotely controlled destruction charges from the shuttle, did not believe the reliability figures cited by NASA. He had collected data for all of the 2,900 previous launches using solid rocket boosters. Of this total, 121 had failed. This data provide a very crude estimate of $P_f = (121)/(2900) = 0.042 = 4.2\%$ or about one chance for failure in every 24 launches. The safety officer considered this high risk of failure to be an upper bound, because improvements made since the early launches had improved the reliability of the solid booster motors. Also, the pre-launch inspections on the shuttle were more thorough. His estimate of risk accounting for taking these improvements was $P_f = 1/100$. Dr. Feynman [3], after extensive interviews with engineers, reliability experts, managers from NASA and many subcontractors, concluded that the shuttle "flies in a relatively unsafe condition with a chance of failure of the order of 1%."

After the Challenger accident NASA began a series of comprehensive assessments of risk of flying the shuttle fleet in space. In 1995 the probability of a catastrophic failure was assessed at 1 in every 145 flights. After a series of safety related upgrades to the shuttle system, a new assessment in 1998 placed the probability at one in 245 flights. In October of 2002 NASA most recent assessment placed the probability at one in 265 flights or about 0.39%. How accurate are these assessments? The disintegration of the shuttle Columbia over Texas as it reentered space on February 1, 2003 was the second shuttle failure in 112 flights. This relatively high probability of failure $(2/112) = 0.01786 = 1.786\%$ raises questions about the validity of NASA's safety assessments [4, 5].

Clearly, risk assessment is a difficult task. Each component in a complete system must be evaluated to ascertain its failure rate. Then the components are placed in either a series or parallel arrangement, or some combination of the two, to model the system prior to determining its reliability. The system reliability may be lower or higher than the component reliability depending on the number and the arrangement of the components. Testing to determine component reliability is possible in some instances when the components are relatively inexpensive. However, establishing reliability estimates by testing requires a very large number of trials that often destroy the component. If you want to show a probability of failure less than $P_f = 1/1,000$, you must test considerably more than 1000 components to failure. Obviously, it would not be possible to test more than 1,000 booster rockets when the cost of a single test is several million dollars.

Often the probability of failure of a component must be estimated based on previous experience with similar applications. In some instances, it is possible to compute the probability of failure; however, these calculations require considerable knowledge of the spectrum of loading and the ability of the material to resist fracture. When analytical methods are inadequate and engineering judgment is required to assess the probability of failure, the estimate should be made by a senior engineer with considerable experience and expertise. Even then, the estimate should be pessimistic rather than optimistic. A frank and honest estimate of P_f, based on all of the data and knowledge available, is much better than unrealistic appraisals that give an unwarranted feeling of safety.

16.6 HAZARDS

When designing a product, there are usually risks involved in either the production of the product or in its use in the marketplace. The public-at-large is exposed. What can you do to minimize the risk? In previous sections, the concept of safety factor, component and system reliability and evaluating the risk were briefly discussed. There is another approach to minimize risk. Namely, developing an ability to recognize the hazards involved. You must clearly recognize potential hazards before taking the necessary precautions in your designs to minimize the risks associated with them.

A Listing of Hazards

Mowrer describes a complete list of hazards and provides an extended discussion of each in reference [6]. You are encouraged to read this chapter and to use the extensive checklists incorporated in his coverage. A very brief excerpt from Mower's reference is provided in this section to assist you in recognizing the many hazards that should be considered when designing a product. This list of hazards includes:

Dangerous chemicals and chemical reactions
Exposure to electrical circuits
Exposure to high forces or accelerations
Explosives and explosive mixtures
Fires and excessive temperature
Pressure
Mechanical hazards
Radiation
Noise

Dangerous Chemicals

Chemicals can be nasty and you must appreciate the extreme dangers of exposures to certain toxic substances. Have you read about the leak that developed in a storage tank in Bhopal, India, in 1984? The storage tank contained the toxic chemical methyl isocyanate used in the manufacture of pesticides. It was a significant leak with about 80,000 pounds of the chemical released to the environment. Three thousand nearby inhabitants were killed, 10,000 permanently disabled and another 100,000 injured. The cause of the failure was not in the design of the tank, but in the training of the individuals responsible for the plant maintenance and operation. Nevertheless, a catastrophic accident occurred because several workers and managers, in positions of responsibility, did not adequately understand the dangers of this very toxic chemical.

Exposure to Voltage and Current

What about exposure to electrical circuits? Almost everyone has been shocked by the standard 120 volt, 60 cycle electrical power supplied by the local utility company. Why worry? You should worry because electrical shocks are **dangerous**. **Yes!** Even the 120-volt supply can cause major problems. Your body acts like a resistor and limits the current flowing from the voltage source through your hands, arms, legs, etc. The problem is that the resistance of your body is variable. It depends on the moisture on your hands, the type of soles on your shoes and even the moisture on the ground. If your hands are dry, your shoes have rubber soles and you are standing on a dry floor when you touch one wire of the circuit with only one hand, then you will probably not be harmed because you have arranged a very high resistance path to ground. Consequently, the current flow through your body will be very small. However, if your hands are wet and you touch both of the wires (the white and the black) from the electrical supply, one with the left hand and the other with the right hand, you have placed 120 volts across your heart. You can be electrocuted with only 120 volts under these conditions. The moisture on your hands greatly reduces the effective resistance of the body.

Do not take chances with electricity. Insulate the operator from the circuits preferably with two independent layers of insulation. High voltages are even more serious than low voltages because the currents flowing through one's body increase dramatically. When the current flow through the human body increases to about 10 to 100 mA (milli-amperes), there are very serious consequences to the respiratory muscles. Higher currents of 75-300 mA produce problems with a person's heart function.

The electrical current I flowing through the body can increase in two ways: first, by increasing the applied voltage V, and second, by decreasing the resistance R offered by the body. Ohm's law gives the relation among the voltage V, current I, and the resistance R as:

$$I = V/R. \tag{16.18}$$

Using adequate electrical insulation in the design of products with electrical power, markedly increases the resistance R and decreases the current I to negligible amounts.

Some people think that low voltages (5 or 10 volts) are safe because they barely feel a tingle when touching a low voltage circuit. However, some low voltage circuits particularly on high performance computers carry substantial (100 or more ampere) currents. If you short a circuit with high current flow, an arc occurs which generates significant amounts of heat. Also, the flash of the arc can damage a person's eyes and the heat may produce serious burns. In your designs, insulate and/or shield even low voltage circuits if they carry significant currents.

The Effects of High Forces and Accelerations

High forces and high accelerations (or decelerations) go hand in hand. Newton's second law requires the connection ($\mathbf{F = ma}$). If the brakes are suddenly applied in an automobile, the car decelerates quickly and the passenger (without seat belt or air bags) is thrown into the windshield. You must always be concerned with acceleration or deceleration because of their effects on the human body. Military pilots are trained to withstand high accelerations (high Gs). A good pilot with a well-designed flight suit can pull 7 or perhaps 8 Gs before losing consciousness. However, a civilian will become irritated at less than 2 Gs. If you want to feel an acceleration thrill, go to an amusement park and ride the roller coaster. There is no need to incorporate high accelerations into the design of most new products. A reasonably hot sports car that accelerates to 60 MPH in six seconds requires an acceleration of only 0.46 Gs. You should be careful to keep both acceleration and decelerations low when you design products that move and accelerate.

Explosives and Explosive Mixtures

With the bombing of the U. S. S. Cole, the disaster at the Federal building in Okalahoma City and terrorists bombings at many locations, most people have become aware of the disastrous effects of large explosions. The gas pressures generated by the blast destroy very substantial buildings, break glass over a very large region and kill or injure many people. The population is generally aware of the characteristics of common explosives like dynamite and ANFO (ammonia nitrate and fuel oil). In most designs, you do not encounter a need to accommodate these traditional explosives. What you must recognize are other less apparent agents that act like explosives under special circumstances. Fuels such as natural gas, propane, butane, etc. can leak and combine with air to produce an explosion when ignited. Boating accidents are common when gasoline leaks in an engine compartment. The resulting mixture of gasoline fumes and air explode when subjected to a small spark, destroying the boat and killing or injuring the passengers. Still another unusual source of fuel for an explosion is dust. When handling large quantities of a combustible solid, dust (fine particles) is generated. If these particles are suspended in air, the resulting mixture will certainly explode when ignited. A grain elevator exploded in Westwego, Louisiana, in 1977 killing 35 people when an explosive mixture of combustible particles (dust) from the grain and air was ignited. More recently on June 8, 1998, the De Bruce grain elevator in Haysville, Kansas exploded killing several workers.

Fires

Over a million fires occur in the U.S. every year. Some are vehicle fires (400,000) and others are structural fires (650,000). Many people die in the fires (4,700) and many more are injured (28,700). Clearly, fire is a serious problem. People are killed, injured or traumatized and property is lost (eight or nine billion dollars per year). Some of these fires are, simply put, stupid. For example, he or she who smokes in bed may some night fall asleep and set the house on fire. A surprisingly large number of fires (about 100,000 per year) are deliberately set—apparently to collect the fire insurance, to take revenge or as a sick kind of diversion.

Engineers have a responsibility to decrease the number of fires resulting from the products that they design. Did these fires start with an appliance or a motor that overheated for some reason? Determine the reason for the overheating and redesign the product so that excessive heat will not be generated. Why did the automobile burst into flames? Did a fitting on the gasoline line leak? Redesign to eliminate the fittings or specify a fitting that will not fail under the prevailing conditions. There are many solutions to the problem of fires in America. As a nation, we are far too casual about fires. Carelessness in personal practices, poor design in products intended for the home and business and fraud (arson) to collect fire insurance reimbursement for lost property are tolerated.

Pressure and Pressure Vessels

Pressurized fluids are used for many good reasons, and in most cases, pressure vessels (the containers that hold the fluids) perform very well. It was not always the case. In Boston during the winter of 1919, a huge tank about 90 feet in diameter and 50 feet high fractured. It contained two million gallons of molasses that flooded the local area. Twelve people died and another 40 were injured in this accident.

It is relatively easy to design a pressure vessel today. In fact, the American Society for Mechanical Engineers (ASME) has developed a code that engineers follow in their design to produce pressure vessels that are certified as safe for service. However, on occasion, tanks fail in service. The difficulty is usually with the steel plates that are welded together to manufacture the tank. The steel employed in both the plates and the welds have to exhibit high fracture toughness at low temperatures, and the welds must be free of large flaws. If you have the responsibility of designing a pressure vessel, follow the ASME code, make certain the welding procedures followed in manufacturing produce crack-free welds, and specify steel for the plate and welding rods that is sufficiently strong and fracture resistant. If you intend to become a mechanical engineer, you will have the opportunity to learn how to design pressure vessels, select suitable materials for their construction and specify welding procedures in courses presented later in the curriculum.

Mechanical Hazards

Mechanical hazards are features, which exist on products that may cause injury to someone nearby. Consider as an example an electric fan used to cool a room. Is the fan blade adequately guarded, or is it possible for someone to stick his or her finger into the rotating blades? Suppose the cabinet you recently designed to hold a special tool has a sharp corner at hip level. Can someone walk into the cabinet and break skin or bruise a hip on that corner? You have designed a center post crane to lift material and move it over a 25-foot diameter area. Will the center post be stable under all conditions of loading or will it collapse? You have designed a new pizza machine that rolls the dough into sheets exactly 2 mm thick. Have you provided protection for the entrance to the rolls that prevent the operator from inserting his or her fingers into the rollers? You have designed a wonderful guard that prevents a person from exposing their hands and arms in operating a punch press. However, the guard is attached to the machine with two small screws. Have you used locknuts and/or safety wires to insure that the screws will not loosen during the operation of the press? Are the screws large enough to resist failure by shear?

There are many mechanical hazards encountered in designing equipment and products. Always examine each component, and look for sharp points, cutting edges, pinch points, rotating parts, etc. Do not count on peoples' good judgment. If it is possible to insert a finger into the machinery, even if it is a foolish act, you can be certain that sooner or later someone will do so.

Radiation Hazards

Radiation hazards are due to exposure to electromagnetic waves. The damage produced depends on many different factors such as:

* The strength of the source.
* The degree to which the emitting radiation is focused.
* The distance from the source.
* The shielding between the source and the object being radiated.
* The time of exposure to the radiation.

The radiation spectrum is divided into several different regions—very short wave length, visible light, infrared and microwave radiation. The short wave length radiation is the most dangerous (x-rays, gamma rays, neutrons, etc.) with serious health risks (cancer) for overexposures. The hazards due to UV, visible light and IR are usually due to excessive exposure where serious burns to the skin occur. For very intense radiation, even short exposures are detrimental to the eyes. Microwave radiation is absorbed into the body and may result in the heating of one's internal organs. Because the means used by organs to dissipate this heat is not known nor is the effect of the localized increase in temperature known, it is prudent to avoid exposure to microwaves. In America, the Offices of Safety and Health Administration (OSHA) has issued a regulation limiting the power density of microwave energy to 10 mW/cm² for an exposure of 6 minutes or more. In designing products, where operators may be exposed to microwave radiation, shielding should be employed to reduce the power density well below the regulatory levels.

Noise

Noise levels that occur in the environment may be damaging to your hearing, may interfere with your work or play and may degrade the quality of your life-style. Most noise is man-made, although occasionally a storm and Mother Nature provides the sounds of thunder and wind. If you listen occasionally to a band playing rock and roll music, there is a temporary shift in the threshold of audibility to a higher level of pressure. However, if you play in the rock and roll band almost every night for an extended period, the shift in the threshold of audibility becomes permanent and your hearing becomes impaired.

Noise is a pressure disturbance that propagates through some medium such as air. The velocity of propagation through air is 344 m/s at room temperature. The pressure disturbance is oscillatory usually with many different frequencies present. The frequency content of the pressure waves depends on the source of the sound. A note from a violin will have much higher frequencies that a note from a tuba. The intensity of the noise is measured with a sound level meter that consists of a microphone, amplifier and a display meter that provides a reading in decibels (dB).

As a general guideline, the threshold for audibility is less than 25 dB[2] before a person is considered handicapped. In addition, there is a threshold for feeling noise-generated pressure at about 120 dB and another threshold for pain between 135 and 140 dB. When designing a product, the noise level is a serious consideration. Clearly, the feeling and pain levels of noise intensity must be avoided, but what levels are satisfactory? The U.S. Environmental Protection Agency (EPA) has established standards, which provide guidance to engineers in the design of products. For example, the results presented in Fig. 16.6 show the relation among the sound pressure level, the communicating distance and the degree of speech intelligibility.

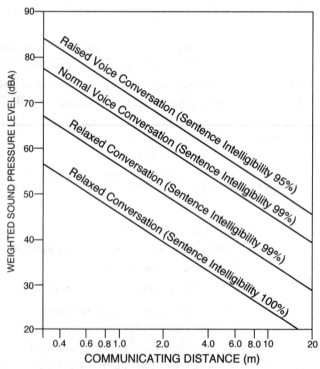

Fig. 16.6 Outdoor distances for intelligible conversation with a background of steady noise [8].

16.7 SUMMARY

When a product is introduced in the marketplace, there is usually some risk involved to the workers making the product, the customers and society-at-large. Hopefully, the risks will be small and acceptable to all concerned because of the significant benefits produced for the customers and/or the sociotechnical system. In determining if the risk is acceptable, the public needs an accurate and honest assessment of the consequences of a failure and the probability of the occurrence of a serious accident. It is the responsibility of the engineering profession to assist business administrators and governmental agencies in these assessments.

---

[2] What constitutes normal hearing differs from one authority to another. A reference by the American Academy of Ophthalmology and Otolaryngology was used in writing this section.

While the public must accept risk, engineers must make every attempt to minimize the probability of harm within reasonable constraints on cost and performance. There are many excellent engineering methods for ensuring safe design. The concepts of stress, strength, safety factor and margin of safety illustrate an approach to minimizing risk due to failure by fracture. Recommendations for a range of commonly accepted safety factors have been given. Issues to consider in selecting the safety factor to employ when sizing components have been described.

Sometimes it is understood that failures will occur in service. In these cases, it is necessary to determine the probability of the failure event. To illustrate methods for determining probability of failure, the concepts of failure rate and mean time between failures were introduced. It was shown that the failure rate could be determined simply by keeping records of the failures of components as a function of time after they were placed in service. These concepts are important, but they should not be confused with the probability of failure or survival.

Component reliability was introduced, and a method for computing reliability from the failure rate was shown and demonstrated. In developing this reliability equation, Eq. (16.12), the failure rate was assumed to be constant over the service life. It is important to observe that the probability of survival decreases as the service life is increased and the probability of failure increases with time in service. For reliable service for a long period of time, the failure rate must be extremely low.

System reliability is different than component reliability. A system is usually composed of many components. If the components are arranged in series, they must all function for the system to operate correctly. The reliability of a series connected system is determined from Eq. 16.13. It is evident from the data presented in Table 16.2 that the number of components arranged in a series markedly lowers the reliability of a system. Sometimes a system must function all of the time. If this is the case, a system failure cannot be allowed, because of its negative consequences. In these instances, engineers design with a number of redundant components arranged in a parallel system. Redundancy improves system reliability, as shown in Table 16.3, but at increased cost and the requirement for more power, weight and size.

The explosion of the space shuttle Challenger and more recently the disintegration of Columbia were described to illustrate the importance of evaluating the risk. It is often a very difficult problem and an analytical solution for the risk is often not possible. Nevertheless, it is a professional responsibility to prepare an intelligent, accurate, honest and frank estimate of the risk, and to insure that all of the principals involved are aware of the consequences of a failure.

Finally, a number of hazards that cause injury or death has been introduced. Unfortunately the listing is relatively long. It is important that you recognize the hazards and be vigilant in your designs to avoid them. Avoidance often does not require extensive calculation from elaborate formulae, but rather a detailed assessment of each component in the system and a good measure of common sense.

REFERENCES

1. Dally, J. W., Packaging of Electronic Systems: A Mechanical Engineering Approach, McGraw Hill, New York, NY, 1990, pp. 288-296.
2. Lewis, R. S. Challenger: The Final Voyage, Columbia University Press, New York, NY, 1988.
3. Feynman, R. "Personal Observations on the Reliability of the Shuttle," Appendix to the Presidential Commission's Report, Ayer Co., Salem, 1986.
4. Mowrer, F. W., Introduction to Engineering Design: ENES 100, McGraw Hill, New York, NY, 1996, pp. 149-172.
5. Whoriskey, P., "Shuttle Failures Raise a Big Question," Washington Post, February 10, 2003, p. A-9.
6. Gugliotta, G and R. Weiss, "Dangers of Gauging Space Safety," The Washington Post, February 17, 2003, p. A-14.

7. Cunniff, P. F., <u>Environmental Noise Pollution</u>, Wiley, New York, NY, 1977, pp. 101-115.
8. "Information on Levels of Environmental Noise Requisite to Protect health and Public Welfare with an Adequate Margin of Safety," EPA Report, March 1974.
9. Powers D. G. and J. Proctor, Editors, <u>Lockheed L-188 Electra</u>, World Transport Press, April 1999.

EXERCISES

16.1 Have you been involved in one or more automobile accidents since you began to drive? Were you or anyone else injured or worse yet killed? Estimate the total mileage you have driven over this period and calculate your accident rate (number/mile). What were the reasons for the accident or accidents? Comment on your driving behavior and its influence on the accident rate.

16.2 Are the benefits of driving worth the risks of injury or death? Determine your probability of being killed in a fatal accident while driving this year. Hint: Statistics on fatal accidents are always listed in the World Almanac and on the web site for the NHTSA. Make any assumptions necessary for your analysis, but justify each of them.

16.3 Your co-worker designs a tie rod (a tension member) from a 20 mm diameter steel bar with a strength of 400 MPa. If the team leader has indicated that he or she wants to maintain a safety factor of 2.4, determine the maximum load that can be applied to the tie rod.

16.4 If a component is designed with a safety factor of 2.3, what is its margin of safety?

16.5 If engineers are so smart and have all of these great computer programs to determine stresses, why is it necessary to specify a margin of safety or a safety factor when designing a structural member?

16.6 NASA's space shuttle utilizes two booster rockets fueled with solid propellant to provide the thrust required for launch. Is this a redundant system? State your reason for this conclusion.

16.7 If the probability of failure of one of the solid propellant, booster rockets on the space shuttle is 1/1,000, determine the reliability of the solid booster rocket system.

16.8 You are designing a very large computer system to contain the database for a world-wide reservation system. It is estimated that at any instant 500 operators will be accessing the database, and another 4,000 operators will soon be ready with their requests for computer availability. The mainframe computers that you plan to employ each have a MTBF of 8,000 hours. Present a design of the computer system that will insure that all 500 operators have a 99.9 % probability of being served. In your design show all calculations and carefully list your assumptions. Justify the costs involved if each mainframe computer employed in the system is valued at $200,000.

16.9 Janet is an engineer in the transmission department of the Fink Motor Corp. Her job is to record data on the mileage prior to failure of the new lightweight transmission that has been placed in 250,000 of the new 2008 model of the Clunker III. She records the following data:

Mileage (1000 miles)	No. Failures, N_f
0 - 5	65
5 - 10	38
10 - 20	20
20 - 30	18
30 - 40	22
40 - 50	34
50 - 60	90
60 - 70	521
70 - 80	1,364
80 - 90	6,677
90 - 100	15,592

Determine the failure rate as a function of service life expressed in terms of mileage, and prepare a graph showing your results.

16.10 Using the data from the table in Exercise 16.9, determine the probability of survival and the probability of failure of the transmissions as a function of service life measured in terms of mileage. Prepare a graph of your results.

16.11 If the transmissions in the Clunker III were under warranty for 100,000 miles, what would be the consequences for the Fink Motor Corporation?

16.12 Prepare a listing showing the ratio of fatal events per flights for the airlines in the U. S. and Canada since 1980. The data needed to prepare this listing is at http://airsafe.com/airline.htm.

16.13 Based on the results of Exercise 16.12 cite the three airlines with the best safety records. Cite the three airlines with the worst safety records.

16.14 Will the results of Exercise 16.13 affect your planning for air travel in the future?

16.15 Prepare a paper describing the top ten fatal air transport accidents. The Website listed in Exercise 16.12 contains the data necessary for you to write this paper.

16.16 Prepare a paper discussing your reaction to the risk assessment of the space shuttle system by NASA. Give arguments for and against NASA's position and practices regarding safety of the shuttle and its crew both before and after the Challenger and Columbia accidents.

16.17 Describe an incident when you received an electrical shock. Something obviously went wrong to cause this incident. Please indicate the problem. Was there a design deficiency? What could be done to prevent the incident from reoccurring?

16.18 You are designing a cabinet-mounted, self-cleaning oven that requires very high power levels to heat its interior surfaces until they are free of all the splattered, burnt-on grease. As the lead engineer on this design team, what precautions should you take to insure that the oven would not be the source of a fire during its anticipated 15-year life?

16.19 Automobiles are the most commonly used mode of transportation in the U. S. and many other developed countries. Describe a least one-design flaw pertaining to safety of many sports utility vehicles. What government agency is charged with the responsibility of insuring safety of automobiles in America?

16.20 There are many fatal accidents each year involving crashes with large trucks. The Federal Motor Carrier Safety Administration (FMCSA) is the government agency responsible for regulations involving trucks. Write a paper describing possible rules the agency could impose on the trucking industry to reduce the number of fatal truck crashes. Cite reasons for the new rules and predict the benefits that would result if they were implemented. Discuss the economic consequences for your suggestions.

Notes

CHAPTER 17

ETHICS, CHARACTER AND ENGINEERING

17.1 INTRODUCTION

This chapter on ethics was written shortly after the 1996 presidential election, and it has been revised nearly every year since then. During this period the newspapers and TV news and talk shows covered many ethical issues emanating from both Congress and the White House. Ethical lapses occurred on nearly a weekly basis, keeping the news services busy reporting on questionable behavior of our elected officials from both political parties[1]. These officials are the people the voters have entrusted and empowered to write and approve the laws governing the country. The news media has pointed out numerous instances where political leaders have skirted the truth if not the law.

During the period from the initial writing of this chapter to its revision in November of 2006, the ethical situation has deteriorated. Ethical issues continue to be reported by the news media with alarming frequency. For example[2], U. S. District Judge Royce Lambert cited the former Secretary of the Interior, Mr. Bruce Babbitt, and the former Secretary of the Treasury, Robert Rubin, for contempt of court for withholding evidence regarding the Indian trust funds. (The government cannot account for $2.4 billion of these funds). Judge Lambert concluded that the employees of the Interior and Treasury Departments had "engaged in a shocking pattern of deception of the court. I have never seen more egregious conduct by the federal government."

Several years ago President Clinton admitted to having an affair with an intern about half his age. He initially claimed on national television that he did not have sex with **that** woman. However, when the results of tests with that woman's dress were known, he admitted it. Previously, the president, in still another legal proceeding, denied the existence of an intimate relationship under oath. When the facts of the president's behavior became apparent, the House of Representatives impeached Mr. Clinton for lying and obstructing justice and a federal judge held him in contempt of court for lying.

The lax ethical practices by many of our elected officials have led to a very serious skepticism about the merits of many of the government institutions and agencies. The fact that only about 50% of the eligible voters took time to cast their ballots in closely contested presidential elections is testimony to the deep skepticism toward the government and our elected officials.

In more recent years, corporate scandals have overshadowed the unethical behavior of our government officials. Many high executives have been found guilty of serious mismanagement and fraud. In some cases they have been sentenced to long prison terms considering their age. The legal processes are still underway and it is expected that many more CEOs and CFOs and other executives will be found guilty of crimes and sentenced to prison. In August of 2002, Forbes.com published an article titled "The Corporate

---

[1] The situation has not changed much as we prepare this revision shortly before the November 2006 mid term elections.
[2] See the article entitled, "Shredding 162 Boxes," Wall Street Journal, December 9, 1999.

Scandal Sheet" in which they listed Adelphia, AOL Time Warner, Arthur Anderson, Bristol Meyers Squibb, CMS Energy, Duke Energy, Dynegy, El Paso, Enron, Global Crossings, Halliburton, Homestores, Kmart, Merck, Nicor Energy, Peregrine Systems, Quest Communications International, Reliant Energy, Tyco, World Com and Xerox as companies with serious accounting problems. Clearly the ethics of many of the executives in corporate America are troubling.

As engineers, we want to preserve and enhance the confidence of the public. To gain public confidence, it is essential that we all behave in an ethical manner every day of every year. Whether we act as individuals or as professionals, consistent ethical behavior is mandatory. Business, our primary employer, must also act in an ethical manner. Because engineering developments are sponsored, financed and controlled by business, corporate and engineering behaviors are inseparable. Our actions and words always must be above reproach.

17.2   CONFUSION ABOUT ETHICAL BEHAVIOR

Many people are confused about ethical behavior. The results of several polls indicate that Americans are less ethical today than in previous generations. Sixty-four percent of a sample of 5,000 individuals admits to lying if it does not cause real damage. An even larger percentage of those sampled (74%) will steal providing the person or business that is being ripped off does not miss it [1]. Many students (75% high school and 50% college) admit to cheating on an important exam [2]. Most students have not developed a moral code to use as a guide for their behavior. Poor behavior, when the author was in high school, included making too much noise, chewing gum in class, talking, littering, running in the halls, etc. Today, these relatively minor transgressions have been replaced with more serious problems involving mass murder, drug and alcohol abuse, pregnancy, suicide, rape, robbery and assault. These are extremely negative changes. Why?

There are many reasons for the degradation of ethical behavior over the past 50 years. Let's discuss two key reasons. First, institutions (family, church and schools) that once taught ethics are much weaker today than 50 years ago. Many families have been torn apart by an extremely high divorce rate and single parent homes are common. The concept of "love, honor and cherish, in sickness and in health" is often ignored.

The influence of our religious organizations has deteriorated due to serious declines in membership. Unfortunately, their finances often limit their activities to current membership. Many social problems are with people who are not well known to a church, synagogue or mosque. The public has also become much more secular with no religious affiliation. Our public school administrators are so concerned with possible litigation that they have largely removed the teaching of moral values from their curriculum. Given the current constrains on discipline in the public school system, it is often difficult for the administrators and teachers to maintain orderly classrooms. Consequently many students graduating from high school have not been given an adequate opportunity to develop a personal code of ethics. The result of the failure of our institutions to teach ethics is that many of our young people are not aware of what is right and what is wrong. All actions are not gray. Many actions are plainly right, and many are obviously wrong.

The second reason for the deterioration in ethics on the part of our younger population is the current state of ethical standards in society. Everyone is exposed to cynical behavior and selfish attitudes on TV, magazines and newspapers every day. The "bad guy" often walks away from severe crimes with minor penalties. Minor infractions are frequently ignored. Killers and child sex offenders are paroled after remarkably short periods of incarceration. Popular TV programs frequently show examples of lax ethics and self-indulgences to generate a popular but sick kind of humor. The result is to create confusion in our youth as they try to develop some system for judging right from wrong. Our youth mirrors the behavior of society leading to cynicism, selfish attitudes, and dishonesty and irresponsible conduct.

17.3 RIGHT, WRONG OR GRAY

Christina Sommers [2] develops an interesting viewpoint regarding what is considered right, wrong and controversial by society-at-large. She argues that there are some ethical issues that are not clearly defined, and that society has not reached a consensus regarding the correctness of these issues. You can develop cogent arguments for and against a specified viewpoint for controversial issues such as abortion, affirmative action, capital punishment, etc. Sommers refers to these controversial issues as "dilemma" ethics that should be distinguished from "basic" ethics. With "dilemma" ethics, society-at-large is not certain about what is right, wrong or gray. There are several attitudes prevalent in society—sometimes the issue is so important to segments of our population that demonstrations are organized to elevate one viewpoint or another.

No attempt will be made in this textbook to resolve any of the "dilemma" issues. These issues have been unresolved for decades. These issues will require much more time and understanding before our society is willing to reach the consensus necessary for resolution. It is much more productive to consider "basic" ethics where right and wrong or white and black are much more clearly recognized by the majority of society.

Right Versus Wrong

It is easy to define right and wrong. In fact, the understanding of right and wrong dates back to Aristotle [3] and fundamental doctrine has not changed much since then. Let's start with the four classic virtues:

- Prudence
- Justice
- Fortitude
- Temperance

Prudence refers to careful forethought, good judgment and discretion. In other words, think before you act and exercise care and wisdom. If you act, will your action cause problems, now or later, for you or anyone else? Are you careful about your remarks concerning others? One of my rules is not to speak ill of others under any circumstances. Is the action that you are about to undertake in your best interest? Will that action be appreciated by your family, friends, coworkers, managers and peers?

Justice involves fairness, honor, keeping your word, honesty, truthfulness, etc. It is clearly wrong to lie, cheat, steal and break promises. It is also wrong to tolerate those about you who do so[3]. The author's behavior with regard to justice is governed by the golden rule. Do unto others as you would have them do unto you. It is a very simple rule, and it is effective.

Fortitude is about courage and persistence. One usually relates fortitude with warfare and/or battle. Warriors stand their ground and exhibit a very special brand of courage when their lives are at extreme risk. However, there are other brands of courage that must be exhibited in more typical circumstances. Will you stay with an idea, even if it is not popular, if you know that it is the "right" approach? Showing determination and resolution is sometimes difficult when the risks of failure are high or when peers encourage you to abandon your idea.

Temperance, of course, refers to moderation in drinking alcoholic beverages. However, temperance has much wider implications with regard to ethics. Temperance can also mean control of human passions such as anger, lust, hostility, exasperation and lechery. With regard to food and beverages, it implies self-control and moderation in consumption. Clearly, much is to be gained by avoiding overindulgence in food, drink and in restraining your emotional extremes.

---

[3] Many people are willing to tolerate about them those who cheat, steal and break promises. They do not believe in imposing ethical constraints on friends or associates. This no tolerance concept is difficult to establish in many honor codes, and the author may be in the minority on this issue.

There are other virtues such as loyalty and obedience that are less commonly discussed these days. With the restructuring, massive downsizing and outsourcing that has taken place in the business world in the past two decades, the concept of company loyalty to the employee and vice versa has been destroyed. Hopefully, loyalty to the family is an invariant, at least in those situations where a family in the true sense exists. The concept of obedience was seriously damaged—if not destroyed—in this country by the Vietnam conflict. The government ordering young people to fight in a war of questionable value was too much for many to endure. Many young men refused to obey the orders to report for military duty and fled the country to avoid persecution.

Because of these events (many others could also be added), the virtues of loyalty and obedience have shifted from the "basic" to the "dilemma" category, and arguments can be advanced on both sides of these questions. What do you think of being loyal to an employer who will eliminate your position during a period of **increasing** profits? How do you feel about completing your tax return for the Internal Revenue Service? Talk about straining one's patience!!!

Theological Virtues

In addition to the four basic virtues, prudence, justice, fortitude and temperance, there are three religious (theological) virtues, which include:

- Faith
- Hope
- Charity

The theological virtues are well known, so it is not necessary to elaborate on their importance. However, it is stressed that their merits are definite and irrefutable. It is good to be faithful, hopeful and charitable. If you want to be even more complete, add other ethical constants such as humility and respect to the list.

While the emphasis in most college courses on ethics is with social dilemmas where arguments and counter arguments are to be developed, most individual behavior can be judged by very clear and well-understood virtues that are part of "basic" ethics. There is no need to be confused, nor is there any reason to impair your "basic" code of ethics with issues raised in the study of moral relativism.

17.4 LAWS AND ETHICS

The governments (federal, state and local) play a role in ethical behavior, because they generate laws and set public policy. Laws (mostly at the state level) are written to aid people in their pursuit of a morally correct and safe life. The purpose of most of these laws is to prohibit a well-defined set of vices. For example, it is illegal to sell drugs, rob banks, commit burglary, kill or abuse another human, engage in prostitution or pursue deviant sexual practices (rape, among other offenses).

These laws, and many more like them, serve to create a moral ecology in which society exists [4]. The laws require visible or outward conformity to a publicly accepted moral code of behavior, but do nothing to inhibit illegal behavior unless it is observed, reported and prosecuted. An unscrupulous person can sell drugs if he or she is not observed in the act. Even if they are observed, they can avoid penalty if they are not arrested and fully prosecuted for one reason or another. Even if they are observed, detained, prosecuted and convicted, they may receive a suspended sentence. The law does not insure that infractions **will not** occur; the law simply indicates that certain behavior **may** not be tolerated.

In some instances, laws are not a suitable approach for creating a moral ecology. An example of an unsuitable law in this country was prohibition, which was enacted by a constitutional amendment ratified in 1919. When it became evident that the law was causing more problems than it solved, it was repealed in 1933. The government wanted to make it illegal to consume alcoholic beverages, but society-at-large wanted to drink. Bootlegging, racketeers, and speakeasies followed bringing crime and violence to the neighborhoods.

It soon became apparent that the law was not a suitable approach to solve a temperance problem. In such cases, the government uses public policy to discourage people from pursuing a vice. Policies are adapted that limit (but do not preclude) the consumption of alcoholic beverages. Hours of sale are restricted, licenses are required, taxes are imposed, the number of outlets is constrained, etc. Public policies are conveyed by means of rules and regulations—not laws. The policies should be crafted to strengthen families, communities and churches by discouraging, but not prohibiting irresponsible conduct.

17.5   ETHICAL BUSINESS PRACTICES

Corporations are entities that under law are treated as persons. A corporation, acting like a person, has the right to conduct business and the obligation to pay taxes. Just as individuals are judged by their ethical behavior, corporations are judged by their ethical conduct [5]. The chief executive officer (CEO) for a corporation sets the moral and ethical tone for the organization. Any large business is organized with several layers of management. If the CEO operates the business with the highest ethical standards, the middle and lower level managers will follow the example set at the highest level in the corporation. However, if lax ethical standards are permitted, a pattern is set and questionable ethics will become the corporate style and standard.

In an excellent paper, Baker [5] has developed a list of issues that occur daily in a typical corporation. The manner in which a corporation deals with these issues establishes its character. Baker's list is shown below:

1. How people, employees, applicants, customers, shareholders and the families who live near our facilities are treated?
2. Is everyone free of systemic or individual practices of discrimination?
3. How do we spend our shareholders' money on our expense accounts as an institution and as individuals?
4. What is the level of quality that goes into our products? Do we meet our customers' expectations for quality? Do we meet our own standards for quality?
5. What is our concern for safety, not only for our employees, but also for our customers and our neighbors in communities in which we operate?
6. How "ethically" do we compete?
7. How well do we adhere to the laws of the locales, regions and nations in which we do business?
8. What is acceptable gift giving and gift taking?
9. How honest are our communications to our employees and our public advertising?
10. What are our corporate and personal positions on public policy issues, and how do we promote those positions?

This is a long list, but we certainly could add more items to Baker's catalog of concerns—sustainability and the environment for instance. Adding more items is not as important as recognizing that corporate decisions are made on a daily basis by its managers and employees. These day-to-day decisions determine the corporate character. The ethical judgments that serve as the basis for most of these decisions will depend on individual personal values (virtues) and experience. The quality and consistency of the ethics applied in these decisions will markedly affect the success of the corporation, its management and its employees.

17.6 HONOR CODES

We will not lie, steal or cheat, nor tolerate among us anyone who does [6]. This is a fourteen word honor code that cadets learn almost as soon as they step foot onto a military academy or college campus that has strong ties to the military (Virginia Military Institute and the Citadel). The military academies (Navy, Army, Air Force and Coast Guard) are educational institutions that serve to train career officers for the services. Honor and integrity is a very large and important element in their educational program.

Honor codes are much more widespread than those found in military institutions. There is a growing trend to introduce honor codes on campuses where they are absent and to strengthen them on campuses where they already exist [7]. An honor code typically forbids lying, cheating and stealing. When a peer notes an infraction of the code, the code requires that student to bring the case forward. A student committee hears the case. Although faculty members may testify if called upon, they have no control over the proceedings. The penalty imposed on those found guilty of breaking the honor code is determined by the student committees—university administrators serve only to implement the student's decisions.

The honor codes are very well respected by the student body and fully appreciated by the faculty. Students generally support the honor codes because they also are concerned with cheating by their peers. They want to play the game fairly and on a level field. When the honor system is established, the students handle the responsibility for academic integrity with careful consideration. Punishments are handed out only after complete investigations involving representation of both sides of the case. In some instances, the individual being judged by the student committee retains lawyers to represent them in the proceedings.

The faculty also favors the honor system. They never have appreciated proctoring exams. With the increase in cheating over the past 25 years, proctoring has evolved into policing that is very distasteful. The advantage of an honor code to the faculty is that an unproctored examination transfers the responsibility for policing to the students. With an effective honor system, students often are permitted to take an exam at a time and place of their choosing.

Of course honor codes are not perfect. Some students cheat regardless of the code. The military academies, where the code is the strictest and the most rigorously enforced, have suffered the most notorious lapses in standards. Scandals at the U.S. Naval Academy have been the most prevalent and widespread in recent years with as many as 133 midshipmen involved in cheating on an electrical engineering exam several years ago. To make matters worse, the administrators at the academy appear to have interfered with the system, and in doing so, lost the trust of the midshipmen.

The lesson from the academies with regard to the effectiveness of the honor codes is that rigid codes and rules reduce cheating. However, if the students see a way of beating this very rigid system, they will occasionally take advantage of the opportunity. Honor codes are most effective when they produce an ambience of trust between the students, faculty and the administration. Cheating, lying, and stealing will generally be reduced. For honest students, the scoring on exams will be fairer. For faculty the distasteful task of policing is eliminated. The feeling of trust between the faculty and the students is enabling. With time and prolonged success, the honor code and the trust that it engenders becomes an essential part of the educational process.

17.7 CHARACTER

The author had the privilege of teaching for a year (1995-96) at the U.S. Air Force Academy. It was a wonderful experience for me for many reasons—both personal and professional. One of the most important reasons was the opportunity to participate in a well-planned approach to incorporate character building into the educational process. Character was a key educational outcome in every class that was taught at the U. S. Air Force Academy.

It is not possible to transplant character into an individual, and it is not possible to accomplish much in efforts to explicitly "teach" character. Character comes implicitly when individuals develop a personal set of ethical responsibilities. Let's consider some of these attributes that form the foundation for a person's character.

- **Be committed to excellence**. This attribute simply means to do the best that you can do in both your personal and professional endeavors. To make this commitment, you have to assess your capabilities. How good are your skills? How determined are you in achieving your goals? How much time can you commit without endangering your health? If you know yourself, it is possible to measure your achievements on a realistic scale. Engage to the full measure of that scale.
- **Respect the dignity of everyone**. Respect is the foundation for all achievements. We work, live and play in a diverse world. You must appreciate everyone that you encounter in life. Race, gender, ethnicity and religion are not criteria for judgment. Recognize the potential of those with whom you work and study. Support and encourage them and carefully avoid demeaning criticism. Teamwork, essential in today's workplace, requires that you accept the differences inevitable in a diverse population and that you fully embrace fellow team members.
- **Integrity**. A single word describes a vitally important attribute of character. Integrity means that you decide to do the right thing solely because it is the right thing to do. You often face decisions that test your integrity. Do you instinctively make the right (honorable) decision? For example, as you parked your car last night, your bumper scraped the fender of the adjacent car. No one observed your accident. Do you leave a note with your name, address and phone number? Or did you split and drive to another nearby parking lot? How many parking lot scrapes can you find on your car? Did you find a note from the party inflicting the damage claiming responsibility? A person with integrity will consistently make the correct decision, not because it is right for their personal benefit, but because it is clearly the "right" course of action.
- **Be decisive**. When you encounter a problem in engineering or your personal life, there is an information-gathering period followed by a decision. Some folks do not want to make that decision. They prolong the information-gathering period; they procrastinate; they hem and haw; they pass the buck. When you establish the facts, evaluate them and make a timely decision. It will be your decision if you make it in isolation. If you made it within the framework of a team, it should represent a consensus of the members of the team.
- **Take full responsibility**. Decisions are made and actions follow. Often your decisions produce a winner, but sometimes they result in a loser. If the outcome was a loser and you participated in the decision process, it is your responsibility. Step up, accept the responsibility, and lead the effort to fix the problem. Never, under any circumstances, begin to participate in a **finger pointing** exercise. Finger pointing does not resolve problems; it exacerbates them.
- **Be temperate**. Self-discipline is an essential part of character. Eat, drink and be merry in moderation. Overindulgence is disgusting. No one appreciates a drunk or a glutton. Self-discipline insures control of the human passions, sensual pleasures, anger, rage and frustration. With self-control, it is possible to attain consistently high levels of achievement.
- **Exhibit fortitude**. On many occasions, you will encounter significant difficulties in completing the tasks at hand. It is easy to give up and divert your attention to more pleasurable undertakings. You can quickly forget the missing assignment, the incomplete solution, the late review or the unfinished drawing. Who is to know? Stamina, mental toughness and discipline are attributes that keep us on task. Stay with the job, and not only complete the work, but do it well and with dispatch.
- **Understand the significance of spiritual values**. Many, but not all, of us have a faith. Most of us endorse a set of theological beliefs. Your beliefs and your church may be different than mine. Nevertheless, it is essential that I respect your convictions and you should respect mine. We all need to be sensitive to the important role that religion occupies in the mental comfort of many people. We must accommodate a diversity of beliefs by supporting the right of an individual to choose his or her faith and to pursue the ceremonies offered by the church representing this faith.

17.8 ETHICS OF ENGINEERS

Most of this chapter has been devoted to a discussion of issues pertaining to individual behavior, ethics, virtues, morals and character. A personal code of ethics is the cornerstone to achieving a meaningful sense of moral values. Professional ethics are also important, but they must begin only after a person has developed a well-understood personal code of ethics. Hopefully, the preceding sections of this chapter will be helpful in any attempt you make to establish a sense of right and wrong.

There are several codes of ethics for engineers. Most of the founding societies of engineering have developed and distributed codes and guidelines for professional conduct and faith statements. The codes of ethics for the different engineering societies are all similar. The fact that each society has their own code, is more to insure a complete distribution of the code than to pursue unique ethical issues.

Accreditation Board for Engineering and Technology*

CODE OF ETHICS OF ENGINEERS

THE FUNDAMENTAL PRINCIPLES

Engineers uphold and advance the integrity, honor and dignity of the engineering profession by:

I. using their knowledge and skill for the enhancement of human welfare;

II. being honest and impartial, and serving with fidelity the public, their employers and clients;

III. striving to increase the competence and prestige of the engineering profession; and

IV. supporting the professional and technical societies of their disciplines.

THE FUNDAMENTAL CANONS

1. Engineers shall hold paramount the safety, health and welfare of the public in the performance of their professional duties.

2. Engineers shall perform services only in the areas of their competence.

3. Engineers shall issue public statements only in an objective and truthful manner.

4. Engineers shall act in professional matters for each employer or client as faithful agents or trustees, and shall avoid conflicts of interest.

5. Engineers shall build their professional reputation on the merit of their services and shall not compete unfairly with others.

6. Engineers shall act in such a manner as to uphold and enhance the honor, integrity and dignity of the profession.

7. Engineers shall continue their professional development throughout their careers and shall provide opportunities for the professional development of those engineers under their supervision.

111 Market Place, Suite 1050, Baltimore, MD 21202-4012

*Formerly Engineers' Council for Professional Development. (Approved by the ECPD Board of Directors, October 5, 1977)

AB-54 2/85

Fig. 17.1 ABET's code of ethics.

Let's consider the Code of Ethics of Engineers, presented in Fig. 17.1, which is sponsored by the Accreditation Board for Engineering and Technology (ABET). The code is divided into two parts. The first part deals with four fundamental principles, and the second part contains seven fundamental canons. The Engineers' Council for Professional Development first advanced this code in 1977. The fact that the code remains without modification for more than 25 years indicates that professional ethical values are as constant as personal ethical values.

The fundamental principles seek to insure that the engineer will uphold and advance the integrity, honor and dignity of the profession. These goals are common to your personal goals that were discussed previously; however, the approach to achieve the professional goals is different as indicated below:

1. We are selective in the use of our knowledge and skills so as to ensure that our work is of benefit to society.
2. We are honest, impartial and serve our constituents with fidelity.
3. We work hard to improve the profession of our discipline.
4. We support the professional organizations in our engineering discipline.

The seven fundamental canons in the ABET Code of Ethics of Engineers, presented in Fig. 17.1, are explained in considerable detail in guidelines that are used to expand and clarify the relatively brief statements that represent the "regulations" that shape our professional behavior. The detailed guidelines may be obtained from ABET[4].

17.9 ETHICS IN LARGE ENGINEERING SYSTEMS

Engineers design and build many different products each year that are included in large and complex sociotechnical systems. Usually these products and the systems are conservatively designed with adequate safety factors, carefully tested, and perform well in service for extended periods of time. They provide a much needed service without endangering either individual or public safety. However, from time to time mistakes are made in the initial design. These mistakes are usually detected in prototype testing and eliminated prior to releasing the product to the market place. In very rare circumstances, a mistake or several mistakes are overlooked, for a variety of reasons, and the system is released and placed in service with an unacceptably high probability for failure. An undetected mistake often leads to a catastrophic accident. The space shuttle system clearly falls into this category. If you are a space supporter, you may argue that the risks are worth the benefits. But if this is the case, an unprepared and untrained high school teacher should not be invited to ride along to enhance the image of the space program. It is one thing to order a career astronaut into peril, but totally a different proposition to invite an uninformed civilian to participate in a very dangerous project.

In Chapter 15, the probability of failure (or an accident) has been discussed in considerable detail. You must understand that there is always some risk of failure when designing high performance systems. It is important to learn to accept a trade-off between risk, safety and performance—it is an inherent part of the process. In most complex sociotechnical systems (air transportation for example), there is a small but finite risk for failure with a subsequent loss of life and property. The public must know this risk. It is the responsibility of the engineering community, industry and the government to alert the potential customers to the dangers involved. Knowing the risk, the customers can decide whether or not they want to use the system.

Some people worry more than others and place a very high value on safety. They require a probability of failure of nearly zero. Do you know someone who will not travel by flying? (John Madden, the popular professional football announcer, travels from game to game each week on a special bus). Other

---

[4] The Accrediting Board for Engineering and Technology (ABET) is located at 111 Market Place, Suite 1050, Baltimore, MD 21202. Visit their Website at http://www.abet.org.

folks love the thrill of a risk, and they are willing to accept a much higher probability of an accident. They think hang-gliding or skiing on black diamond slopes is great sport.

17.10 THE CHALLENGER ACCIDENT—A CASE STUDY

With this background on risk, safety and performance, let's begin a discussion of the Challenger accident [9]. The Challenger was one of the original four orbiters built by the National Air and Space Administration (NASA) to serve the space shuttle system. The rocket fuel (liquid hydrogen) on the Challenger 51-L mission exploded 73 seconds after launch on Tuesday, January 28, 1986. The crew of six and a civilian passenger were killed, and the space shuttle was lost.

The Challenger accident has been selected as a case study because it illustrates several ethical issues in the engineering and management of large, complex and inherently dangerous systems: Some issues that will be raised are:

1. The design of the shuttle and the selection of the contractors involved many political considerations [9].
2. The lack of communications between key people and organizations was a significant factor in the accident.
3. The interface between the upper-level administrators (business managers) and the engineers was an important element in the decision to launch on that disastrous morning.
4. Public attention and opinion, not safety, markedly affected the decision-making process.
5. The risk potential was not known by the public and not appreciated by top administrators and politicians directly involved in the launch decision.
6. Christa McAuliffe, a high school teacher and mother of two children, was killed in the accident. Why was she invited to fly on such a dangerous mission?

Background Information

To set the stage for the accident, you have to go back to the early 1970s. NASA had been very successful with the Apollo missions (even with the problems of Apollo 13), and looked forward to larger and more aggressive space endeavors. They proposed an integrated space system that would include a space station, space shuttle, space tug and manned bases on both Mars and the moon. This agenda sounded wonderful until the public examined the price tag. The public did a quick look and wanted no part of it. A poll indicated that the public believed that Apollo had been too costly. The politicians, always driven by the polls, took note and reduced NASA's budget. NASA recognized the need for a new, cost-effective project to follow Apollo that the public (and politicians) would buy. Responding to these political pressures, NASA proposed the space shuttle that would serve the military, the scientific community and the rapidly growing commercial business of placing satellites in orbit. The space shuttle system, illustrated in Fig. 17.2, was marketed as a relatively routine space transport system. Even the name "shuttle" connected with the airline shuttle services that routinely fly every hour from one large city to another.

To make the space shuttle system a commercial success, NASA proposed a fleet of four orbiters [10], which would eventually fly on a weekly basis (NASA initially set a goal of 160 hours for the turnaround time for an orbiter). They planned on nearly 600 flights in the period from 1980 to 1991. The early estimate of the cost of a launch was $28 million with a payload delivery cost of $100 to $270 per pound. After some operational experience, the cost estimates proved to be completely unrealistic. Launches actually cost on the order of $280 million, and the cost to place a pound of payload in orbit on the shuttle was in excess of $5,200 [11]. Early experience with the space shuttle system indicated that NASA could not hold their schedules and their costs were running more than a factor of 10 higher than the original estimates. Because of its poor performance, NASA was struggling to improve its image as 1985 ended.

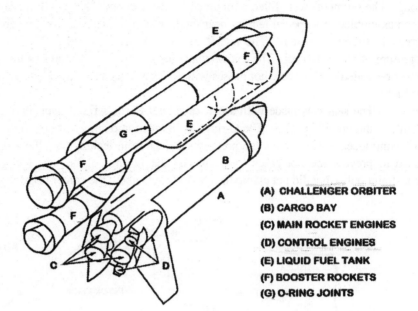

Fig. 17.2 Illustration showing the main components on the NASA space shuttle.

(A) CHALLENGER ORBITER
(B) CARGO BAY
(C) MAIN ROCKET ENGINES
(D) CONTROL ENGINES
(E) LIQUID FUEL TANK
(F) BOOSTER ROCKETS
(G) O-RING JOINTS

Prior to the launch of the Challenger on January 28, 1986, twenty-four shuttle flights had been made. These flights were all successful in that they returned to earth with all crewmembers safe. They also showed the operational capabilities of the shuttle system in placing commercial satellites into orbit, repairing satellites in space and salvaging malfunctioning satellites. However, these flights also indicated that the shuttle system had many serious problems. The three main liquid rocket engines were too fragile with many critical components. Some of the tiles in the heat shield needed repair and/or replacement after every flight. The computers and the inertial navigation systems experienced occasional failures. The brakes and landing gear were stressed to the limit when the orbiter (an 80 ton dead stick glider) landed at speeds ranging from 195 to 240 MPH. Finally, the seals in the solid fuel booster rockets showed distress (sometimes extensive) in 12 of the 24 previous launches.

The record of the shuttle during the 1981-1985 period showed a consecutive series of successful launches, but with many prolonged delays to repair failing components in a very large, highly stressed system. The maintenance records showed so many problems that NASA estimated it took three man-years of work preparing for a launch for every minute of mission flight time. Clearly, the word shuttle to describe such a transportation system is a misnomer.

The Solid Propellant Booster Rockets

The explosion of the main fuel (liquid hydrogen) tank on the Challenger was due to the failure of the O-ring seals on the solid fuel boosters that were adjacent to the hydrogen fuel tank. To understand the seals and their purpose, it is essential that you appreciate the size and function of the two booster rockets used to provide much of the thrust necessary for the launch. They are enormous cylinders—12 feet in diameter and 149 feet tall. Each cylinder is filled with 500 tons of a solid propellant consisting of a rubber mixture filled with aluminum powder and an oxidizer—ammonium perchlorate. When ignited, the propellant burns to produce an internal pressure in the motor case of about 450 psi (lbs/in^2) at a temperature of about 6,000 °F. The expanding gasses exit the rocket motor case through a nozzle, to produce about 2.6 million pounds of thrust from each booster. These booster rockets are essentially very **big** Roman candles!

Big was the nub of the problem. The solid rocket boosters were made by Morton Thiokol in Utah, but launched from the Kennedy Space Center in Florida. They were much too long to ship across the country as a single cylinder. To circumvent this problem, the motor casing was fabricated in segments—each 27 feet long. The segments were filled with propellant and shipped by train to the launch site in Florida. They were then assembled in a special facility near the launch pad. This procedure solved the shipping problem, but it created different problem. The rocket casing is a pressure vessel that must contain very hot gasses at a pressure of about 450 psi. The joints, where the cylindrical segments of the motor case were fitted together, must be sealed so that these hot gasses will not leak and cause damage to adjacent components of the launch vehicle.

The seal was made with a pair of rubber O-rings fitted over the internal finger of a clevis type of joint, as shown in Fig. 17.3. The O-rings, 0.280 inch in diameter, were compressed between the two surfaces affecting a seal that prevents the pressurized fluid from leaking past the joint. A putty like compound was used to prevent the hot gases from eroding the O-rings. The pins locked the segments together (axial constraint only), but did not clamp the clevis fingers about the center finger.

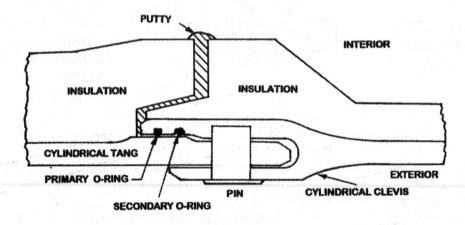

Fig. 17.3 Design of the O-ring seal used at the circumferential joints of the solid rocket motor cases.

The sealing of the segmented cylinders with the O-rings was not a new concept. A single O-ring had been employed previously on the Titan III rocket effectively sealing the circumferential joints on a smaller diameter motor case. Morton Thiokol, the contractor building the boosters, introduced the second O-ring to provide a margin of safety in the event the first O-ring failed. It all sounded good, and the initial test firings of the booster in Utah were apparently successful.

Unfortunately, one test does not validate the safety of even a simple system. To insure high reliability of a component, many tests are required. Moreover, this test firing had little or no bearing on problems associated with the reuse of the solid rocket boosters (up to 20 times). After splash down following a launch and recovery from the Atlantic Ocean, the cylinders were shipped back and forth between Florida and Utah.

Failure of the O-ring Seal

Motion photographs taken at the final launch of the Challenger indicated that the O-ring seals failed almost immediately after ignition of the booster. The hot gasses cut through the O-rings and the joint of the booster. A hole was formed in the wall of the booster at the joint, and a flaming jet of white-hot gasses escaping from the booster rocket cut through the wall of the adjacent tank containing the liquid hydrogen. The launch was effectively over. While many things happened in the last five seconds before the devastating explosion—all bad—the penetration of the adjacent tank, which contained liquid oxygen and liquid hydrogen, spelled the disastrous end of the mission.

It was a terrible day, and we as a nation were in shock while we mourned the loss of the crew and the schoolteacher/mother. It was sometime later, when the facts were brought to the public about the space shuttle program that we learned the shuttle should not have been permitted to fly that day. The on-site engineers understood the very high probability of failure of the O-ring seals. Moreover, knowledgeable engineers at Morton Thiokol tried without success to prevent the launch.

The story containing all of the facts about the failure of the O-ring seal is too long to be covered here, but you are encouraged to read references [9 – 14], where very complete and well-written accounts are given. The essential elements leading to the catastrophic failure are listed below:

1. The circumferential joint changed shape when the motor case was pressurized, and the gap, which the O-rings filled, increased markedly in size. A schematic illustration of the new gap geometry is presented in Fig. 17.4.

2. The new gap opening was so large that the back-up O-ring probably could not seal the joint. When the booster case was pressurized, the seal depended on a single O-ring—not two.

3. The hot exhaust gases had eroded the O-rings on 12 of the previous 24 launches indicating some leakage about half of the time, and on a few occasions, significant erosion of the rings occurred. There was also clear evidence that the putty failed to keep the hot gasses from attacking the rubber O-rings.

4. The joints moved during the early launch sequence as the orbiter engines and then the booster engines were ignited. This motion requires the O-ring seals to be flexible and to reseat and reseal continuously during these movements.

5. The temperature the night before the launch dropped to 22°F and had only increased to about 28°F by the time of the launch.

6. The O-ring seals were not certified to operate below a temperature of 53°F by the contractor responsible for the solid rocket boosters.

7. Tests conducted by the contractor Morton Thiokol in 1985, six months before the accident, indicated that the O-ring seal was not effective at low temperatures. (At 50°F the O-rings would not expand and follow the movement in the joint during the period of operation of the booster.) In fact, at room temperature 75°F, it required 2.4 seconds for the O-rings to reseat and seal pressurized gasses after a gap in the joint was opened.

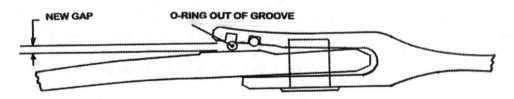

Fig. 17.4 Rotation of the joint due to cylinder pressurization opens the gap in the seal region.

The seven facts listed above show very clear evidence of two problems. First, the design of the seal in the circumferential joint was marginal and should have been fixed much earlier in the program. You do not inspect a piece of burnt and eroded O-ring and walk away from the problem.

Second, the launch should have been postponed due to cold weather for several different reasons. The very low temperatures were well below the limits to which the system had been certified. The rubber in the O-rings becomes very stiff at these low temperatures and cannot respond quickly enough to accommodate joint movement. The O-rings were almost guaranteed to leak. Also, there was considerable ice on the rocket motors and the fuel tank. Pieces of this ice might damaged the rocket motors, the tiles on the heat shield or the wing's leading edges of the orbiter when it separates and falls during the violent vibrations that occur just after ignition but before lift-off.

Ignoring the Problem

Problems usually do not go away when they are ignored. They persist and sooner or later failure will occur. Although the seals had failed on 12 of the previous 24 flights, the failures were not catastrophic because the leaks were small and at locations which did not endanger the adjacent components. The boosters on the shuttle operated for only a few minutes before they are cut loose to fall into the Atlantic Ocean for recovery at a later time. So those concerned with the launch schedule tolerated the small leaks for short periods of time. The trouble with the leak on the Challenger was that it was large, and the jet of hot gas issuing from the leak was pointed directly at the liquid hydrogen fuel tank. The jet of hot gas acted exactly like a cutting torch, and the huge thin-walled fuel tank was penetrated within seconds.

The O-ring seals were inadequate and the joint needed to be redesigned and modified. In fact, during the development of the boosters in 1977–79, several memos were written by engineers at the Marshall Space Flight Center indicating their concern over the design of the joint. They recommended that the joint be redesigned because "the adequacy of the clevis joint was completely unacceptable." The suppliers of the O-rings stated: "The O-ring was being required to perform beyond its intended design and that a different type of seal should be considered."

Apparently there was a major breakdown in communication; none of these memos and reports from the Marshall Space Flight Center was forwarded to Morton Thiokol, the designers and builders of the boosters. A problem, properly detected by engineering at the NASA Center in charge of monitoring the technical aspects of the development of the boosters, was not pursued to its logical conclusion. Instead of insisting on a redesign of the joint, Marshall Space Flight Center approved the boosters in September of 1980. The pressures of cost overruns and repeated schedule slippage sometimes make managers (and engineers) accept flawed designs. It is very shortsighted and not in the best interest of safe design to follow this practice.

NASA accepted a booster rocket with inadequate seals, but the bird flew. On the first launch, with Columbia, there was no reported damage to the seals. The O-ring seals failed on the second flight. One of the O-rings was burnt with about 20% of the thickness of the ring vaporized. The ring failure did not affect the mission, but it was clear that the putty was not protecting the rings from the hot gas. Marshall Space Flight Center reacted to the field experience by reclassifying the joint to a "Criticality 1 category." This was the official recognition on the part of NASA that a seal failure could result in "loss of mission, vehicle and crew due to metal erosion, burn through and probable case burst resulting in fire and deflagration." The failure of the O-rings by erosion continued with high frequency (50% of the missions), but NASA decided to accept the situation and did not recommend remedial action.

It appears that NASA decided to live with the seal problem at least until a new lighter booster motor could be designed with improved joints. These new boosters were to be ready about six months after the Challenger exploded. A very real lesson—in too-little-too-late.

Recognizing the Influence of Temperature

From the initial development phase of the boosters, the O-ring seals had been recognized as a problem. A problem that NASA classified as very serious, but one that they must have erroneously believed did not elevate the risk to unacceptable limits. In January of 1985, a year before the accident, the Challenger was launched on a cold day at a temperature of 51°F. An examination of the joints after the recovery of the boosters showed that a number of O-rings were severely damaged and evidence of extensive gas leakage was observed. The contractor, Morton Thiokol, and Marshall Space Flight Center reviewed the O-ring seal problems. During this review, Morton Thiokol noted that low temperatures were a contributing factor to the inability of the O-rings to seal properly. The condition was considered undesirable, but was acceptable. However, the O-ring problems persisted and NASA eventually placed a launch constraint on the space shuttle.

The launch constraint would prohibit launches until the problem had been resolved or reviewed in detail prior to each launch. Unfortunately, NASA routinely provided launch waivers for each subsequent flight to avoid delays in an unrealistic launching schedule. The problem of O-ring erosion was common; one or more joints on 50% of the launches exhibited significant erosion. Fortunately, the erosion and attendant leaks had not caused any serious difficulty until the Challenger accident. The problem was expected and accepted as routine.

A Management Decision

The very cold weather on the night before the fatal launch was no surprise. The weatherman (or woman) was on the mark in predicting the temperature and the ice that formed on structures that night. The engineers at Morton Thiokol, in Utah, were very concerned about the effects of the very low temperatures on the ability of the O-rings to function properly. Teleconferences took place between the Morton Thiokol engineers and the NASA program managers. The engineers argued to postpone the launch until the temperature increased to at least 53°F.

The NASA program managers were very unhappy about the engineer's concerns. They did not want the extended delay required to wait for warmer weather. Senior vice presidents from Morton Thiokol sensed the displeasure from the customer and took over the decision process. Senior managers decided to keep the customer happy and reversed the decision of the Vice President of Engineering not to launch at these very low temperatures. Senior management at Morton Thiokol signed off on the launch ignoring the fact that the temperature at launch time was expected to be 25°F lower than the lowest temperature which engineering believed the seals would be effective (53°F).

The senior management at Morton Thiokol yielded to client pressure. The engineers were unanimous in their opposition to launch. Senior management asked the engineers to prove that the O-ring seals would fail. Of course, they could not state with 100% certainty that the rings would fail. Failure analysis is performed in terms of probabilities. The risk had elevated as the temperature decreased, but the engineers could not prove conclusively that the seals would fail in a manner that would detrimentally affect the mission. Senior managers at Morton Thiokol and program managers at NASA concluded erroneously that the risks were low enough to proceed with the launch.

Approvals at the Top

NASA has a four-level approval procedure to control the launch of the shuttle. This sounds good, but for a multilevel approval process to be of any value, information must flow freely from the bottom of the organization to the top of the chain of command. The technical discussions concerning the ability of the O-rings to function at the very low temperatures took place between NASA administrators and engineers at level 4 and Morton Thiokol. Program managers at Marshall Space Flight Center, level 3 administrators, were also included in these discussions. While the discussion was extended, with clear polarization between engineering and management, not a word of these concerns was conveyed to the top two levels of management at NASA.

Approvals for the launch were given by the level 2 administrator for the National STS Program in Houston, TX and the level 1 administrator at NASA's Headquarters in Washington D.C. These approvals were essentially automatic because the administrators in charge had been isolated. They were not privy to the management decision to fly with very cold O-ring seals. They were not informed of the engineering recommendation to postpone the launch and to wait until the temperature increased to at least 53°F.

The lesson here is clear—approvals by the very high level managers are worthwhile only if they are informed decisions. If complete information is not presented to the executives for their evaluation, then their approval is meaningless. These administrators are not knowledgeable, and they add no value in an informed decision making process.

17.11 THE COLUMBIA FAILURE

On February 1, 2003 the Columbia shuttle disintegrated as it reentered the atmosphere at the end of its 16-day mission. In Mission Control, Columbia's re-entry appeared normal until 8:54:24 AM[5], when engineers monitoring the sensors informed the flight director that four hydraulic sensors in the left wing were indicating—off-scale low—a reading that falls below the minimum capability of the sensor. At 8:55:00 a.m. nearly 11 minutes after Columbia had re-entered the atmosphere, its wing leading edge temperatures reached their normal level of nearly 3,000 °F. At 8:55:32 a.m. Columbia crossed from Nevada into Utah while traveling at Mach 21.8 at an altitude of 223,400 ft. At 8:58:20 a.m. as Columbia crossed from New Mexico into Texas, a thermal protection tile was lost. At 8:59:15 a.m. the engineers monitoring the sensors informed the flight director that the sensors monitoring the pressure on both left main landing gear tires were lost. The flight director then informed the Columbia's crew that Mission Control was evaluating these sensor readings, and added that the crew's last transmission was not clear. At 8:59:32 a.m. a broken response from Columbia was recorded but it was cut off in mid-word. Videos

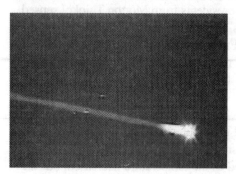

made by observers on the ground less than a minute later, presented in Fig. 17.5, revealed that Columbia was disintegrating.

Fig. 17.5 A video photograph of Columbia disintegrating as it is passing over Texas while traveling at about Mach 19.5 at an altitude of about 210,000 ft.

Columbia's History

Columbia was the first space-rated Orbiter, and it made the Space Shuttle Program's first four orbital test flights. Because it was the first of its kind, Columbia differed slightly from the other four Orbiters—Challenger, Discovery, Atlantis, and Endeavour. Built to earlier engineering specifications, Columbia was slightly heavier, and was not able to carry sufficient cargo to conduct cost effective missions to the International Space Station. For this reason, Columbia was not equipped with a Space Station docking system that gave it more space in its cargo bay for longer apparatus. Consequently, Columbia usually flew science missions and serviced the Hubble Space Telescope.

The final flight of Columbia was designated STS-107. This was the Space Shuttle program's 113th flight and Columbia's 28th. The flight was nearly trouble-free. Unfortunately, there were no clear indications to either the crew onboard Columbia or to personnel in Mission Control that the flight was in trouble as a result an impact by at least one large piece of foam insulation that occurred shortly after launch. Mission management failed to note a few clues revealed during launch that the Orbiter was in trouble and take corrective action.

---

[5] Times are given as Eastern Standard Time (EST.).

STS-107 was an intense science mission that required the seven-member crew to form two teams, enabling round-the-clock shifts. Because the extensive science cargo and its extra power sources required additional checkout time, the launch sequence and countdown were about 24 hours longer than normal. Nevertheless, the countdown proceeded as planned, and Columbia was launched on January 16, 2003. At 81.7 seconds after launch, when the Shuttle was at an altitude of about 65,600 feet and traveling at Mach 2.46 (1,650 mph), a large piece of insulating foam came off an area where the Orbiter attaches to the external tank. This piece of foam impacted the leading edge of Columbia's left wing at a velocity of about 500 MPH. The foam impact was not detected by the flight crew or observed by ground personnel. However, the next day, during detailed reviews of launch camera photography revealed a large piece of foam striking the Orbiter. This foam impact and resulting damage to the left wing of Columbia had no effect on the daily activities performed by the astronauts during the 16-day mission. This mission had met all its objectives prior to reentry.

The Accident Investigation

Immediately following the failure of the Orbiter, NASA established a Columbia Accident Investigation Board to identify the chain of events that caused the Columbia accident. Evidence the Board considered included:

- Film and video during launch.
- Radar images of Columbia in orbit.
- Amateur video of debris shedding during reentry.
- Onboard sensor data from the onboard recorder recovered after the accident.
- Analysis of the debris recovered.
- Computer modeling.
- Impact and wind tunnel tests.

Fig. 17.6 The upper circle shows the foam covered bipod on the external tank and the lower circle shows the impact point on the leading edge of the Columbia's left wing.

The reason for the loss of Columbia and its crew was a breach in the thermal protection system on the leading edge of the left wing. The breach was initiated by a piece of insulating foam that separated from the left bipod ramp of the external tank and struck the wing in the vicinity of the lower half of a reinforced carbon-carbon panel 81.9 seconds after launch. During re-entry, this breach in the thermal protection system allowed superheated air to penetrate the leading-edge insulation. This hot air (plasma) heated and softened the aluminum structure of the left wing and weakened the structure until aerodynamic forces caused loss of control, collapse of the wing and subsequent disintegration of the Orbiter. A photograph of Columbia prior to launch, showing the area of the foam released from the external tank and the location of its impact on the left wing of the Orbiter is presented in Fig 17.6.

The external tank is the largest component of the Space Shuttle. It serves as the main structural component during assembly, launch and supports both the solid rocket boosters and the Orbiter. It also serves as the cryogenic, propellant tank for the space shuttle main engines. It contains 143,351 gallons of liquid oxygen at $-297°$ F in its upper tank and 385,265 gallons of liquid hydrogen at -423 ° F in its lower tank.

The Orbiter is attached to the external tank by two umbilical fittings at the bottom and by a "bipod" at the top. The bipod is attached to the external tank by fittings to the right and left of the external tank centerline. The bipod fittings, which are bolted to the external tank, are located near the flange joint at the inter tank as shown in Fig. 17.7.

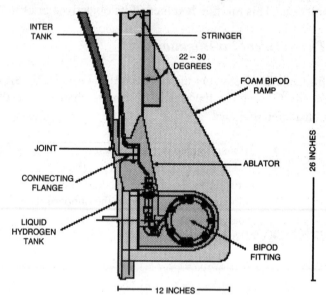

Fig. 17.7 Drawing of the bipod showing the foam insulation that covered the bipod fitting.

The external tank is coated with two materials that serve as its thermal protection system. The inner coating is a dense composite ablator for dissipating heat, and the outer coating is low-density closed-cell foam for high insulation efficiency. The external tank thermal protection system is designed to:

- Reduce heat flow into the cryogenic tanks to minimize the vaporization of the oxygen and hydrogen prior to launch.
- Maintain the temperature of the external surfaces sufficiently high to prevent the formation of frost or ice.

A combination of factors led to the loss of the left bipod foam ramp (see Fig. 17.7) during the ascent of Columbia shortly after launch. NASA personnel believe that pre-existing defects in the foam were a major factor and that these defects were necessary to induce the foam to fail. However, analysis indicated that pre-existing defects were not the only factor responsible for foam loss. Other factors are involved such as the technique for producing and applying the foam, particularly in the bipod region. The foam used to insulate the external tank consists of two chemical components that must be mixed in an exact ratio and is then sprayed according to strict specifications. The foam is applied to the bipod fitting in a manual operation to form the ramp illustrated in Fig. 17.7. This process is probably the primary source of defects in the foam. Dissection of these foam ramps, conducted after the accident, revealed defects such as voids, pockets and debris. These defects are due to a lack of control of several parameters in the manual spraying used to apply the foam. The quality control of the foam is exacerbated by the complexity of the underlying hardware configuration. The dissection studies showed that the defects usually occurred at the boundaries between each layer as the ramp is formed by the repeated application of thin layers.

It is important to recognize that the failure of the foam insulation on previous shuttle missions was not a rare event. Foam loss has occurred on more than 80 percent of the 79 previous Shuttle missions for which imagery is available. Foam was lost from the left bipod ramp on nearly 10 percent of missions analyzed where the left bipod ramp was visible following external tank separation. For many of the missions, it was not possible to determine if foam was lost because the launches were at night or the external tank bipod ramp areas were not in view at the time when the photographs of the external tank were taken.

It is believed that the impact of a large piece of foam damaged the leading edge of Columbia's left wing. The leading edge of an Orbiter's wing is comprised of 22 panels fabricated from a reinforced carbon-carbon (RCC) composite. To prevent oxidation, the outer layers of the carbon substrate are converted into a 0.02 to 0.04 inch thick layer of silicon carbide. The components of the Orbiter's wing leading edge provide the aerodynamic load bearing, structural and thermal control capability with temperatures exceeding 2,300 °F. The design requirements for the leading edge of the wing included 100 missions with minimal rework, limiting the temperature of the aluminum wing structure to less than 350 °F, and withstanding a kinetic energy impact of only 0.006 foot-pounds. The engineering specification clearly indicate that the wing leading edge would not be required to withstand impact from debris or ice, because these objects would not pose a threat during the launch sequence.

The risk of micrometeoroid or debris damage to the RCC panels has been evaluated several times. Hypervelocity impact testing, using a variety of projectiles, as well as low-velocity impact testing with projectiles of ice and other materials, resulted in a design change that improved the resistance of the leading edges to impact damage. Analysis of this design change predicted that an Orbiter could survive re-entry with a ¼ to 1-inch diameter hole in the lower surfaces of the RCC panels depending on panel location.

Analysis of maintenance reports indicated that the RCC leading edge panels from all of the Orbiters had been struck by objects throughout their operational life. However, none of the panels had been completely penetrated. A sampling of 21 post-flight reports noted 43 hypervelocity impacts. The largest damage zone was 0.2 inch in diameter. The most significant low-velocity impact was to one of Atlantis' panels. The damaged area was 1.9 inches by 1.6 inches on the exterior surface and 0.5 inches by 0.1 inches in the interior surface. The substrate of the panel was exposed and oxidized. After inspection the severely damaged panel was replaced. A study concluded that the damage was caused by a strike by a man-made object, possibly during ascent. This panel damage to Atlantis was a clear warning of an impending disaster.

Upon completion of its exhaustive study, the Columbia Accident Investigation Board concluded that the cause of the accident was a breach in the thermal protection system on the leading edge of the left wing. The breach was initiated by a piece of foam that separated from the left bipod ramp of the external tank and impacted the left wing in the vicinity of the lower half of RCC panel 8. The conclusion that foam separated from the external tank bipod ramp and struck the wing in the vicinity of panel 8 is clearly documented by photographic evidence. Sensor data and the aerodynamic and thermodynamic analyses confirmed that the breach was in the vicinity of panel 8. This data and subsequent analysis also indicated subsequent melting of the supporting structure, the spar, and the wiring behind the spar. The detailed examination of the debris also pointed to panel 8 as the breach site. Impact tests established that a large piece of foam could breach an RCC leading edge panel. These tests and the physical evidence collected over two states convinced those individuals who were discounting of the analytical evidence of the cause of the failure.

Hard Lessons to Learn

Dr. Diane Vaughan, an authority on risk assessment, testified before the Columbia Accident Investigation Board and stated: "What we find out from a comparison between Columbia and Challenger is that NASA is an organization did not learn from its previous mistakes and it did not properly address all of the factors that the presidential commission identified."

Organizational failures always occur usually with higher frequency in larger organizations with several layers of management. Failures happen regardless of the safeguards and systems that an organization employs. Failures are even more common in high-risk organizations like NASA— missions fail and astronauts are killed. NASA uses several risk-avoidance systems that are designed to insure that the instruments and astronauts sent into space complete their missions and return safely. However, NASA has failed in three cases to achieve this objective when sending astronauts in space or when preparing space systems to do so. The Space Shuttles Challenger and Columbia tragedies as well as the Apollo launch pad fire in 1967 are examples NASA's failures that resulted in the deaths of 17 astronauts. Organizational failures played an important role in all three cases.

As was the case with O-ring damage in the years before the Challenger accident[6], the failure of Columbia did not represent the first time that foam had detached from the external tank and caused damage to an Orbiter. There had been many reports of impacts on previous flights by pieces of foam that were much smaller than that one which severely damaged Columbia. Orbiters would often return with many damage sites due to both hypervelocity and low velocity impact. Some sites were larger than 1 inch in diameter. There are numerous reports in the Problem Reporting and Corrective Action (PRACA) system that showed some of the thermal barrier damage was probably due to foam failure during launch and assent. Not only was foam impact not considered in the design of the thermal protection system, but the engineering specification for the Orbiter clearly states that nothing should impact the shuttle during launch.

However, foam had been failing and striking the Orbiter with high frequency and a team of chemists and engineers were working on improving the foam and the techniques used for applying it. The studies of impacts with foam concluded that they did not pose a "flight safety" risk because experience showed that the small pieces of foam that had been coming off the external tank only caused small pits and craters on the Orbiters. Unfortunately, the engineers did not consider a large piece of foam traveling at high velocity[7] when they declared foam failure not to be a "flight safety" issue. The impact of such a large piece of foam on a vulnerable area of the Orbiter was unprecedented. (So was the three days of abnormal cold before the Challenger launch.)

Foam failure that regularly damages the thermal protection system on a large fraction of the flights was not considered to be dangerous because the orbiters were returning safely with damage that could be repaired or ignored. What the engineers or management did not considered was the possibility of a larger or heavier piece of foam hitting the Shuttle in a particularly vulnerable area like the leading wing edge. This is a classic example of Vaughan's "normalization of deviance" [16] where an unpredicted anomaly becomes routine.

---

[6] See Section 17.10 for a complete description of the Challenger launch.
[7] Photographic analysis revealed that one large piece and at least two smaller pieces of foam had separated from the bipod ramp area of the external tank. The large piece of foam was about 24 in. long and 15 inches wide. It was rotating at 18 RPS and moving with a velocity of about 500 MPH when it impacted the Columbia's wing.

Normalization of Deviance or Accepting Problems Instead of Pursuing Solutions

Vaughan [16] has developed the concept of **normalization of deviance** to explain the way technical flaws escape the scrutiny of the various safety boards within a large organization such as NASA. Frequently these boards observe unanticipated problems that continue to occur on a regular basis without leading to an accident. People begin to believe the problems are benign and are lead to the pragmatic notion of "acceptable" deviance. The leadership at NASA found it very expensive and time consuming to establish the cause of some problems and to incorporate added inspections in the regular maintenance cycle of the Shuttle system without strong evidence of "flight safety " risk. Under the pressures of the Shuttle's flight schedule and limited budget, NASA did not commit significant resources on problems that its management did not classify as "flight safety" risks that could cause loss of an Orbiter. This practice resulted in disincentives for the engineers to determine the source of problems, even though scenarios could be envisioned that would cause significant "flight safety" risks. NASA frequently cleared flights as operational based on previously successful flights that had completed their missions while exhibiting clear evidence of a design problem. This reasoning prompted Richard Feynman [13] in his analysis of the Challenger accident to remark: "When playing Russian roulette the fact that the first shot got off safely is little comfort for the next."

The Role of NASA's Management Structure in Columbia's Accident

NASA's organizational and physical infrastructure was developed in the days of Apollo program when it had a large budget and a sharply focused mission—to land a man on the moon before the Russians. A layered bureaucratic group of middle and upper level managers functioned well and the Apollo program accomplished its goals with only one serious mishap in 1967 and a near catastrophic mission on Apollo 13. After more than 30 years, NASA's organizational structure is nearly the same although its missions have changed markedly. There is reason to question if their organizational structure is sufficiently flexible considering its plans for the future and its ongoing projects.

For the Shuttle program, there should be a mechanism for engineers to bypass the bureaucracy and hierarchy, especially in the pre-launch and launch processes. Suppose the engineers had succeeded in calling off the launch of the Challenger due to the effect of cold weather on the O-ring seals. The flight would have been cancelled, Challenger would have been taken off of the launch pad and the solid rocket booster disassembled to replace the damaged O-rings. This action would have been expensive but not nearly as costly as the loss of crew and the Orbiter. Similarly, if an engineer needs a certain type of data, there should be a way to bypass the formal bureaucratic procedures and to obtain the data with dispatch. Engineers have many intuitions and hunches that take time and resources to translate into analysis and data. These intuitions need to be respected, given credence, explored and welcomed by upper management [17].

Probability of Failure

Richard Feynman served on the Presidential commission that investigated the Challenger accident that occurred on January 28, 1986. It was near the end of his life, and the famous physicist devoted significant amounts of time to the investigation. After the investigation, he prepared a paper describing his personal observations on the reliability of the Shuttle [18]. This paper is available on the Website given in Reference 19. We suggest you read it, because we are only reviewing a small portion of this reference. The following paragraphs are direct quotes from Feynman's personal observations.

1. It appears that there are enormous differences of opinion as to the probability of a failure with loss of vehicle and of human life. The estimates range from roughly 1 in 100 to 1 in 100,000. The higher figures come from the working engineers, and the very low figures from

management. What are the causes and consequences of this lack of agreement? Since 1 part in 100,000 would imply that one could put a Shuttle up each day for 300 years expecting to lose only one, we could properly ask—"What is the cause of management's fantastic faith in the machinery?"

2. If a reasonable launch schedule is to be maintained, engineering often cannot be done fast enough to keep up with the expectations of originally conservative certification criteria designed to guarantee a very safe vehicle. In these situations, subtly, and often with apparently logical arguments, the criteria are altered so that flights may still be certified in time. They therefore fly in a relatively unsafe condition, with a chance of failure of the order of a percent (it is difficult to be more accurate).

3. Official management, on the other hand, claims to believe the probability of failure is a thousand times less. One reason for this may be an attempt to assure the government of NASA perfection and success in order to ensure its annual funding from the government. The other may be that they sincerely believed it to be true, demonstrating an almost incredible lack of communication between themselves and their working engineers.

4. In any event this has had very unfortunate consequences, the most serious of which is to encourage ordinary citizens to fly in such a dangerous machine, as if it had attained the safety of an ordinary airliner. The astronauts, like test pilots, should know their risks, and we honor them for their courage. Who can doubt that McAuliffe was equally a person of great courage, who was closer to an awareness of the true risk than NASA management would have us believe?

5. Let us make recommendations to ensure that NASA officials deal in a world of reality in understanding technological weaknesses and imperfections well enough to be actively trying to eliminate them. They must live in reality in comparing the costs and utility of the Shuttle to other methods of entering space. They must be realistic in making contracts, in estimating costs and the difficulty of the projects. Only realistic flight schedules should be proposed, schedules that have a reasonable chance of being met. If in this way the government would not support them, then so be it. NASA owes it to the citizens from whom it asks support to be frank, honest and informative, so that these citizens can make the wisest decisions for the use of their limited resources.

6. For a successful technology, reality must take precedence over public relations, for nature cannot be fooled.

As one studies the Columbia's fate, we appreciate the significant value of the personal observations made by Richard Feynman nearly 30 years ago. It appears that his message was lost on NASA as they continue to ignore significant operational problems in their efforts to met schedule with a constrained budget. Dr. Feynman estimated that the Shuttle flies in a relatively unsafe condition, with a chance of failure of the order of a percent. With the additional Shuttle flights since the Challenger accident it is possible to test his estimate with the additional flight experience. The Columbia accident was the second in 113 flights; thus the estimate of the probability of failure P_f of future Shuttle flights is:

$$P_f = N_f / N = 2/121 = 1.65\%$$

This estimate is the same order as Dr. Feynman predicted after the Challenger accident and 1,700 times higher than the probability of failure of 0.001% (1/100,000) predicted by NASA managers.

17.12 SUMMARY

The degradation of ethical behavior during the past two or three generations has been described. Some of the probable reasons for these behavioral changes were discussed. While there are many controversial issues that can be classified as "dilemma" ethics, there are several "basic" characteristics where the difference between right and wrong is clearly defined. The list of basic virtues was described that include:

- Prudence
- Justice
- Fortitude
- Temperance
- Faith
- Hope
- Charity
- Humility
- Respect

Laws are written to prohibit, by punishment, a well-defined set of vices. These laws control behavior, but only affect the extremely repulsive actions on the part of individuals or corporations.

Individual behavior is important, but as we learned in the Challenger explosion, it is not sufficient. Corporations (business) must operate with the highest ethical standards. The chief executive officer (CEO) establishes the standards and the lower level managers act to establish the corporate character.

Honor codes exist on some college campuses to guide ethical behavior (lying, cheating and stealing) and the students and the instructors respect them. Character development and honor codes go together in producing a meaningful educational experience. Some of the more important attributes in character development are listed below:

1. Excellence
2. Respect
3. Integrity
4. Decisiveness
5. Responsible
6. Temperance
7. Fortitude
8. Understanding

Professional ethics was described by introducing the Code of Ethics of Engineers sponsored by the Accreditation Board for Engineering and Technology (ABET). The code is short with four fundamental principles and seven basic canons.

The fatal accidents of the Challenger and Columbia Space Shuttles, which failed during their missions in 1986 and 2003, were discussed. Many ethical issues dealing with individual and corporate behavior were apparent when the events leading to this fatal accident were reviewed. This example should aid you in developing both your individual and professional character.

REFERENCES

1. Patterson, J. and P. Kim, The Day America Told the Truth: What People Really Believe about Everything that Really Matters, Prentice Hall, New York, NY, 1991.
2. Sommers, C. H., "Teaching the Virtues," *Public Interest,* No. 111, Spring 1993, pp. 3-13.
3. Woodward, K. L. "What is Virtue," *Newsweek*, June 13, 1994.
4. George, R. P., Making Men Moral: Civil Liberties and Public Morality, Oxford University Press, New York, NY, 1994.
5. Baker, D. F. "Ethical Issues and Decision Making in Business," Vital Speeches of the Day, 1993
6. United States Air Force Character Development Manual, USAF Academy, CO, December 1994.
7. Abramson, R., "A Matter of Honor," *Los Angeles Times*, April 3, 1994.
8. Lickona, T. A. Educating for Character, Bantam Book, New York, NY, 1991.
9. Jensen, C., No Down Link: A Dramatic Narrative about the Challenger Accident, Farrar, Straus and Gitoux, New York, NY, 1996.
10. *Time,* February 10, 1986
11. Lewis, R. S., The Voyages of Columbia: The First True Space Ship, Columbia University Press, New York, NY, 1984.
12. *The New York Times* April 23, 1986.
13. Feynman, R. P. What Do You Care What Other People Think? Bantam, New York, NY, 1988.
14. Report to the President by the Presidential Commission on the Space Shuttle Challenger Accident, Ayer Co., Salem, MA, 1986.
15. Anon, Report of Columbia Accident Investigation Board, Volume I, August 26, 2003, See http://www.nasa.gov/columbia/home/CAIB_Vol 1.html.
16. Hall, J. L., "Columbia and Challenger: Organizational Failure at NASA," Space Policy Vol. 19, 2003, pp. 239-247.
17. Vaughan, D., The Challenger Launch Decision: Risky Technology, Culture and Deviance at NASA, The University of Chicago Press, Chicago, 1996.
18. Feynman, R. P., Personal Observations on the Reliability of the Shuttle, http://www.fotuva.org/feynman/challenger-appendix.html.

EXERCISES

17.1 Political leaders are often attacked by the media for lax ethical behavior. Please write a short paper describing three recent lapses of ethical behavior on the part of the leadership in either a State or the Federal government. Did these officials break the law? Is it important that they did or did not break the law?

17.2 Many corporate executives have been involved in serious crimes involving hundreds of millions of dollars. Identify one of these executives and write a paper about his misfeasance and the consequences.

17.3 List what you consider poor behavior in a college classroom.

17.4 Have you ever cheated on an important exam in high school? What about a college exam? Do you know of someone who cheated? What was your attitude when you observed this cheating?

17.5 Is there an honor code at the University where you are pursuing your studies? Have you read it? Do you abide by the rules?

17.6 Write a paragraph or two explaining your position regarding:

- Prudence
- Justice

- Fortitude
- Temperance

17.7 Write a paragraph or two describing your feelings about:

- Faith
- Hope
- Charity

17.8 If you could write a law that would go on the books tomorrow and be strictly enforced, what behavior would it require or prohibit?

17.9 Why does the CEO of a corporation set the standard for ethical behavior? What are some of the issues that arise on a daily basis that develop corporate character? Is it possible for a corporation with tens of thousands of employees to develop a character like an individual?

17.10 Write an essay describing your character. Include a discussion of your weaknesses and strengths.

17.11 Examine the code of ethics for engineers given in Fig. 17.1, and describe your opinion of the fundamental principles or canons. Can you live with these expectations if you become a professional engineer? Do you believe ABET's code is adequate, or should it be revised to reflect a more modern viewpoint of what is right and wrong?

17.12 Suppose that you were a lead engineer working for Morton Thiokol on the evening of January 27, 1986 involved in the discussion of the O-rings. What would you have done?

- Early in the teleconference.
- At the critical stage of the teleconference.
- After management had taken over the decision process.

There is no right or wrong answer to these questions. The purpose of the exercise is to place you in a professional dilemma. Someday you may be placed in a similar situation where there is a trade-off between safety and corporate business interests. Think about your response in advance and be prepared to deal with such a dilemma and its consequences.

17.13 Suppose you were the engineer responsible for maintaining the foam on the external tank of the Space Shuttle. What would you do to prevent the failure of foam during the launch and ascent of a Shuttle?

17.14 Suppose you were the engineer responsible maintaining the RCC composite panels that form the leading edge of the three remaining Orbiters. What changes in your maintenance plan would you make following the Columbia accident?

17.15 Write a review of Chapter 2 "Columbia's Final Flight" from volume I of the Report of the Columbia Accident Investigation Board. Reference 15 provides the URL for the Website.

17.16 Write a review of Chapter 3 "Accident Analysis" from volume I of the Report of the Columbia Accident Investigation Board. Reference 15 provides the URL for the Website.

17.17 Write a review of Chapter 5 "From Challenger to Columbia" from volume I of the Report of the Columbia Accident Investigation Board. Reference 15 provides the URL for the Website.

17.18 Are there government agencies that are concerned with product safety? Name them and describe the way they function.

Notes

APPENDIX A

GUIDES AND FORMS
FOR DESIGN TEAMS

TEAM MEMBERS' AGREEMENT

- The information discussed by our team members will remain confidential.
- We will acknowledge problems and deal with them.
- We will be supportive rather than judgmental.
- We will respect differences.
- We will provide responses directly and openly in a timely fashion.
- We will provide information that is specific and focused on the task and not on personalities.
- We will be open, but will respect the right of privacy.
- We will not discount the ideas of others.
- We are each responsible for the success of the team experience.
- The team has all of the resources needed to solve the problems that arise during the development process.
- Every member of the team will contribute to its success.
- We will try to know and understand our fellow team members so as to identify ways for enhancing everyone's professional development.
- We recognize the importance of the team meeting to the success of the group, and accordingly we agree to:
- Always attend the scheduled team meetings.
- When attendance is impossible, notify the team leader and as many other members as possible in advance.
- Use meeting time wisely
- Start on time.
- Limit our breaks and return from them on time.
- Keep focused on our goals.
- Avoid side issues, personality conflicts and hidden agenda.
- Share responsibility for briefing members when they miss a meeting.
- Avoid making phone calls or engaging in side conversations that interrupt the team.
- _____.
- _____.
- _____.
- _____.
- _____.
-

SIGNATURES:

_____   _____

_____   _____

_____   _____

Date_____

Agenda
Weekly Meeting
Date _____ Time _____

Team Member Responsible

1. Weekly status report   _____

2. Review of outstanding action items

   - Action item #_____   _____
   - Action item #_____   _____
   - Action item #_____   _____
   - Action item #_____   _____
   - Action item #_____   _____

3. Report on progress

   - Subsystem _____   _____
   - Subsystem _____   _____
   - Subsystem _____   _____
   - Subsystem _____   _____
   - Subsystem _____   _____

4. Identify new problems

   - Problem #_____   _____
   - Problem #_____   _____
   - Problem #_____   _____
   - Problem #_____   _____
   - Problem #_____   _____

5. Assignment of action items

   - Action item #_____   _____
   - Action item #_____   _____
   - Action item #_____   _____
   - Action item #_____   _____
   - Action item #_____   _____

6. New business
7. Summary
8. Adjourn

Agenda
Weekly Meeting
Date _____ Time _____

Team Member Responsible

1. Weekly status report _____

2. Review of outstanding action items

   - Action item #_____   _____
   - Action item #_____   _____
   - Action item #_____   _____
   - Action item #_____   _____
   - Action item #_____   _____

3. Report on progress

   - Subsystem _____   _____
   - Subsystem _____   _____
   - Subsystem _____   _____
   - Subsystem _____   _____
   - Subsystem _____   _____

4. Identify new problems

   - Problem #_____   _____
   - Problem #_____   _____
   - Problem #_____   _____
   - Problem #_____   _____
   - Problem #_____   _____

5. Assignment of action items

   - Action item #_____   _____
   - Action item #_____   _____
   - Action item #_____   _____
   - Action item #_____   _____
   - Action item #_____   _____

6. New business
7. Summary
8. Adjourn

Action Item Record

Team Meeting of _____

Decisions made	Follow-up required	Responsible member	Date complete
1.	1.	1.	1.
2.	2.	2.	2.
3.	3.	3.	3.
4.	4.	4.	4.
Action at next meeting	**Preparation required**	**Responsible member**	**Date complete**
1.	1.	1.	1.
2.	2.	2.	2.
3.	3.	3.	3.
4.	4.	4.	4.

Team Meeting of _____

Decisions made	Follow-up required	Responsible member	Date complete
1.	1.	1.	1.
2.	2.	2.	2.
3.	3.	3.	3.
4.	4.	4.	4.
Action at next meeting	**Preparation required**	**Responsible member**	**Date complete**
1.	1.	1.	1.
2.	2.	2.	2.
3.	3.	3.	3.
4.	4.	4.	4.

Team Meeting of _____

Decisions made	Follow-up required	Responsible member	Date complete
1.	1.	1.	1.
2.	2.	2.	2.
3.	3.	3.	3.
4.	4.	4.	4.
Action at next meeting	**Preparation required**	**Responsible member**	**Date complete**
1.	1.	1.	1.
2.	2.	2.	2.
3.	3.	3.	3.
4.	4.	4.	4.

CONCEPT SELECTION
PUGH CHART

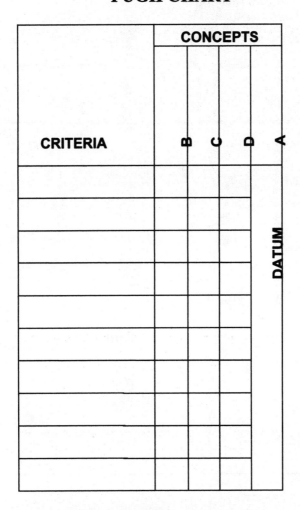

S - CONCEPT EQUAL TO DATUM

+ CONCEPT SUPERIOR TO DATUM

– CONCEPT INFERIOR TO DATUM

Notes: